16.45   8/81

ANALYTIC GEOMETRY

SECOND EDITION

MURRAY H. PROTTER
University of California, Berkeley

CHARLES B. MORREY, JR.
University of California, Berkeley

ANALYTIC GEOMETRY

SECOND EDITION

ADDISON-WESLEY PUBLISHING COMPANY

Reading, Massachusetts
Menlo Park, California • London • Amsterdam • Don Mills, Ontario • Sydney

ISBN 0-201-05997-5
FGHIJKLMNO-MA-89876543210

Preface
to the
Second Edition

This text in analytic geometry is designed to prepare students thoroughly for a course in calculus. Most of the material in the book can be covered in a three-credit course for one semester or, equivalently, in a four- or five-credit course for one quarter. The presentation is intended for a typical first course at the college level for students who have had three years of high school mathematics. However, we believe that both the text and the material are suitable also for the twelfth-grade student. After completing such a course, the high school senior should have sufficient background to make the transition from those materials written primarily for high school to those written primarily for college.

Chapter 1 is concerned with inequalities; the rules governing them are given as theorems derived from an appropriate axiom. Set notation is introduced here, and the student is encouraged to learn both the traditional and the modern ways of describing solutions of inequalities. Precise definitions of function and relation are presented in Chapter 2, where functional notation is introduced. Although the book is devoted to analytic geometry, we present in Appendices 1 and 2 an indication of how the rules of algebra can be derived from the axioms for a field. This supplementary material should encourage able students to develop a serious interest in the logical structure of mathematics.

In Chapter 4 we define a vector in a geometric plane as the "equivalence class" of all directed line segments having a given magnitude and direction. In this way we avoid certain difficulties which arise when a vector is identified with a directed line segment. The operations of addition and scalar multiplication and the inner product of vectors are defined geometrically. Although we use coordinate systems in the proofs of some theorems, it is apparent that many of the theorems are independent of the particular coordinate system employed. Some problems from Euclidean geometry are included to emphasize this fact. In Chapter 10 we introduce solid analytic geometry. In Chapter 11 we develop the corresponding theory for vectors in space. In general we emphasize the logical structure of each of the topics.

It is possible that a student taking this course will have studied most of the topics in Chapters 1, 2, and 3 in a high school algebra course. Also, the student may have had practice in plotting the curves of various equations, including conics. In such cases the instructor may wish to review these topics briefly, concentrating on those topics which are completely new to the student. However, an entering student who is poorly prepared should study the early material intensively in order to develop a proper basis for understanding calculus. For this reason we have also included a brief review of trigonometry in Appendix 3 and a review of determinants of the second and third order in Appendix 4.

Other special features are the inclusion in Chapter 3 of a section on the parametric equations of a line in the plane and a section on systems of linear inequalities in the plane. We have included more material on vectors than can be found in most texts on analytic geometry because we believe that the mathematical concepts encountered in the study of vectors are important in higher mathematics; moreover for those students interested in science, a knowledge of vectors is indispensible in the study of physics. We have not attempted to develop analytic geometry by means of vectors because we believe that the logical development of vectors can be presented most successfully when the student has obtained some previous knowledge of analytic geometry by coordinate methods.

Some of the more difficult aspects of curve plotting, namely those concerned with finding the domain, range, and asymptotes, are deferred until Chapter 7, after the study of conics. This chapter also contains a section on nonlinear inequalities in the plane; the study of the solution of such inequalities gives the student additional practice in dealing with sets, plotting curves, and describing regions bounded by them. Chapter 8 begins with a section on the graphs of trigonometric functions in which periodicity is emphasized. We follow this with a special section on inverse functions and relations in general. Then there are sections on inverse trigonometric functions, as well as on exponential and logarithmic functions.

Chapter 12 is new to this second edition. We show that the basic formulas in analytic geometry of two and three dimensions can be extended by analogy to yield corresponding formulas for analytic geometry in four and more dimensions. We retain all the geometric terminology and we show how lines, planes, hyperplanes, and other geometric figures in higher dimensions may be defined and their properties explored in a logical manner. The outstanding beginning student who is seriously interested in studying mathematics should be attracted to this material.

As a result of our experience with the first edition of the text, we have clarified and improved the exposition of various sections. We have also introduced a large number of new and challenging problems; furthermore, most of the exercises in the first edition have been changed or modified.

Berkeley, California   M.H.P.
January 1975   C.B.M., Jr.

Contents

Chapter 1   INEQUALITIES

  1. The axioms of algebra 1

  2. Inequalities 3

  3. Proofs of the theorems in Section 2 11

  4. Absolute value 12

  5. Inequalities by factoring 16

Chapter 2   RELATIONS. FUNCTIONS. GRAPHS

  1. The number plane. Loci 21

  2. Functions. Functional notation 25

  3. Relations. Intercepts. Symmetry 31

  4. Intersections of loci 37

Chapter 3   THE LINE

  1. The distance formula. The midpoint formula 40

  2. The equations of certain loci 43

  3. Slope of a line 45

  4. Parallel and perpendicular lines 48

  5. The straight line 51

  6. Parametric equations of a line. Directed lines. Point of
     division formula 58

  7. Distance from a point to a line. Projections 65

  8. Families of lines 70

  9. Angle between two lines. Bisectors of angles 72

  10. Geometric theorems 76

  11. Linear inequalities in two unknowns 79

  12. Systems of linear inequalities; relation to linear
      programming 82

Chapter 4   VECTORS IN A PLANE

1. Directed line segments and vectors 85
2. Operations with vectors 88
3. Operations with vectors, continued 93

Chapter 5   THE CIRCLE

1. The equation of a circle 99
2. Tangents. Families of circles 102

Chapter 6   THE CONICS

1. The parabola 108
2. Tangents. Geometrical properties of a parabola 114
3. The ellipse 120
4. Geometrical properties of an ellipse. Directrices. Tangents 125
5. The hyperbola 130
6. The hyperbola: directrices, tangents, asymptotes 135
7. Translation of axes 140
8. Rotation of axes. The general equation of the second
   degree..145

Chapter 7   GRAPHS OF ALGEBRAIC RELATIONS

1. Loci of algebraic relations 155
2. Nonlinear inequalities in the plane 160

Chapter 8   THE TRANSCENDENTAL FUNCTIONS

1. Graphs of trigonometric functions 165
2. Inverse relations and functions 169
3. The inverse trigonometric functions 173
4. Exponential and logarithmic curves 176

Chapter 9   PARAMETRIC EQUATIONS, POLAR COORDINATES

1. Parametric equations 182
2. Polar coordinates 187
3. Graphs in polar coordinates....................................190
4. Equations in Cartesian and polar coordinates 195
5. Straight lines, circles, and conics 197

Chapter 10   SOLID ANALYTIC GEOMETRY

   1.   The number space R_3. Coordinates. The distance formula 202

   2.   Direction cosines and numbers 206

   3.   Equations of a line 212

   4.   The plane 216

   5.   Angles. Distance from a point to a line 220

   6.   The sphere. Cylinders 227

   7.   Quadric surfaces 230

   8.   Translation of axes 237

   9.   Other coordinate systems 241

  10.   Linear inequalities 244

Chapter 11   VECTORS IN THREE DIMENSIONS

   1.   Operations with vectors 248

   2.   Linear dependence and independence 253

   3.   The inner (scalar or dot) product................................257

   4.   The vector or cross product 261

   5.   Products of three vectors 268

Chapter 12   ANALYTIC GEOMETRY IN FOUR DIMENSIONS

   1.   The space R_4. The distance formula. Straight lines 272

   2.   Equations of a line 278

   3.   Two dimensional planes in Euclidean four space 285

   4.   Hyperplanes in four dimensional space 296

Appendices

   1.   The axioms of algebra 309

   2.   Natural numbers. Sequences. Extensions 316

   3.   Trigonometry review 322

   4.   Determinants 329

   5.   Proofs of Theorems 6, 12, and 13 of Chapter 11 338

   6.   Proofs of Theorems 8, 13(ii), and Lemma 2 of Chapter 16 342

   Answers to odd-numbered problems 347

   Index 365

Inequalities 1

▶1. **THE AXIOMS OF ALGEBRA**

Almost all high school students learn plane geometry as a single logical development in which theorems are proved on the basis of a system of axioms or postulates. Unlike plane geometry, however, algebra has traditionally been taught in high school without the aid of a formal logical system. In this method the student simply learns a few rules—or many—for manipulating algebraic quantities; these rules lead to success in solving problems but do not shed any light on the structure of algebra. In recent years, however, mathematicians have developed a number of new programs which present algebra in a logical manner analogous to the one first developed by Euclid for plane geometry.

Since many of the students taking this course in analytic geometry have learned algebra in the traditional way, we feel it is important to state a set of axioms upon which algebra (and consequently, analytic geometry) is based. It is a fact that the usual rules of algebra are logical consequences of the axioms given below, and we suggest that all students become familiar with these axioms. To establish the results given in the remainder of the text by a direct use of the axioms would make the proofs of the theorems cumbersome and unwieldy. Therefore in most proofs we shall employ the usual rules of manipulation which the student already knows. Those readers interested in the derivation of the rules of algebra from the axioms will find this development in the first two appendices (pp. 309–316).

Throughout this book we shall denote real numbers by letters or numerals. The word **equals** or its symbol $=$ will stand for the words "is the same as." The student should compare this with other uses for the symbol $=$ such as that in plane geometry where, for example, two line segments are said to be equal if they have the same length.

Although the axioms of algebra as given below apply to many different kinds of number systems, we shall understand in the statement of the axioms the word number to be *real* number.

Axioms of addition and subtraction

A–1. *Closure property.* *If a and b are numbers, there is one and only one number, denoted by a $+$ b, called their* **sum.**

1

A–2. *Commutative law.* *For every two numbers a and b, we have*

$$b + a = a + b.$$

A–3. *Associative law.* *For all numbers a, b, and c, we have*

$$(a + b) + c = a + (b + c).$$

A–4. *Existence of a zero.* *There is one and only one number 0, called **zero**, such that $a + 0 = a$ for every number a.*

A–5. *Existence of a negative.* *If a is any number, there is one and only one number x such that $a + x = 0$. This number is called the **negative** of a and is denoted by $-a$.*

Axioms of multiplication and division

M–1. *Closure property.* *If a and b are numbers, there is one and only one number, denoted by ab (or $a \times b$ or $a \cdot b$), called their **product.***

M–2. *Commutative law.* *For every two numbers a and b, we have*

$$ba = ab.$$

M–3. *Associative law.* *For all numbers a, b, and c, we have*

$$(ab) \cdot c = a \cdot (bc).$$

M–4. *Existence of a unit.* *There is one and only one number u, different from zero, such that $au = a$ for every number a. This number u is called the **unit** and (as is customary) will be denoted by 1.*

M–5. *Existence of a reciprocal.* *For each number a, different from zero, there is one and only one number x such that $ax = 1$. This number x is called the **reciprocal** of a and is denoted by a^{-1} (or $1/a$).*

In addition to the above axioms, we include one which relates both to addition and multiplication.

Axiom D. *Distributive law.* *For all numbers a, b, and c, we have*

$$a(b + c) = ab + ac.$$

Axioms M–1 through M–5 are obtained from the corresponding axioms A–1 through A–5 by replacing addition by multiplication and 0 by 1. The need for the restriction $a \neq 0$ in M–5 is explained in Appendix 1.

As we stated earlier, it is understood that the axioms hold for all real numbers, positive, negative, and zero. However, it is true that the above axioms hold as well for the system of all complex numbers, the system of rational numbers, and many other number systems. The axioms A, M, and D deal only with those laws of algebra concerned with addition, subtraction, multiplication, and division, including such operations for polynomials. These axioms do not, for example, imply the *existence* of a number whose square is 2. Since the rational numbers satisfy all these axioms and there is no rational number whose square is 2, Axioms A, M, and D are not sufficient to distinguish the real number system from the rational number system. An additional axiom, usually introduced in calculus, is required to guarantee the existence of irrational numbers and, in particular, a real number whose square is 2.

▶ 2. INEQUALITIES

In elementary algebra and geometry we study equalities almost exclusively. The solution of linear and quadratic equations, the congruence of geometric figures, and relationships among various trigonometric functions are topics concerned with equality. As we progress in the study of mathematics, however, we shall see that a knowledge of *inequalities* is both interesting and useful. (Moreover, facility in working with inequalities becomes indispensable in the study of calculus.)

We begin the topic of inequalities with the statement of an axiom and a number of fundamental theorems concerning inequalities. These theorems, which will be proved in the next section, enable us to find solutions to linear inequalities in one unknown as well as to linear systems of inequalities.

Axiom I. *Axiom of inequality. Among the real numbers there is a set called the* **positive numbers** *which satisfies the conditions:* (i) *for any number a exactly one of the three alternatives holds: a is positive or $a = 0$ or $-a$ is positive;* (ii) *any sum or product of positive numbers is positive.*

When Axiom I is added to those of the previous section, the resulting system of axioms is applicable only to those number systems which have a linear order. For example, the system of complex numbers does not satisfy Axiom I. However, both the real number system and the rational number system satisfy all the axioms given thus far.

DEFINITIONS. We say that a number a is **negative** whenever $-a$ is positive. If a and b are any real numbers, we say that $a > b$ (read a **is greater than** b) whenever $a - b$ is positive; we write $a < b$ (read a **is less than** b) whenever $a - b$ is negative.

Notation. The expression "if and only if," a technical one used frequently in mathematics, requires some explanation. Suppose A and B stand for propositions which may be true or false. To say that A *is true* **if** B *is true* means that the truth of B implies the truth of A. The statement A *is true* **only if** B *is true* means that the truth of A implies the truth of B. Thus the shorthand statement, "A is true if and only if B is true," is equivalent to the *double implication:* the truth of A implies and is implied by the truth of B. As a further shorthand notation we use the symbol $\leftrightarrow$ to represent "if and only if," and we write

$$A \leftrightarrow B$$

for the two implications stated above.*

The following theorem is an immediate consequence of Axiom I and the definitions based on that axiom.

Theorem 1. (i) $a > 0$ if and only if a is positive; (ii) $a < 0$ if and only if a is negative; (iii) $a > 0$ if and only if $-a < 0$; (iv) $a < 0$ if and only if $-a > 0$; (v) *if a and b are any numbers, then exactly one of the following three alternatives holds: $a > b$ or $a = b$ or $a < b$;* (vi) $a < b$ *if and only if* $b > a$.

The next four theorems give rules for manipulating inequalities. These theorems, often given as fundamental laws of manipulation in high school algebra courses, will be proved in the next section.

Theorem 2. *If $a > b$ and c is any real number, then*

$$a + c > b + c$$

and

$$a - c > b - c.$$

Theorem 3. (i) *If $a > 0$ and $b < 0$, then $a \cdot b < 0$;* (ii) *if $a < 0$ and $b < 0$, then $a \cdot b > 0$;* (iii) $1 > 0$.

Theorem 4. (i) *a and b are both positive or both negative if and only if $a \cdot b > 0$; they have opposite signs if $a \cdot b < 0$.* (ii) *If $a \neq 0$ and $b \neq 0$, then $a \cdot b$ and a/b have the same sign.*

Theorem 5. (i) *If $a > b$ and $c > 0$, then*

$$ac > bc \quad and \quad a/c > b/c.$$

(ii) *If $a > b$ and $c < 0$, then*

$$ac < bc \quad and \quad a/c < b/c.$$

(iii) *If a and b have the same sign and $a > b$, then*

$$1/a < 1/b.$$

(iv) *If $a > b$ and $b > c$, then $a > c$.*

---

* The term "necessary and sufficient" is frequently used as a synonym for "if and only if."

From Theorems 2 and 5, we obtain the following statements in a form familiar to many readers.

(1) *The direction of an inequality is unchanged if* (i) *the same number is added to both sides or* (ii) *both sides are multiplied or divided by a positive number.*
(2) *The direction of an inequality is reversed if both sides are multiplied or divided by a negative number.*

$$-3 \quad -2 \quad -1 \quad 0 \quad 1 \quad 2 \quad 3$$ **Figure 1–1**

From the geometric point of view we associate a horizontal axis with the totality of real numbers. The origin may be selected at any convenient point, with positive numbers to the right and negative numbers to the left (Fig. 1–1). For every real number there will be a corresponding point on the line and, conversely, every point will represent a real number. Then the inequality $a < b$ could be read: *a is to the left of b.* This geometric way of looking at inequalities is frequently of help in solving problems. It is also helpful to introduce the notion of an **interval** *of numbers* or *points.* If a and b are numbers (as shown in Fig. 1–2), then the **open interval from** a **to** b is the collection of all numbers which are both larger than a and smaller than b.

$$\overset{(\qquad\qquad)}{\underset{a \qquad\qquad b}{\rule{3cm}{0.4pt}}}$$ **Figure 1–2**

That is, an open interval consists of all numbers *between a and b.* A number x is between a and b if *both* inequalities $a < x$ and $x < b$ are true. A compact way of writing this is

$$a < x < b.$$

The **closed interval** from a to b consists of all the points between a and b, *including a and b* (Fig. 1–3).

$$\overset{[\qquad\qquad]}{\underset{a \qquad\qquad b}{\rule{3cm}{0.4pt}}}$$ **Figure 1–3**

Suppose a number x is either equal to a or larger than a, but we don't know which. We write this conveniently as $x \geq a$, which is read: x is **greater than or equal to** a. Similarly, $x \leq b$ is read: x **is less than or equal to** b, and means that x may be either smaller than b or may be b itself. A compact way of designating a closed interval from a to b is to state that it consists of all points x such that

$$a \leq x \leq b.$$

An interval which contains the endpoint b but not a is said to be **half-open on the left.** That is, it consists of all points x such that

$$a < x \leq b.$$

Similarly, an interval containing a but not b is called **half-open on the right,** and we write

$$a \leq x < b.$$

Parentheses and brackets are used as symbols for intervals in the following way:

> (a, b) for the open interval: $a < x < b$,
> $[a, b]$ for the closed interval: $a \leq x \leq b$,
> $(a, b]$ for the interval half-open on the left: $a < x \leq b$,
> $[a, b)$ for the interval half-open on the right: $a \leq x < b$.

We can extend the idea of an interval of points to cover some unusual cases. Suppose we wish to consider *all* numbers larger than 7. This may be thought of as an interval extending to infinity to the right. (See Fig. 1–4.) Of course, infinity is not a number, but we use the symbol $(7, \infty)$ to represent all numbers larger than 7. We could also write: all numbers x such that

$$7 < x < \infty.$$

In a similar way, the symbol $(-\infty, 12)$ will stand for all numbers less than 12. The double inequality

$$-\infty < x < 12$$

is an equivalent way of representing all numbers x less than 12.

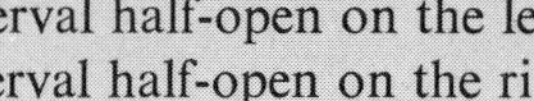

Figure 1–4

The first-degree equation $3x + 7 = 19$ has a unique solution, $x = 4$. The quadratic equation $x^2 - x - 2 = 0$ has two solutions, $x = -1$ and $x = 2$. The trigonometric equation $\sin x = \frac{1}{2}$ has an infinite number of solutions: $x = 30°, 150°, 390°, 510°, \ldots$ *The **solution** of an inequality involving a single unknown, say x, is the collection of all numbers which make the inequality a true statement.* Sometimes this is called the **solution set.** For example, the inequality

$$3x - 7 < 8$$

has as its solution set *all* numbers less than 5. To demonstrate this we argue in the following way. If x is a number which satisfies the above inequality we can, by Theorem 2, add 7 to both sides of the inequality and obtain a true statement. That is, we have

$$3x - 7 + 7 < 8 + 7, \qquad \text{or} \qquad 3x < 15.$$

Now, dividing both sides by 3 (Theorem 5), we obtain

$$x < 5,$$

and we observe that *if* x is a solution, *then* it is less than 5. Strictly speaking, however, we have not *proved* that every number which is less than 5 is a solution. In an actual proof we would begin by supposing that x is any number less than 5; that is,

$$x < 5.$$

We multiply both sides by 3 (Theorem 5) and then subtract 7 (Theorem 2) to get

$$3x - 7 < 8,$$

the original inequality. Since the condition that x is less than 5 implies the original inequality, we have proved the result. The important thing to notice is that the proof consisted of *reversing* the steps of the original argument which led to the solution $x < 5$ in the first place. So long as each of the steps we take is *reversible*, the above procedure is completely satisfactory so far as obtaining solutions is concerned. The step going from $3x - 7 < 8$ to $3x < 15$ is reversible, since these two inequalities are equivalent. Similarly, the inequalities $3x < 15$ and $x < 5$ are equivalent. Finally, note that we can say that the solution set consists of all numbers in the interval $(-\infty, 5)$.

By use of the symbol $\Leftrightarrow$ which stands for "if and only if," a solution to this example can be given in a compact form. We write

$$3x - 7 < 8 \Leftrightarrow 3x < 15 \qquad \text{(adding 7 to both sides)}$$

and

$$3x < 15 \Leftrightarrow x < 5 \qquad \text{(dividing both sides by 3).}$$

The solution set is the interval $(-\infty, 5)$.

We present a second illustration of the same technique.

EXAMPLE 1. Find the solution set of the inequality:

$$-7 - 3x < 5x + 29.$$

Solution.

$$-7 - 3x < 5x + 29 \Leftrightarrow -36 < 8x \qquad \text{(adding } 3x - 29 \text{ to both sides)}$$
$$\Leftrightarrow 8x > -36 \qquad \text{(Theorem 1)}$$
$$\Leftrightarrow x > -\tfrac{9}{2}.$$

The solution set is the interval $(-\tfrac{9}{2}, \infty)$.

Notation. It is convenient to introduce some terminology and symbols concerning sets. In general, a **set** is a collection of objects. The objects may have any character (numbers, points, lines, etc.) so long as we know which objects are in a given set and which are not. If S is a set and P is an object in it, we write $P \in S$ and say that P **belongs to** S or that P **is an element of** S. If S_1 and S_2 are two sets, their **union**, denoted by $S_1 \cup S_2$, consists of all objects each of which is in at least one of the two sets. The **intersection of S_1 and S_2**, denoted by $S_1 \cap S_2$, consists of all objects each of which is in both sets. Schematically, if S_1 is the horizontally shaded set (Fig. 1–5) and S_2 the vertically shaded set, then $S_1 \cup S_2$ consists of the entire shaded area and $S_1 \cap S_2$ consists of the doubly shaded area. Similarly, we may form the union and intersection of any number of sets. When we write $S_1 \cup S_2 \cup \cdots \cup S_7$ for the union of the seven sets $S_1, S_2, \ldots, S_7$, this union consists of all elements each of which is in at least one of the 7 sets. The intersection of these 7 sets is written $S_1 \cap S_2 \cap \cdots \cap S_7$. It may happen that two sets S_1 and S_2 have no elements in common. In such a case we say that their intersection is empty, and we use the term **empty set** for the set which has no members.

Most often we will deal with sets each of which is specified by some property or properties of its elements. For example, we may speak of the set of all even integers or the set of all rational numbers between 0 and 1. We employ the special symbol

$$\{x: x = 2n \quad \text{and} \quad n \text{ is an integer}\}$$

to represent the set of all even integers. In this notation the letter x stands for a generic element of the set, and the properties which determine membership in the set are listed to the right of the colon. The symbol

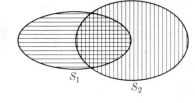

Figure 1–5

$$\{x: x \in (0, 1) \quad \text{and} \quad x \text{ is rational}\}$$

represents the rational numbers in the open interval $(0, 1)$. If a set has only a few elements, we may specify it by listing its members between braces. Thus the symbol $\{-2, 0, 1\}$ denotes the set whose elements are the numbers -2, 0, and 1. A set may be specified by any number of properties and we may use a variety of notations to determine these properties. If a set of objects has properties A, B, and C, we may denote this set by

$$\{P: P \text{ has the properties } A, B, \text{ and } C\}.$$

Other examples are

$$(0, 2) = \{x: 0 < x < 2\}, \qquad [2, 14) = \{t: 2 \le t < 14\}.$$

The last set is read as "the set of all numbers t such that t is greater than or equal to 2 and less than fourteen."

To illustrate the use of the symbols for set union and set intersection, we observe that

$$[1, 3] = [1, 2\tfrac{1}{2}] \cup [2, 3], \qquad (0, 1) = (0, \infty) \cap (-\infty, 1).$$

The words *and* and *or* have precise meanings when used in connection with sets and their properties. The set consisting of elements which have property A *or* property B is the *union* of the set having property A and the set having property B.

Symbolically,

$$\{x: x \text{ has property } A \text{ or property } B\}$$
$$= \{x: x \text{ has property } A\} \cup \{x: x \text{ has property } B\}.$$

The set consisting of elements which have property A *and* property B is the *intersection* of the set having property A with the set having property B. In set notation,

$$\{x: x \text{ has property } A \text{ and property } B\}$$
$$= \{x: x \text{ has property } A\} \cap \{x: x \text{ has property } B\}.$$

The next example illustrates the utility of this terminology.

EXAMPLE 2. Solve for x:

$$\frac{3}{x} < 5 \qquad (x \neq 0).$$

Solution. We have an immediate inclination to multiply both sides by x. However, since we don't know in advance whether x is positive or negative, we must proceed cautiously. We do this by considering two cases: (i) x is positive, and (ii) x is negative. Then the desired solution can be written

$$\left\{x: \frac{3}{x} < 5 \quad \text{and} \quad x > 0\right\} \cup \left\{x: \frac{3}{x} < 5 \quad \text{and} \quad x < 0\right\}.$$

Denoting

$$S_1 = \left\{x: \frac{3}{x} < 5 \quad \text{and} \quad x > 0\right\}, \qquad S_2 = \left\{x: \frac{3}{x} < 5 \quad \text{and} \quad x < 0\right\}$$

we see that the solution set is $S_1 \cup S_2$. Now

$$x \in S_1 \Leftrightarrow 3 < 5x \quad \text{and} \quad x > 0$$
$$\Leftrightarrow x > \tfrac{3}{5} \quad \text{and} \quad x > 0$$
$$\Leftrightarrow x > \tfrac{3}{5}.$$

Similarly,

$$x \in S_2 \Leftrightarrow 3 > 5x \quad \text{and} \quad x < 0$$
$$\Leftrightarrow x < \tfrac{3}{5} \quad \text{and} \quad x < 0$$
$$\Leftrightarrow x < 0.$$

The solution set (see Fig. 1–6) is $(\tfrac{3}{5}, \infty) \cup (-\infty, 0)$.

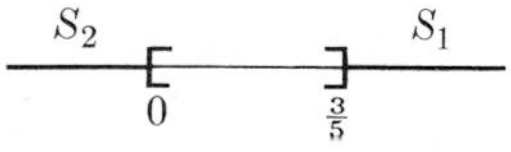

Figure 1–6

EXAMPLE 3. Solve for x:

$$\frac{2x - 3}{x + 2} < \frac{1}{3} \qquad (x \neq -2).$$

Solution. As in Example 2, the solution set is the union $S_1 \cup S_2$, where

$$S_1 = \left\{ x: \frac{2x - 3}{x + 2} < \frac{1}{3} \quad \text{and} \quad x + 2 > 0 \right\},$$

$$S_2 = \left\{ x: \frac{2x - 3}{x + 2} < \frac{1}{3} \quad \text{and} \quad x + 2 < 0 \right\}.$$

For numbers in S_1 we may multiply the inequality by $x + 2$ and, since $x + 2$ is positive, the direction of the inequality is preserved. Hence

$$\begin{aligned}
x \in S_1 &\Leftrightarrow 3(2x - 3) < x + 2 \quad \text{and} \quad x + 2 > 0 \\
&\Leftrightarrow 5x < 11 \quad \text{and} \quad x + 2 > 0 \\
&\Leftrightarrow x < \tfrac{11}{5} \quad \text{and} \quad x > -2 \\
&\Leftrightarrow x \in (-2, \tfrac{11}{5}).
\end{aligned}$$

For numbers in S_2, multiplication of the inequality by the negative quantity $x + 2$ reverses the direction. We have

$$\begin{aligned}
x \in S_2 &\Leftrightarrow 3(2x - 3) > x + 2 \quad \text{and} \quad x + 2 < 0 \\
&\Leftrightarrow 5x > 11 \quad \text{and} \quad x + 2 < 0 \\
&\Leftrightarrow x > \tfrac{11}{5} \quad \text{and} \quad x < -2.
\end{aligned}$$

Since there are no numbers x satisfying *both* conditions $x > \tfrac{11}{5}$ and $x < -2$, the set S_2 is empty.

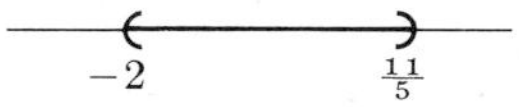

Figure 1–7

The solution set (Fig. 1–7) consists of $S_1 = (-2, \tfrac{11}{5})$.

PROBLEMS

In problems 1 through 6, find in each case the solution set, express this set as an interval, and plot.

1. $6x < 27 - 3x$

2. $9x - 17 < 8 + 4x$

3. $3(x + 2) > 11 - (x + 1)$

4. $3 + x < 7 - 3x$

5. $\dfrac{2x - 1}{6} < \dfrac{x + 2}{5}$

6. $\dfrac{3x + 5}{4} - 2 + \dfrac{x}{2} > \dfrac{3 - x}{4}$

In problems 7 through 10, find the solution of each pair of simultaneous inequalities.

7. $2x + 3 > 0$ and $3 - 2x > 0$

8. $x - \tfrac{3}{2} < 0$ and $2x + 5 > 4$

9. $2x + 1 > 0$ and $3x + 6 < 0$

10. $3x - 2 < 1$ and $2x + 4 < 0$

In problems 11 through 14, express each given combination of intervals as an interval. Plot a graph in each case.

11. $[-3, \infty) \cap (-\infty, -1)$

12. $(2, \infty) \cap (4, \infty)$

13. $[-2, 1) \cap (-1, 4)$

14. $(-1, 1) \cup (-2, 3)$

In problems 15 through 20, find the solution set.

15. $\dfrac{2}{x} < \dfrac{3}{4}$ 16. $\dfrac{x-1}{x} < 2$ 17. $\dfrac{2x+3}{2-x} < 3$

*18. $\dfrac{x+1}{2x-3} < \dfrac{1}{2}$ 19. $x^2 - 1 > 0$ 20. $x^2 + x - 6 > 0$

Using the theorems of this section, prove the following statements.

21. If $a > b$ and $c > d$, then $a + c > b + d$.
22. If $a > b > 0$ and $c > d > 0$, then $a \cdot c > b \cdot d$.
23. If n is a positive integer and $a > 1$, then $a^n \geq a$.
24. If $0 < b < 1$, then $b^n \leq b$.

▶3.   PROOFS OF THE THEOREMS IN SECTION 2

We now restate and prove Theorems 2 through 5 of Section 2.

Theorem 2. *If $a > b$ and c is any real number, then*

$$a + c > b + c,$$
$$a - c > b - c.$$

Proof. Since $a > b$, it follows from the definition that $a - b > 0$. However

$$(a + c) - (b + c) = a - b$$

and so $(a + c) - (b + c) > 0$; we conclude that $a + c > b + c$. Similarly, $(a - c) - (b - c) = a - b > 0$; therefore $a - c > b - c$.

Theorem 3. (i) *If $a > 0$ and $b < 0$, then $a \cdot b < 0$;* (ii) *if $a < 0$ and $b < 0$, then $a \cdot b > 0$;* (iii) $1 > 0$.

Proof. (i) The hypothesis implies that $-b > 0$. Then from the laws of signs† we see that $a \cdot (-b) = -(a \cdot b) > 0$. Hence, from Theorem 1, we conclude that $a \cdot b < 0$.
(ii) The hypothesis implies that $(-a) > 0$ and $(-b) > 0$, so that by Axiom I, $(-a) \cdot (-b) = a \cdot b > 0$.
(iii) By Axiom M–4, we note that $1 \neq 0$. If 1 were negative, then $1 \cdot 1$ would equal 1 and be positive by part (ii). Thus 1 is positive and so $1 > 0$.

---

* Throughout the text an asterisk indicates an unusually difficult problem.
† The laws of signs $a \cdot (-b) = (-a) \cdot b$ and $(-a) \cdot (-b) = a \cdot b$ are proved in Appendix 1 on the basis of the axioms. Therefore, these laws are true for all real numbers, whether positive, negative, or zero.

Theorem 4. (i) *a and b are both positive or both negative if and only if* $a \cdot b > 0$; *they have opposite signs if* $a \cdot b < 0$. (ii) *If* $a \neq 0$ *and* $b \neq 0$, *then* $a \cdot b$ *and* a/b *have the same sign.*

Proof. (i) This result follows from Axiom I and part (ii) of Theorem 3. (ii) We observe that $(a/b) \cdot (ab) = a^2 > 0$, and so by (i) a/b and ab must have the same sign.

Theorem 5. (i) *If* $a > b$ *and* $c > 0$, *then* $ac > bc$ *and* $a/c > b/c$. (ii) *If* $a > b$ *and* $c < 0$, *then* $ac < bc$ *and* $a/c < b/c$.

Proof. We note that $a > b$ is equivalent to the statement $a - b > 0$. If $c > 0$, we make use of (i) of Theorem 4. The remainder of the proof is left to the student (see problem 23, Section 4).

▶ **4. ABSOLUTE VALUE**

If a is a real number, we define the **absolute value** of a, denoted by $|a|$, by the conditions

$$|a| = a \quad \text{if} \quad a > 0,$$
$$|a| = -a \quad \text{if} \quad a < 0,$$
$$|0| = 0.$$

For example,

$$|7| = 7, \qquad |-13| = 13, \qquad |2 - 5| = |-3| = 3.$$

Algebraic manipulations with absolute values are facilitated by the results of the next theorem.

Theorem 6

(i) $|a| \geq 0, |-a| = |a|,$ and $|a|^2 = a^2$.
(ii) $|a \cdot b| = |a| \cdot |b|$ and, if $b \neq 0$, $|a/b| = |a|/|b|$.
(iii) $|a| = |b| \Leftrightarrow a = \pm b$.
(iv) *For positive numbers b, we have*

$$|a| < b \Leftrightarrow -b < a < b.$$

Proof. Parts (i) and (ii) are simple consequences of the definition of absolute value. To prove (iii) we first observe that if $a = \pm b$, then it follows from (i) that $|a| = |b|$. Also, if $|a| = |b|$, then $|a|^2 = |b|^2$ so that from (i) again $a^2 = b^2$ and $a = \pm b$. To establish (iv), we note that the solution set of the inequality

$|x| < b$ is the union of the sets S_1 and S_2, where

$$S_1 = \{x: |x| < b \text{ and } x \geq 0\}, \qquad S_2 = \{x: |x| < b \text{ and } x < 0\}.$$

Then

$$x \in S_1 \Leftrightarrow |x| < b \text{ and } x \geq 0$$
$$\Leftrightarrow x < b \text{ and } x \geq 0 \qquad (\text{since } |x| = x \text{ if } x \geq 0)$$
$$\Leftrightarrow x \in [0, b).$$

Similarly,

$$x \in S_2 \Leftrightarrow |x| < b \text{ and } x < 0$$
$$\Leftrightarrow -x < b \text{ and } x < 0 \qquad (\text{since } |x| = -x \text{ if } x < 0)$$
$$\Leftrightarrow x \in (-b, 0).$$

The result of (iv) follows since $S_1 \cup S_2 = (-b, b)$.

EXAMPLE 1. Find the solution set of the equation

$$\left| \frac{x + 2}{2x - 5} \right| = 3.$$

Solution. Using the result of (ii) in Theorem 6, we have

$$\left| \frac{x + 2}{2x - 5} \right| = 3 \Leftrightarrow \frac{|x + 2|}{|2x - 5|} = 3 \qquad (x \neq \tfrac{5}{2})$$
$$\Leftrightarrow |x + 2| = 3|2x - 5|.$$

Now we employ (iii) of Theorem 6 to get

$$\left| \frac{x + 2}{2x - 5} \right| = 3 \Leftrightarrow x + 2 = \pm 3(2x - 5)$$
$$\Leftrightarrow x + 2 = 3(2x - 5) \quad \text{or} \quad x + 2 = -3(2x - 5)$$
$$\Leftrightarrow -5x = -17 \quad \text{or} \quad 7x = 13$$
$$\Leftrightarrow x = \tfrac{17}{5} \quad \text{or} \quad \tfrac{13}{7}.$$

The solution set consists of the two numbers $\tfrac{17}{5}$, $\tfrac{13}{7}$ which, written in set notation, is $\{\tfrac{17}{5}, \tfrac{13}{7}\}$.

EXAMPLE 2. Find the solution set of the inequality $|3x - 4| \leq 7$.

Solution. Using parts (iii) and (iv) of Theorem 6, we have

$$|3x - 4| \leq 7 \Leftrightarrow -7 \leq 3x - 4 \leq 7.$$

Adding 4 to each portion of this double inequality, we find

$$|3x - 4| \leq 7 \Leftrightarrow -3 \leq 3x \leq 11.$$

Dividing by 3, we obtain

$$|3x - 4| \leq 7 \Leftrightarrow -1 \leq x \leq \tfrac{11}{3}.$$

The solution set is the interval $[-1, \tfrac{11}{3}]$.

EXAMPLE 3. Solve for x:

$$\left| \frac{2x - 5}{x - 6} \right| < 3.$$

Solution. Proceeding as in Example 2, we see that

$$\left| \frac{2x - 5}{x - 6} \right| < 3 \Leftrightarrow -3 < \frac{2x - 5}{x - 6} < 3 \qquad (x \neq 6).$$

The solution set consists of the union of S_1 and S_2, where

$$S_1 = \left\{ x : -3 < \frac{2x - 5}{x - 6} < 3 \quad \text{and} \quad x - 6 > 0 \right\},$$

$$S_2 = \left\{ x : -3 < \frac{2x - 5}{x - 6} < 3 \quad \text{and} \quad x - 6 < 0 \right\}.$$

For numbers in S_1, we find

$$x \in S_1 \Leftrightarrow -3(x - 6) < 2x - 5 < 3(x - 6) \quad \text{and} \quad x - 6 > 0.$$

Considering the three inequalities separately, we may write

$$x \in S_1 \Leftrightarrow -3x + 18 < 2x - 5 \quad \text{and} \quad 2x - 5 < 3x - 18 \quad \text{and} \quad x - 6 > 0$$
$$\Leftrightarrow 23 < 5x \quad \text{and} \quad 13 < x \quad \text{and} \quad x > 6$$
$$\Leftrightarrow 13 < x.$$

Thus $x \in S_1 \Leftrightarrow x \in (13, \infty)$. Similarly, for $x \in S_2$ we see that multiplication of an inequality by the negative quantity $x - 6$ reverses the direction. Therefore,

$$x \in S_2 \Leftrightarrow -3(x - 6) > 2x - 5 > 3(x - 6) \quad \text{and} \quad x - 6 < 0$$
$$\Leftrightarrow 23 > 5x \quad \text{and} \quad 13 > x \quad \text{and} \quad x - 6 < 0$$
$$\Leftrightarrow x < \tfrac{23}{5}.$$

Hence $x \in S_2 \Leftrightarrow x \in (-\infty, \tfrac{23}{5})$. The solution set consists of $S_1 \cup S_2$ (see Fig. 1–8).

Figure 1–8

We now establish the following important theorem and corollary.

Theorem 7. *If a and b are any numbers, then*

$$|a + b| \leq |a| + |b|.$$

Proof. * We consider three cases.

Case 1. The numbers a and b are both positive. Then $|a| = a$, $|b| = b$, and $|a + b| = a + b$. The conclusion of the theorem is satisfied, since $|a + b| = |a| + |b|$.

Case 2. The numbers a and b are both negative. Then $|a| = -a$, $|b| = -b$, and $|a + b| = -(a + b)$. As in Case 1, we have $|a + b| = |a| + |b|$.

Case 3. One number, say a, is positive, and the other, b, is negative. Then $|a| = a$, $|b| = -b$, and we have $|a + b|$ as either $a + b$ or $-(a + b)$, depending on whether $a + b$ is larger or smaller than zero. But we know that

$$a + b < a - b = |a| + |b|,$$

and

$$-(a + b) < a - b = |a| + |b|.$$

Therefore, we have shown that in all possible circumstances

$$|a + b| \leq |a| + |b|.$$

Corollary. *If a and b are any numbers, then*

$$|a - b| \leq |a| + |b|.$$

Proof. We write $a - b$ as $a + (-b)$, and apply Theorem 1 to obtain

$$|a - b| = |a + (-b)| \leq |a| + |-b| = |a| + |b|.$$

The final equality holds since, from (i) of Theorem 6, it is always true that $|-b| = |b|$.

PROBLEMS

In problems 1 through 22, find the solution set.

1. $|x - 2| = \frac{1}{2}$
2. $|x + 1| = 1$
3. $|3x + 1| = 8$
4. $|2x - 3| = 5$
5. $|x - 1| = |2x + 6|$
6. $|4 - 3x| = |x|$
7. $\left| \dfrac{2x + 3}{3x + 2} \right| = 2$
8. $|2x - 3| = |2x + 5|$

---

* A shorter although less intuitive proof is given by the following argument. Since $|a| = a$ or $-a$, we may write $-|a| \leq a \leq |a|$, and similarly, $-|b| \leq b \leq |b|$. Adding these inequalities (see problem 21, Section 2), we get $-(|a| + |b|) \leq a + b \leq |a| + |b|$. The conclusion of Theorem 7 is equivalent to this double inequality.

9. $|x|^2 = |x|$

10. $|x^2 - 6| = |x|$

11. $|x - 1| < \frac{1}{2}$

12. $|x + 3| < \frac{1}{4}$

13. $\left|\dfrac{x + 2}{2x - 1}\right| < 1$

14. $\left|\dfrac{3 - x}{1 - 2x}\right| < 3$

15. $\left|\dfrac{2x - 1}{x + 3}\right| < 2$

16. $\left|\dfrac{3 - 2x}{x + 1}\right| < 2$

17. $|x - 2| < |2 + x|$

18. $\left|\dfrac{2x - 3}{x - 6}\right| < 4$

19. $\left|\dfrac{x + 4}{1 - x}\right| \le 3$

20. $|1 - 2x| < |x + 5|$

21. $|x + 2| \le |2x - 8|$

22. $|2x + 3| < 2|x - 1|$

23. Prove Theorem 5.

24. Prove that for any numbers a and b, $|a| - |b| \le |a - b|$.

25. Given that a and b are positive and c and d are negative, and $a > b, c > d$. Show that

$$\frac{a}{c} < \frac{b}{d}.$$

26. If $a_1, a_2, \ldots, a_n$ are any numbers, show that $|a_1 + a_2 + \cdots + a_n| \le |a_1| + |a_2| + \cdots + |a_n|$.

27. If a, b, c, d are any numbers, show that $|a + b| - |c| - |d| \le |a - c + b - d|$.

▶5. INEQUALITIES BY FACTORING

We wish to determine the solution set of inequalities involving polynomial expressions. For example, we consider the inequality

$$(x - 2)(x - 1)(x + \tfrac{1}{2}) > 0;$$

we want to find those values of x which make the inequality valid. First, we observe that the values $x = 2, 1, -\frac{1}{2}$ are *not* in the solution set since they make the left-hand side of the inequality zero (Fig. 1–9). For all other values of x the left-hand side of the above inequality is either positive or negative.

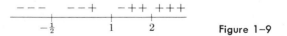

Figure 1–9

It is convenient to proceed geometrically and examine the behavior of $(x - 2)(x - 1)(x + \frac{1}{2})$ in each of the intervals which separate the zeros of this expression. If $x > 2$, we see that $(x - 2)$ is positive, as are $(x - 1)$ and $(x + \frac{1}{2})$.

Thus we have the product of three positive quantities, which is positive. This fact is shown in Fig. 1–9 by the three plus signs above the interval $(2, \infty)$. We conclude that the inequality $(x - 2)(x - 1)(x + \frac{1}{2}) > 0$ holds for $(2, \infty)$. In the interval $1 < x < 2$, we note that $(x - 2)$ is negative, $(x - 1)$ is positive, and $(x + \frac{1}{2})$ is positive. We indicate this fact by placing two positive signs and one negative sign above the interval $(1, 2)$, as shown in Fig. 1–9. The law of signs states that $(x - 2)(x - 1)(x + \frac{1}{2})$ is negative in this interval. Proceeding to the intervals $(-\frac{1}{2}, 1)$ and $(-\infty, -\frac{1}{2})$, we get the signs $- - +$ and $- - -$, respectively. The solution set of the inequality

$$(x - 2)(x - 1)(x + \tfrac{1}{2}) > 0$$

is the set $S = (-\frac{1}{2}, 1) \cup (2, \infty)$.

More generally, suppose we have an inequality of the form

$$A(x - a_1)(x - a_2)(x - a_3) \cdots (x - a_n) > 0,$$

where A and $a_1, a_2, \ldots, a_n$ are numbers. For convenience we place the a_i, $i = 1, 2, \ldots, n$, in decreasing order (see Fig. 1–10), and we allow two or more of the a_i to coincide.

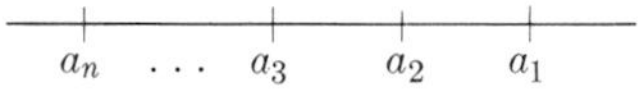

Figure 1–10

The number A may be positive or negative. It is clear that $x = a_1, x = a_2, \ldots, x = a_n$ are values for which the inequality does *not* hold since each of these numbers makes the polynomial expression zero. For each of the intervals in between the numbers $a_1, a_2, \ldots, a_n$, we examine the sign of each factor $(x - a_i)$, use the law of signs, and determine whether or not the inequality is valid. The solution set is the union of the intervals in which the inequality holds. We illustrate with another example.

EXAMPLE 1. Solve the inequality

$$2x^2 - x > 6.$$

Solution. We rearrange the inequality so that a polynomial is on the left and zero is on the right. We get

$$2x^2 - x > 6 \Leftrightarrow 2x^2 - x - 6 > 0.$$

We factor the polynomial, getting

$$2x^2 - x > 6 \Leftrightarrow (2x + 3)(x - 2) > 0.$$

Next we write the factor $2x + 3$ in the form $2(x + \frac{3}{2})$ and rearrange the terms so that the a_i are in decreasing order. We get

$$2x^2 - x > 6 \Leftrightarrow 2(x - 2)(x - (-\tfrac{3}{2})) > 0.$$

We draw a line indicating the values 2, $-\frac{3}{2}$ as shown in Fig. 1–11 and we use the law of signs to determine the validity of the inequality in each interval.

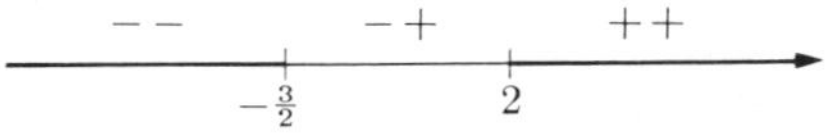

Figure 1–11

We conclude that the solution set is

$$(-\infty, -\tfrac{3}{2}) \cup (2, \infty).$$

It is a fact, although we have not proved it, that in each of the intervals separating the zeros of a polynomial, the polynomial maintains one sign. Also for values of x above a_1, the polynomial cannot change sign. The same is true for values of x below a_n. We now work another example.

EXAMPLE 2. Determine the solution set of the inequality $2x^5 - 3x^4 > -x^3$.

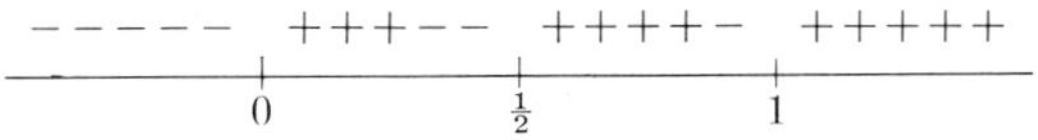

Figure 1–12

Solution. We have

$$2x^5 - 3x^4 > -x^3 \Leftrightarrow 2x^5 - 3x^4 + x^3 > 0$$
$$\Leftrightarrow x^3(2x - 1)(x - 1) > 0$$
$$\Leftrightarrow 2(x - 1)(x - \tfrac{1}{2})x^3 > 0.$$

Here $a_1 = 1$, $a_2 = \frac{1}{2}$, $a_3 = a_4 = a_5 = 0$. The numbers 1, $\frac{1}{2}$, 0 are marked off as shown in Fig. 1–12, and the signs of the factors in each of the intervals are shown. The solution set is $(0, \frac{1}{2}) \cup (1, \infty)$.

If $P(x)$ and $Q(x)$ are polynomials, then the quotient $P(x)/Q(x)$ and the product $P(x)Q(x)$ are always both positive or both negative. This fact is a restatement of Theorem 4, part (ii), which asserts that if $a \neq 0$, $b \neq 0$, then ab and a/b always have the same sign. We conclude that

$$\left\{ \text{the solution set of } \frac{P(x)}{Q(x)} > 0 \right\} = \left\{ \text{the solution set of } P(x)Q(x) > 0 \right\}.$$

In this way, the solution of inequalities involving the division of polynomials can always be reduced to a problem in polynomial inequalities.

EXAMPLE 3. Solve for x: $\dfrac{210}{3x - 2} < \dfrac{50}{x}$.

Solution. We have

$$\frac{210}{3x-2} < \frac{50}{x} \Leftrightarrow \frac{210}{3x-2} - \frac{50}{x} < 0$$

$$\Leftrightarrow \frac{60x+100}{x(3x-2)} < 0$$

$$\Leftrightarrow \frac{20(3x+5)}{x(3x-2)} < 0$$

$$\Leftrightarrow 20(3x+5)x(3x-2) < 0 \qquad \text{(changing the quotient of poly-}$$
$$\text{nomials to the product of poly-}$$
$$\text{nomials)}$$

$$\Leftrightarrow 180(x - \tfrac{2}{3})x(x - (-\tfrac{5}{3})) < 0.$$

$$\begin{array}{cccc} --- & +-- & ++- & +++ \\ \hline & -\frac{5}{3} & 0 & \frac{2}{3} \end{array}$$

Figure 1–13

Here we have $a_1 = \tfrac{2}{3}$, $a_2 = 0$, $a_3 = -\tfrac{5}{3}$. The signs of the factors are indicated in Fig. 1–13. The solution set is $(-\infty, -\tfrac{5}{3}) \cup (0, \tfrac{2}{3})$.

PROBLEMS

Determine the solution sets of the following inequalities.

1. $x^2 - 2x - 3 < 0$

2. $x^2 + x - 6 < 0$

3. $2x^2 + 3x + 1 < 0$

4. $7x^2 - 6x - 1 > 0$

5. $3x^5 - 2x^4 < x^3$

6. $3x^6 - 2x^5 < x^4$

7. $x^2 < 9$

8. $x^2 > 9$

9. $x^2 < 10$

10. $x^2 + 4x + 1 < 0$ (complete the square)

11. $x^2 + 4x + 5 < 0$

12. $(x + 2)(x - 1)^2 > 0$

13. $(x^2 - 4)(x^2 - 9) < 0$

14. $x^3 + 3x^2 > 2x + 6$

15. $\dfrac{2x - 1}{3x} < \dfrac{4x - 2}{5x}$

16. $\dfrac{x}{2} - \dfrac{2x^2 - 2}{3x} + \dfrac{1}{3} < \dfrac{5x}{6}$

17. $\dfrac{4}{x - 1} + \dfrac{x - 2}{5} < \dfrac{3}{x - 1} + 1$

18. $\dfrac{2x - 3}{2x - 5} < \dfrac{3x - 5}{3x}$

19. $\dfrac{x + 1}{x^2 - 5x + 4} > 0$

20. $\dfrac{x + 1}{x^2 + 5x} > 0$

21. $x^2 + \dfrac{2}{x} < 3$

22. $\dfrac{2x - 3}{x^2 - 4} - \dfrac{1}{x} < 0$

23. $\dfrac{4x + 4}{2x - 1} - \dfrac{1}{2} > \dfrac{3x + 3}{2x + 5}$

24. $\dfrac{1}{x^2 - 2x - 3} < \dfrac{1}{12}$

25. $\dfrac{x + 1}{x^2 + 2x + 2} < \dfrac{2}{5}$

26. If $P(x)$, $Q(x)$, $R(x)$, and $S(x)$ are polynomials, prove that the solution set of the inequality

$$\frac{P(x)}{Q(x)} + \frac{R(x)}{S(x)} > 0$$

is the same as the solution set of the inequality

$$P(x)Q(x)S^2(x) + Q^2(x)R(x)S(x) > 0.$$

Relations. Functions. Graphs. 2

▶ 1. THE NUMBER PLANE. LOCI

In Chapter 1, we saw that the study of inequalities is greatly illuminated when we represent real numbers by means of points on a straight line. In the study of inequalities involving two unknowns, we shall see that a corresponding geometric interpretation is both useful and important.

The determination of solution sets of equations and inequalities in two or more unknowns is an important topic in mathematics, especially from the point of view of applications. The geometric representation of a solution set, called the *graph*, is particularly helpful in the study of analytic geometry.

We now introduce some terminology which will be used throughout the text. It is essential that the student become familiar with these terms as quickly as possible. *The set of all real numbers is denoted by R_1.* Any two real numbers a and b form a **pair.** For example, 2 and 5 are a pair of numbers. When the *order* of the pair is prescribed, we say that we have an **ordered pair.** The ordered pair 2 and 5 is different from the ordered pair 5 and 2. We usually designate an ordered pair in parentheses with a comma separating the first and second elements of the pair: (a, b) is used for the ordered pair consisting of the numbers a (first) and b (second).

DEFINITIONS. The set of all ordered pairs of real numbers is called the **number plane** and is denoted by R_2. Each individual ordered pair is called a **point** in the number plane. The two elements in a number pair are called its **coordinates.**

The number plane can be represented on a geometric or Euclidean plane. It is important that we keep separate the concept of the number plane, R_2, which is an abstract system of ordered pairs, from the concept of the geometric plane, which is the two-dimensional object we studied in Euclidean geometry.

In a Euclidean plane we draw a horizontal and a vertical line, denote them as the x and y axes, respectively, and label their point of intersection O (Fig. 2–1). The point O is called the **origin.** We select a convenient unit of length and, starting from the origin as zero, mark off a number scale on the horizontal axis, positive to the right and negative to the left. Similarly, we insert a scale along the vertical

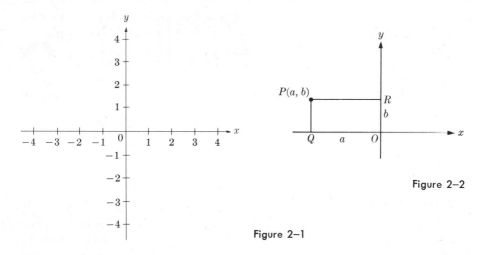

Figure 2–2

Figure 2–1

axis, with positive numbers extending upward and negative ones downward. It is not necessary that the units along the horizontal axis have the same length as the units along the vertical axis.

We now set up a one-to-one correspondence between the points of the number plane, R_2, and the points of the Euclidean (geometric) plane. For each point P in the Euclidean plane we construct perpendiculars from P to the coordinate axes, as shown in Fig. 2–2. The intersection with the x axis is at the point Q, and the intersection with the y axis is at the point R. The distance from the origin to Q (measured positive if Q is to the right of O and negative if Q is to the left of O) is denoted by a. Similarly, the distance OR is denoted by b. Then the point in the number plane corresponding to P is (a, b). It is easy to see that, conversely, to each point in the number plane there corresponds exactly one point in the Euclidean plane. Because of the one-to-one correspondence between the number plane and the plane of Euclidean geometry, we will frequently find it convenient to use geometric terms for the number plane. For example, a "line" in the number plane is used for the set of points corresponding to a line in the geometric plane.

The description above suggests the method to be used in plotting (that is, representing geometrically) points of the number plane. When plotting points, the coordinates are enclosed in parentheses adjacent to the point, as illustrated in Fig. 2–3.

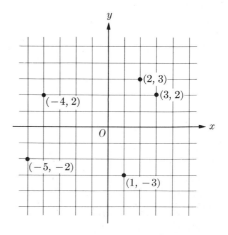

Figure 2–3

DEFINITIONS. The **solution set,** or **locus,** of an equation in two unknowns consists of all points in the number plane, R_2, whose coordinates satisfy the equation. A geometric representation of the locus (that is, the actual drawing) is called the **graph of the equation.**

The solution set of an equation in two unknowns is, of course, a set in R_2. It is a simple matter to extend the set notation already introduced to include sets in the number plane. If S denotes the set of all (x, y) in R_2 having certain properties, which, for example, we call A and B, we describe S as follows:

$$S = \{(x, y): (x, y) \text{ has properties } A \text{ and } B\}.$$

As an illustration, if S is the solution set of the equation

$$2x^2 - 3y^2 = 6,$$

we write

$$S = \{(x, y): 2x^2 - 3y^2 = 6\}.$$

To construct the exact graph of the solution set of some equation is generally impossible, since it would require the plotting of infinitely many points. Usually, in drawing a graph, we select enough points to exhibit the general nature of the locus, plot these points, and then approximate the remaining points by drawing a "smooth curve" through the points already plotted.

It is useful to have a systematic method of choosing points on the locus. We can do this for an equation in two unknowns when we can solve for one of the unknowns in terms of the other. Then, from the formula obtained, we can assign values to the unknown in the formula and obtain values of the unknown for which we solved. The corresponding values are then tabulated as shown in the examples below.

EXAMPLE 1. Sketch a graph of the solution set of the equation $2x + 3y = 5$. Use set notation to describe this set.

Solution. The solution set S in set notation is

$$S = \{(x, y): 2x + 3y = 5\}.$$

To sketch a graph, we first solve for one of the unknowns, say x. We find $x = \frac{5}{2} - \frac{3}{2}y$. Assigning values to y, we obtain the table:

x	7	$\frac{11}{2}$	4	$\frac{5}{2}$	1	$-\frac{1}{2}$	-2
y	-3	-2	-1	0	1	2	3

Plotting these points, we see that they appear to lie on the straight line shown in Fig. 2–4.

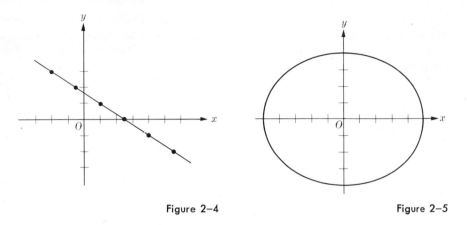

Figure 2–4   Figure 2–5

EXAMPLE 2. Sketch a graph of the locus of the equation

$$\frac{x^2}{25} + \frac{y^2}{16} = 1.$$

Use set notation to describe this set.

Solution. $S = \{(x, y): (x^2/25) + (y^2/16) = 1\}$. Solving for y in terms of x, we obtain

$$y = \pm\tfrac{4}{5}\sqrt{25 - x^2}.$$

Since x enters into the equation only in the term involving x^2, we may abbreviate the table as indicated below.

x	0	±1	±2	±3	±4	±5
y	±4	$\pm\tfrac{4}{5}\sqrt{24}$	$\pm\tfrac{4}{5}\sqrt{21}$	$\pm\tfrac{16}{5}$	$\pm\tfrac{12}{5}$	0
y (approx.)	±4	±3.92	±3.67	±3.2	±2.4	0

Plotting these points, we see that they appear to lie on the oval curve of Fig. 2–5. Note that each column except the first and last really represents four points. What happens if we choose $x > 5$ or $x < -5$?

PROBLEMS

In problems 1 through 28, sketch a graph of the locus of each of the equations. Describe each locus using set notation as in the examples.

1. $y = 2x$ 2. $2x + y = 3$ 3. $2y - x = 5$
4. $2x + 3y = 6$ 5. $x = -2$ 6. $y = 3$
7. $2x + 3y = 0$ 8. $3x + 4y = 12$ 9. $y = \tfrac{1}{3}x^2$
10. $x = \tfrac{1}{3}y^2$ 11. $y = x^2 - 1$ 12. $x = y^2 - 4$

13. $y = x^2 + 2x - 3$ 14. $y = x^2 - 4x + 5$

15. $x^2 + y^2 = 16$ 16. $x^2 + y^2 = 4$

17. $9x^2 + 16y^2 = 144$ 18. $y^2 - x^2 = 4$

19. $3x^2 - 2y^2 = 6$ 20. $y^2 - x^2 + 4 = 0$

21. $y = -x^2 - 2x + 3$ 22. $y = -\frac{1}{8}x^3$

23. $y = x^3 - x$ 24. $y = -x^3 + 3x + 2$

25. $3x^2 - 2xy - y^2 = 0$ [*Hint*: Write as a quadratic in y.]

26. $3x^2 + 2xy + y^2 = 6$ 27. $x^2 + 2xy + y^2 + 2x - 2y = 3$

28. $x^2 - 3x + 2 = 0$

29. Show that the locus of the equation $x^2 - 2x + y^2 + 9 = 0$ is the empty set.

30. Describe the set of points which lie on the loci of both of the equations

$$x^2 + y^2 = 16 \quad \text{and} \quad y = 2x + 4.$$

▶2. FUNCTIONS. FUNCTIONAL NOTATION

In mathematics and many of the physical sciences, simple formulas occur repeatedly. For example, if r is the radius of a circle and A is its area, then

$$A = \pi r^2.$$

If heat is added to an ideal gas in a container of fixed volume, the pressure p and the temperature T satisfy the relation

$$p = a + cT$$

where a and c are fixed numbers with values depending on the properties of the gas, the units used, and so forth.

The relationships expressed by these formulas are simple examples of the concept of function, to be defined precisely later. However, it is not essential that a function be associated with a particular formula. As an example, we consider the cost C in cents of mailing a first-class letter which weighs x ounces. Since postal regulations state that the cost is "10c per ounce or fraction thereof," we can construct the table:

Weight x in ounces	$0 < x \le 1$	$1 < x \le 2$	$2 < x \le 3$	$3 < x \le 4$	$4 < x \le 5$
Cost C in cents	10	20	30	40	50

This table could be continued until $x = 320$, the maximum weight permitted by postal regulations. To each value of x between 0 and 320 there corresponds a precise cost C. We have here an example of a function relating x and C.

It frequently happens that an experimenter finds by measurement that the numerical value y of some quantity depends in a *unique* way upon the measured value x of some other quantity. It is usually the case that no known formula expresses the relationship between x and y. All we have is the set of ordered pairs (x, y). In such circumstances, the entire interconnection between x and y is determined by the ordered pairs. We are led to the following definition.

DEFINITION. A **function** is a set of ordered pairs (x, y) of real numbers in which no two pairs have the same first element. In other words, to each value of x (the first member of the pair) there corresponds exactly one value of y (the second member). The set of all values of x which occur is called the **domain** of the function, and the set of all y which occur is called the **range** of the function.

An example of a function is given by the set of all pairs (r, A) obtained from the formula $A = \pi r^2$ when $r > 0$. The domain of this function is the half-infinite interval $(0, \infty)$. The range is $(0, \infty)$.

The set of ordered pairs (T, p) obtained from the formula $p = a + cT$ is also an example of a function. The domain of T and the range of p will depend on the particular conditions of the gas, the container, and so forth. Although no simple formula is available, the set of ordered pairs (x, C) as given in Table 1 determines a function, the so-called "postage function."

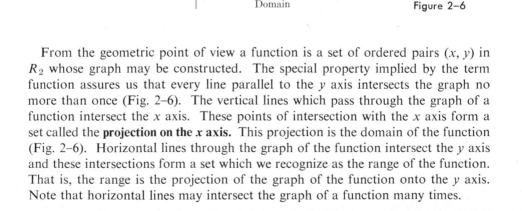

Figure 2–6

From the geometric point of view a function is a set of ordered pairs (x, y) in R_2 whose graph may be constructed. The special property implied by the term function assures us that every line parallel to the y axis intersects the graph no more than once (Fig. 2–6). The vertical lines which pass through the graph of a function intersect the x axis. These points of intersection with the x axis form a set called the **projection on the x axis.** This projection is the domain of the function (Fig. 2–6). Horizontal lines through the graph of the function intersect the y axis and these intersections form a set which we recognize as the range of the function. That is, the range is the projection of the graph of the function onto the y axis. Note that horizontal lines may intersect the graph of a function many times.

It is important to be able to discuss functions and their properties without actually specifying the particular ones we have in mind. For this purpose we introduce a symbol, a letter of the alphabet, to stand for a function. The letters most often used are f, g, F, G, φ, Φ. Sometimes, if a problem concerns many different functions, subscripts are employed, so that, for example, f_1, f_2, and f_3 would stand for three different functions.

Suppose that f is a function and x is a number in its domain. To each number x the function f associates a unique number which we denote by y. We also use the symbol $f(x)$ (to be read f of x) as a synonym for the number y. That is, $y = f(x)$. For example, if f is the postage function and the weight of a letter is $3\frac{1}{2}$ ounces then, from Table 1, $f(3\frac{1}{2}) = 40$. In set notation,* we describe a function f by the statement

$$f = \{(x, y): x \text{ is in the domain of } f \text{ and } y = f(x)\}.$$

In other words, a function f is just the solution set of the equation $y = f(x)$, and a graph of f is the graph of this equation.

When specifying a function f, we must give its domain and a precise rule for determining the value of $f(x)$ for each x in the domain. For the most part, we shall give functions by means of formulas such as

$$f(x) = x^2 - x + 2.$$

Such a formula, by itself, does not give the domain of x. Both in this case and in general, we shall take it for granted that if the domain is not specified, then any value of x may be inserted in the formula so long as the result makes sense. The domain shall consist of the set of all such values x.

In prescribing a function by means of a formula, the particular letter used is usually of no importance. The function F determined by the formula

$$F(x) = x^3 - 2x^2 + 5$$

is identical with the function determined by

$$F(t) = t^3 - 2t^2 + 5.$$

The difference is one of notation only.

EXAMPLE 1. Suppose that f is the function defined by the equation

$$f(x) = x^2 - 2x - 3.$$

Find $f(0)$, $f(-1)$, $f(-2)$, $f(2)$, $f(3)$, $f(t)$, and $f(f(x))$. Plot a graph of f for the portion of the domain in $-2 \leq x \leq 3$.

---

* Several symbols for function are now in common use. One such is $f : x \rightarrow y$ where x is a generic element of the domain of f and y is the element of the range which is the *image* of x. A second notation for function is $f : D_1 \rightarrow D_2$ where D_1 is the set forming the domain of f and D_2 is the set forming the range.

Solution. We have

$$f(0) = 0^2 - 2 \cdot 0 - 3 = -3,$$

$$f(-1) = (-1)^2 - 2(-1) - 3 = 0,$$

$$f(-2) = (-2)^2 - 2 \cdot (-2) - 3 = 5,$$

$$f(2) = 2^2 - 2 \cdot 2 - 3 = -3,$$

$$f(3) = 3^2 - 2 \cdot 3 - 3 = 0,$$

$$f(t) = t^2 - 2t - 3.$$

The difficult part is finding $f(f(x))$, and here a clear understanding of the meaning of the symbolism is needed. The formula defining f means that whatever is in the parenthesis in $f(\ )$ is substituted in the right side. That is

$$f(f(x)) = (f(x))^2 - 2 \cdot (f(x)) - 3.$$

However, the right-hand side again has $f(x)$ in it, and we can substitute to get

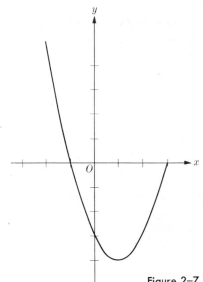

Figure 2–7

$$f(f(x)) = (x^2 - 2x - 3)^2 - 2(x^2 - 2x - 3) - 3$$

$$= x^4 - 4x^3 - 4x^2 + 16x + 12.$$

To plot the graph we compute

$$f(1) = 1^2 - 2 \cdot 1 - 3 = -4$$

and assemble all the results above to obtain the table

x	-2	-1	0	1	2	3
$f(x) = y$	5	0	-3	-4	-3	0

The graph is plotted in Fig. 2–7.

Sometimes successions of parentheses become unwieldy, and we may use brackets or braces with the same meaning: $f[f(x)]$ is the same as $f(f(x))$.

If we write $g(x) = x^3 + 2x - 6$, $-2 \le x \le 3$, this means that the domain of the function g is the interval $[-2, 3]$. If, in the same formula, the portion $-2 \le x \le 3$ were omitted, we would have assumed that g is defined by that formula for *all* x; the domain would be $(-\infty, \infty)$. This opens up many possibilities. For example, we could define a function F by the following conditions (the function F defined in this way has the interval $[-1, 5]$ for its domain):

$$F(x) = \begin{cases} x^2 + 3x - 2, & -1 \le x < 2, \\ x^3 - 2x + 4, & 2 \le x \le 5. \end{cases}$$

A formula may fail to have meaning for certain values of x. In such a case the domain of the function given by this formula cannot include these particular values of x. The student should be able to find such values and note that they are excluded from the domain. For example, if f_1 is defined by the formula

$$f_1(x) = \frac{1}{x - 3},$$

then $f_1(x)$ is not defined for $x = 3$ since division by zero is always excluded. If f_2 is defined by

$$f_2(x) = \sqrt{4 - x^2},$$

its domain is the interval $[-2, 2]$ since complex numbers are excluded.

EXAMPLE 2. Given that
$$f(x) = x^2,$$
show that
$$f(x^2 + y^2) = f[f(x)] + f[f(y)] + 2f(x)f(y).$$

Solution
$$f(x^2 + y^2) = (x^2 + y^2)^2 = x^4 + 2x^2y^2 + y^4,$$
$$f[f(x)] = f[x^2] = (x^2)^2 = x^4,$$
$$f[f(y)] = f[y^2] = (y^2)^2 = y^4,$$
$$2f(x)f(y) = 2x^2y^2.$$

Adding the last three lines, we obtain
$$f[f(x)] + f[f(y)] + 2f(x)f(y) = x^4 + y^4 + 2x^2y^2,$$
which is just $f(x^2 + y^2)$.

PROBLEMS

1. Given that $f(x) = x^2 + 3x - 3$, find $f(-4)$, $f(-3)$, $f(-2)$, $f(-1)$, $f(0)$, $f(1)$, $f(2)$, $f(a + 2)$. Plot a graph of the equation $y = f(x)$ for $-4 \leq x \leq 2$.

2. Given that $f(x) = \frac{1}{3}x^3 + 2x - 3$, find $f(-3), f(-2), f(-1), f(0), f(1), f(2), f(3)$, and $f(a - 1)$. Plot a graph of the equation $y = f(x)$ for $-3 \leq x \leq 3$.

3. Given that
$$f(x) = \frac{x - 1}{2x - 3},$$
find $f(-4)$, $f(-3)$, $f(-2)$, $f(-1)$, $f(0)$, $f(1)$, $f(-1000)$, $f(+1000)$. Is $\frac{3}{2}$ in the domain of f? Plot a graph of f for x on $[-4, 1]$ using the values above and additional values of x near $\frac{3}{2}$.

4. Given that

$$f(x) = \frac{x - 2}{3x + 1},$$

find $f(x)$ for

$$x = -1000, -3, -2, -1, 0, 1, 2, 3, 1000$$

and find $f[f(x)]$. Is $-\frac{1}{3}$ in the domain of f? Plot a graph of f for x on $[-3, 3]$ using the values above and additional values near $-\frac{1}{3}$.

5. Given that

$$f(x) = \frac{4x}{4x^2 + 1},$$

find $f(x)$ for

$$x = -1000, -3, -2, -1, 0, 1, 2, 3, 1000.$$

Show that $f(-x) = -f(x)$ for all values of x. Plot a graph of f for x on $[-3, 3]$.

6. Given that

$$f(x) = \frac{2 - x}{2 + x},$$

show that

$$\frac{f(x) - f(y)}{1 + f(x)f(y)} = \frac{2(y - x)}{4 + xy}$$

whenever both are defined.

7. Given $f(x) = \sqrt{x + 3}$, find $f(-3), f(-2), f(-1), f(0), f(1)$. What is the domain of f? Plot its graph.

8. Given $f(x) = |x + 1|$, what is the domain of f? Find $f(x)$ for $x = -3, -2, -1, 0, 1, 2, 3$ and plot the graph of f on $[-3, 3]$.

9. Given $f(x) = 2 - |x - 1|$, find $f(x)$ for $x = -2, -1, 0, 1, 2, 3, 4$ and plot a graph of f for x on $[-2, 4]$.

10. Given

$$f(x) = \frac{|x - 1|}{|x + 1|},$$

what is the domain of f? Find $f(0), f(-2), f(1), f(2), f(3)$. Plot the graph of f.

11. Given

$$f(x) = \frac{x + 3}{|x^2 - 4|},$$

what is the domain of f? Find $f(x)$ for $x = -3, -1, 0, 1, 3$. Plot the graph.

12. Given $f(x) = x^2 + x - 2$, $g(y) = y^3 - 1$, find the values of

$$f[g(0)], f[g(1)], f[g(t)].$$

13. Given $f(x) = \sqrt{(x - 2)(x - 4)}$, what is the domain of f? Plot a graph of f for x on $[-2, 6]$, using several values of x near 2 and 4.

14. Given $f(x) = \sqrt{1 - 4x - x^2}$, what is the domain of f? [*Hint*: Complete the square under the radical.] Plot a graph of f.

In problems 15 through 20, find and simplify the value of

$$\frac{f(x + h) - f(x)}{h}$$

assuming $h \neq 0$.

15. $f(x) = \frac{1}{2}x^2$ 16. $f(x) = x^2 + 2x + 3$

17. $f(x) = 2x^3 - 2$ 18. $f(x) = \dfrac{1}{(x + 1)}$

19. $f(x) = \dfrac{1}{(x - 1)^2}$ 20. $f(x) = \sqrt{x + 1},\ x > -1$

▶3. RELATIONS. INTERCEPTS. SYMMETRY

Many of the loci discussed in Section 1 are not functions. It can be seen in Example 2 on page 24 that the locus of $(x^2/25) + (y^2/16) = 1$ is not the graph of a function. We simply observe in Fig. 2–5 that each vertical line for $x \in (-5, 5)$ intersects the locus *twice* and not once, as is required for functions. For convenience in discussing loci which are not functions, we introduce the following terminology.

DEFINITIONS. Any set in the number plane, R_2, is called a **relation.** More specifically, a set in R_2 is called a **relation from R_1 to R_1.** The **domain** of a relation is the set of all x such that (x, y) is in the set forming the relation. The set of all y such that (x, y) is in the relation is the **range** of the relation.

As in the case for functions, the domain of a relation is its projection on the x axis. The range of a relation is its projection on the y axis. We observe that, unlike the situation in the case of functions, a vertical line may intersect a relation any number of times. Indeed, such a line may intersect it in infinitely many points. After all, according to the definition above, the entire number plane is a relation.

We shall be interested in relations in which a vertical line intersects the relation in only a finite number of points. Under these circumstances, the relation is com-

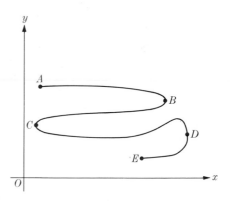

Figure 2–8

posed of the union of a finite number of functions. As Fig. 2–8 shows, the curve going from A to E is the locus of a relation. However, the arc $\overset{\frown}{AB}$ is the locus of a function, as are the arcs $\overset{\frown}{BC}$, $\overset{\frown}{CD}$, and $\overset{\frown}{DE}$. The relation is the union of these four functions.

The solution set of an equation in two unknowns is a relation. A function which forms part of this solution set or relation is said to be **defined implicitly** by this equation.

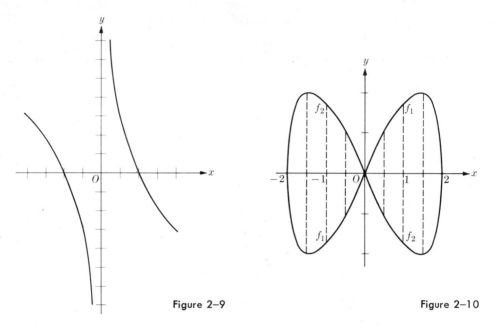

Figure 2–9   Figure 2–10

EXAMPLE 1. Given the equation

$$x^2 + xy - 4 = 0,$$

determine whether or not the solution set S is a function. Plot the graph of S. If S is not a function show how it can be represented as the union of several functions.

Solution. Solving the above equation for y, we find

$$x^2 + xy - 4 = 0 \Leftrightarrow y = \frac{4 - x^2}{x} = \frac{4}{x} - x.$$

The set S is the solution set of the equation $y = (4/x) - x$. Since there is only one value of y for each value of x, S is a function. In fact, S is the function f defined by

$$f(x) = \frac{4 - x^2}{x}.$$

To plot the graph we make the table

x	-4	-3	-2	-1	$-\frac{1}{2}$	$\frac{1}{2}$	1	2	3	4
$f(x) = y$	3	$\frac{5}{3}$	0	-3	$-\frac{15}{2}$	$\frac{15}{2}$	3	0	$-\frac{5}{3}$	-3

A smooth curve is drawn through the points plotted from the table. Since f is not defined for $x = 0$, we computed $f(-\frac{1}{2})$ and $f(\frac{1}{2})$ to get an indication of the behavior of f near $x = 0$. (See Fig. 2–9.)

EXAMPLE 2. Discuss the locus of the equation

$$x^4 - 4x^2 + y^2 = 0.$$

Plot the graph of the solution set S. If S is not a function, show how it can be represented as the union of several functions.

Solution. Solving for y, we obtain

$$x^4 - 4x^2 + y^2 = 0 \Leftrightarrow y = \pm x\sqrt{4 - x^2}.$$

Then S is a relation which is the union of two functions f_1 and f_2 defined by

$$f_1(x) = x\sqrt{4 - x^2}, \qquad f_2(x) = -x\sqrt{4 - x^2}.$$

We construct the table

x	± 2	$\pm\frac{3}{2}$	± 1	$\pm\frac{1}{2}$	0
y	0	$\pm\frac{3}{4}\sqrt{7}$	$\pm\sqrt{3}$	$\pm\frac{1}{4}\sqrt{15}$	0
y approx.	0	± 1.98	± 1.73	± 0.97	0

The graph is shown in Fig. 2–10. We note that f_1 and f_2 are defined implicitly by the equation $x^4 - 4x^2 + y^2 = 0$.

Sketching the locus of an equation by plotting a few of its points and drawing a smooth curve through them is an adequate method if the equation is sufficiently simple. However, if the equation is at all complicated, this method not only may be too laborious but, in some cases, may even lead to incorrect graphs. For example, we may try to plot the graph of the equation

$$y = \frac{1}{2x - 1}$$

by letting x take on a sequence of integer values. That is, we construct the table

x	-3	-2	-1	0	1	2	3
y	$-\frac{1}{7}$	$-\frac{1}{5}$	$-\frac{1}{3}$	-1	1	$\frac{1}{3}$	$\frac{1}{5}$

Plotting these points and drawing a smooth curve through them leads to the *incorrect* graph shown in Fig. 2–11. However, noticing that $x = \frac{1}{2}$ is not in the domain of the function f given by $f(x) = 1/(2x - 1)$ and assigning to x several values near $\frac{1}{2}$ leads to the correct graph shown in Fig. 2–12.

In this section we discuss several facts about loci which are easily obtained from the equation and which are substantial aids in constructing quick, accurate graphs. Other helpful devices are discussed in Chapter 7.

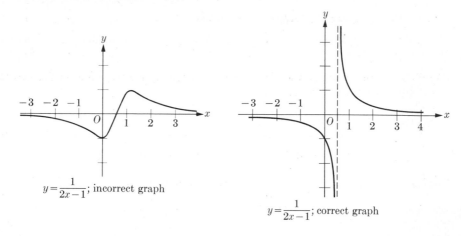

Figure 2–11   Figure 2–12

First, it is useful to know where the locus crosses the axes. A point at which a locus crosses the x axis is called an **x intercept**; a point where it crosses the y axis is called a **y intercept.** It may happen that a locus does not intersect a given axis or intersects it many times. To locate the intercepts, if any, we have the rule:

to find the x intercepts, set $y = 0$ and solve for x;
to find the y intercepts, set $x = 0$ and solve for y.

EXAMPLE 3. Find the x and y intercepts of the locus of

$$x^2 - 3y^2 + 6x - 3y = 7.$$

Solution. Setting $y = 0$, we get

$$x^2 + 6x - 7 = 0 \quad \text{or} \quad x = -7 \quad \text{or} \quad 1.$$

The x intercepts are $(-7, 0)$ and $(1, 0)$.
    Setting $x = 0$, we get

$$3y^2 + 3y + 7 = 0 \quad \text{or} \quad y = \frac{-3 \pm \sqrt{-75}}{6}.$$

Since the solutions for y are complex, there are no y intercepts and the locus does not intersect the y axis.

In constructing the graphs of the loci discussed in Section 1, we see that some of them have certain kinds of symmetry. For example, Fig. 2–5 exhibits several obvious symmetries. If we had known about these symmetries before drawing the graph, the actual sketching would have been much easier. The statement that a locus is **symmetric with respect to the x axis*** $\Leftrightarrow$ whenever a point (a, b) is on the locus, the point $(a, -b)$ is also on the locus. A locus is **symmetric with respect to the y axis** $\Leftrightarrow$ whenever (a, b) is on the locus, the point $(-a, b)$ is also on the locus. We also consider a third type of symmetry. A locus is **symmetric with respect to the origin** $\Leftrightarrow$ whenever (a, b) is on the locus, $(-a, -b)$ is also on the locus.

Note that a curve may be symmetric with respect to the origin without being symmetric with respect to either axis. Figure 2–9 shows such a locus. In Fig. 2–10, we have an example of a curve which is symmetric with respect to both axes, and hence automatically symmetric with respect to the origin.

To test for various kinds of symmetry, we state the following rules:

(i) *Replace y by $-y$ in the equation of the locus; if the resulting equation is equivalent to the original* (i.e., has the same locus), *then the locus is symmetric with respect to the x axis.*

(ii) *Replace x by $-x$ in the equation of the locus; if the resulting equation is equivalent to the original, the locus is symmetric with respect to the y axis.*

(iii) *Replace x by $-x$ and y by $-y$ in the equation of the locus; if the resulting equation is equivalent to the original, the locus is symmetric with respect to the origin.*

The rules for symmetry reduce the work involved in sketching the graph of an equation. If a curve is symmetric with respect to the x axis, then only the portion above the x axis has to be plotted carefully. The remainder of the graph is obtained as the mirror image in the x axis of the part already sketched. A similar procedure works if the locus is symmetric with respect to the y axis.

EXAMPLE 4. Test the locus of the equation

$$9x^2 + 16y^2 = 144$$

for symmetry, find its intercepts, and determine whether the locus is or is not a function. If not, find the functions of which it is the union. Sketch the graph.

Solution. (i) Replacing y by $-y$ in the equation yields

$$9x^2 + 16(-y)^2 = 144$$
$$\Leftrightarrow 9x^2 + 16y^2 = 144.$$

Therefore the locus is symmetric with respect to the x axis. (ii) Replacing x by $-x$ leads similarly to an equivalent equation, so the locus is symmetric with respect to the y axis. (iii) Replacing x by $-x$ and y by $-y$ leads to an equivalent equation, so the locus is

---

* The symbol $\leftrightarrow$ stands for "if and only if." For ease in reading the student may translate the symbol as "is equivalent to." Also, in a definition, the symbol is used to connect the definition with the term being defined. In these definitions we may read "$\Leftrightarrow$" as "means that."

symmetric with respect to the origin. Setting $y = 0$, we get

$$9x^2 = 144$$

or

$$x = \pm 4,$$

and the x intercepts are $(4, 0)$ and $(-4, 0)$. Setting $x = 0$, we find that the y intercepts are $(0, \pm 3)$. Solving for y, we obtain

$$9x^2 + 16y^2 = 144$$

$$\Leftrightarrow y = \pm\tfrac{3}{4}\sqrt{16 - x^2},$$

so the locus is the union of the two functions f_1 and f_2:

$$f_1(x) = \tfrac{3}{4}\sqrt{16 - x^2}$$

and

$$f_2(x) = -\tfrac{3}{4}\sqrt{16 - x^2}.$$

The graph is sketched in Fig. 2–13.

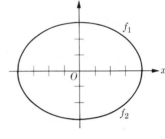

Figure 2–13

EXAMPLE 5. Test the locus of the equation

$$y = \tfrac{1}{8}(x^3 - 4x)$$

for symmetry and find its intercepts. Sketch the graph.

Solution. Replacing y by $-y$ leads to the equation

$$-y = \tfrac{1}{8}(x^3 - 4x)$$

or

$$y = -\tfrac{1}{8}(x^3 - 4x),$$

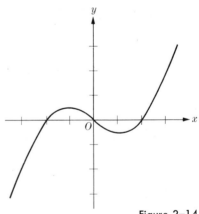

which is not equivalent to the original. There-fore the locus is not symmetric with respect to the x axis. Similarly, replacing x by $-x$, we

Figure 2–14

see that it is not symmetric with respect to the y axis. However, when we replace x by $-x$ and y by $-y$, we get the equation

$$-y = \tfrac{1}{8}[(-x)^3 - 4(-x)] = -\tfrac{1}{8}(x^3 - 4x),$$

which *is* equivalent to the original equation. Therefore the locus is symmetric with respect to the origin. Setting $y = 0$, we find that the x intercepts are $(-2, 0)$, $(0, 0)$, and $(2, 0)$. Letting $x = 0$, we see that the y intercept is $(0, 0)$. A graph of a portion of the locus is shown in Fig. 2–14.

PROBLEMS

In problems 1 through 25, discuss each locus with respect to intercepts and symmetry. When the locus is not a function, express it as the union of several functions. Sketch the graph.

1. $y = x^2 - 9$

2. $y = 9 - x^2$

3. $y^2 = 4x$

4. $y^2 = -3x$

5. $y^2 = 2x - 2$

6. $y^2 + 3x = 2$

7. $x = y^2 - 4$

8. $9x^2 + 4y^2 = 36$

9. $3x^2 + 4y^2 = 12$

10. $xy + 2 = 0$

11. $y^2 + 4xy = 4$

12. $y^2 + 2x = 3$

13. $y^2 = x^2 + 9$

14. $y^2 = 4x^2 - 9$

15. $2x^2 + 2xy + y^2 = 5$

16. $y = 3x + x^3$

17. $y = \frac{1}{2}x^3 + 1$

18. $x^2 = y^3$

19. $y^2 + 2y + 2 = 3x$

20. $y^3 + 2x = 2$

21. $y^2 = 4x^4$

22. $y^2 + 2y = 2x + 5$

23. $y^4 + 2y^2 = x^2 - 2x$

24. $x^2 + 4y^2 = 16$

25. $x^4 + 4x^2 - 4y^2 = 0$

26. Discuss for intercepts and plot the locus of the equation $2x = y + \frac{1}{2}y^3$. Show that if (x_1, y_1) and (x_2, y_2) are on the locus and $y_1 < y_2$, then $x_1 < x_2$. [*Hint*: Consider the cases $0 \le y_1 < y_2$, then $y_1 < 0 < y_2$, and finally $y_1 < y_2 \le 0$.] Would you say that the locus is a function?

27. Suppose that a locus is symmetric with respect to the origin and symmetric with respect to the x axis. Is it symmetric with respect to the y axis? Justify your answer.

28. Discuss the locus of $|y - 1| = |x - 2|$ for symmetry and intercepts. Sketch the graph.

29. Draw the graph of $y = \dfrac{x - 3}{|x - 3|}$. Find the intercepts, if any.

30. Discuss the locus of $y = \dfrac{|x - 2|}{|x + 3|}$ for intercepts and symmetry. Sketch the graph.

▶ 4. INTERSECTIONS OF LOCI

Suppose that we have two equations, each of which has a locus. In many problems we seek the point or points of intersection of these loci. If a point is on both loci, its coordinates must satisfy *both* equations. Consequently, *the points of intersection may be found by solving the equations simultaneously*. We illustrate the method with two examples.

EXAMPLE 1. Find the points of intersection of the solution sets of the equations

$$x + 2y = 4 \quad \text{and} \quad y^2 = x + 4.$$

Sketch the two loci on the same graph.

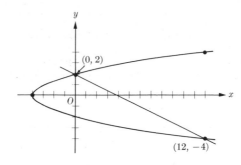

Figure 2-15

Solution. We seek the set S determined by the two conditions

$$S = \{(x, y): x + 2y = 4 \quad \text{and} \quad y^2 = x + 4\}.$$

The first equation may be rewritten as $x = 4 - 2y$. Substituting this value of x in the second equation, we obtain

$$y^2 = 8 - 2y \quad \text{or} \quad y^2 + 2y - 8 = 0.$$

We conclude that $y = 2$ or -4. Since $x = 4 - 2y$, the corresponding values of x are 0 and 12. The student easily verifies that S consists of the points $(0, 2)$ and $(12, -4)$. Graphs of the loci are sketched in Fig. 2–15.

EXAMPLE 2. Find the solution set S determined by

$$S = \{(x, y): 2x^2 - 3y^2 = 5 \quad \text{and} \quad 4x^2 + 9y^2 = 25\}.$$

Sketch the loci on the same graph.

Solution. Suppose that (x, y) is a point of S. Multiplying the equation $2x^2 - 3y^2 = 5$ by 3 and adding the result to $4x^2 + 9y^2 = 25$, we get

$$10x^2 = 40,$$

so that

$$x = \pm 2.$$

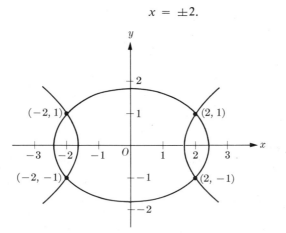

Figure 2-16

Next, substituting the values $x = 2, -2$ in either of the equations, we get the corresponding values for y. For example, letting $x = 2$ in the first equation, we find

$$-3y^2 = -3 \quad \text{or} \quad y = \pm 1.$$

Similarly, $x = -2$ yields the values $y = \pm 1$. We find

$$S = \{(2, 1), (2, -1), (-2, 1), (-2, -1)\}.$$

Both loci are plotted in Fig. 2–16.

PROBLEMS

In each of the following problems, find all the points of intersection of the solution sets of the given equations. Then sketch both loci on the same graph, identifying the points of intersection.

1. $y = 2x$
   $y = x^2$

2. $y = 2x$
   $2y = 4x^2 - 3$

3. $y = 2x - 2$
   $y^2 + 4x = 12$

4. $3x - y = 5$
   $x^2 + y^2 = 25$

5. $x + 2y = 5$
   $y^2 = 3x + 1$

6. $2x + y = 5$
   $y^2 + x = 3$

7. $2x + y = 5$
   $xy = 2$

8. $x - y + 1 = 0$
   $y = x^2 - 1$

9. $x^2 + y^2 = 13$
   $2x^2 + y^2 = 17$

10. $x + 3y + 1 = 0$
    $2x^2 + 3y^2 = 11$

11. $3x^2 + 2y^2 = 59$
    $2x^2 - 3y^2 = -30$

12. $x^2 - 2xy = 1$
    $x + y = 1$

13. $2x^2 + 3y^2 = 12$
    $x^2 - 2y^2 = -1$

14. $2x + y = 3$
    $x^2 + y^2 + 5y = 0$

15. $y^2 = 6x$
    $y^4 + 18x^2 = y^2$

16. $y = |x - 1|$

    $2y = x + 4$

*17. $y = \dfrac{x + 2}{|x + 2|}$

    $2y = 2x + 5$

3 The Line

▶ 1. THE DISTANCE FORMULA. THE MIDPOINT FORMULA

In Chapter 2 we showed how to establish a one-to-one correspondence between the points of the number plane, R_2, and the points of the Euclidean plane. Suppose that the unit of distance in the Euclidean plane is given. If, in setting up the one-to-one correspondence, we choose the unit distance along both the x and y axes to be equal to the unit distance in the geometric plane, we say that we have constructed a **Cartesian coordinate system.**

Suppose we have a Cartesian coordinate system and that P_1 with coordinates (x_1, y_1) and P_2 with coordinates (x_2, y_2) are two points in the plane (Fig. 3–1). The distance between any two points in the plane is the length of the line segment joining them. We shall derive a formula for this distance in terms of the coordinates of the two points. For this purpose, construct the lines P_1Q_1 and P_2Q_2 parallel to the y axis as shown in Fig. 3–1. Similarly, let P_1R_1 and P_2R_2 be the lines parallel to the x axis through P_1 and P_2, respectively. The point Q_1, the intersection of P_1Q_1 with the x axis, has coordinates $(x_1, 0)$. Similarly, Q_2 has coordinates $(x_2, 0)$. We denote by S the point of intersection of P_1R_1 and P_2Q_2. The coordinates of S are (x_2, y_1).

In general, *we shall denote the distance between two points A and B by the symbol* $|AB|$. Then from plane geometry we have

$$|P_1S| = |Q_1Q_2|,$$
$$|SP_2| = |R_1R_2|.$$

Also, from the Pythagorean Theorem, we find

$$|P_1S|^2 + |SP_2|^2 = |P_1P_2|^2,$$

and so

$$|P_1P_2|^2 = |Q_1Q_2|^2 + |R_1R_2|^2.$$

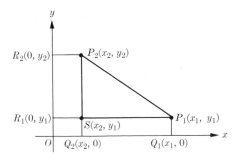

Figure 3–1

Referring to Fig. 3–1, we see that $|Q_1Q_2| = |x_2 - x_1|$ and $|R_1R_2| = |y_2 - y_1|$. We conclude that

$$|P_1P_2|^2 = (x_2 - x_1)^2 + (y_2 - y_1)^2.$$

Since the distance between two points is always positive, we take the positive square root and obtain for the length of the segment P_1P_2,

$$|P_1P_2| = \sqrt{(x_2 - x_1)^2 + (y_2 - y_1)^2}.$$

EXAMPLE 1. Find the distance between the points $P_1(3, 1)$ and $P_2(5, -2)$.

Solution. We substitute in the distance formula and get

$$d = \sqrt{(5 - 3)^2 + (-2 - 1)^2} = \sqrt{4 + 9} = \sqrt{13}.$$

EXAMPLE 2. The point $P_1(5, -2)$ is 4 units away from a second point P_2, whose y coordinate is 1. Locate the point P_2.

Solution. The point P_2 will have coordinates $(x_2, 1)$. From the distance formula we have the equation

$$4 = \sqrt{(x_2 - 5)^2 + (1 - (-2))^2}.$$

To solve for x_2, we square both sides and obtain

$$16 = (x_2 - 5)^2 + 9 \quad \text{or} \quad x_2 - 5 = \pm\sqrt{7}.$$

There are two possibilities for x_2:

$$x_2 = 5 + \sqrt{7}, \quad 5 - \sqrt{7}.$$

In other words, there are two points P_2, one at $(5 + \sqrt{7}, 1)$ and the other at $(5 - \sqrt{7}, 1)$, which have y coordinate 1 and which are 4 units from P_1.

Let $P_1(x_1, y_1)$ and $P_2(x_2, y_2)$ be any two points in the plane. We shall show how to find the coordinates $(\bar{x}, \bar{y})$ of the midpoint $\bar{P}$ of the line segment joining P_1 and P_2. We drop perpendiculars to the x axis from $P_1, \bar{P}$, and P_2 as shown in Fig. 3–2. The intersections of these perpendiculars with the x axis are labeled $Q_1, \bar{Q}$, and Q_2 respectively. The coordinates of these points are shown in Fig. 3–2. From plane geometry we know that $\bar{Q}$ is the midpoint of the segment joining Q_1 and Q_2; that is,

$$|Q_1\bar{Q}| = |\bar{Q}Q_2|$$

or

$$|\bar{x} - x_1| = |x_2 - \bar{x}|.$$

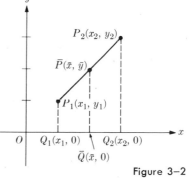

Figure 3–2

Now we have

$$|\bar{x} - x_1| = |x_2 - \bar{x}| \Leftrightarrow (\bar{x} - x_1)^2 = (x_2 - \bar{x})^2$$
$$\Leftrightarrow \bar{x}^2 - 2x_1\bar{x} + x_1^2 = x_2^2 - 2x_2\bar{x} + \bar{x}^2$$
$$\Leftrightarrow 2(x_2 - x_1)\bar{x} = x_2^2 - x_1^2.$$

When $x_1 \neq x_2$, the last equation holds if and only if

$$\bar{x} = \frac{x_1 + x_2}{2}.$$

The case $x_1 = x_2$ leads to the same formula since the points P_1 and P_2 are on a vertical line and $\bar{x} = x_1 = x_2$. Repeating the argument using y coordinates, we obtain

$$\bar{y} = \frac{y_1 + y_2}{2}.$$

EXAMPLE 3. Locate the midpoint of the line segment joining the points $P(3, -2)$ and $Q(-4, 5)$.

Solution. From the above formula,

$$\bar{x} = \frac{3 - 4}{2} = -\frac{1}{2} \quad \text{and} \quad \bar{y} = \frac{-2 + 5}{2} = \frac{3}{2}.$$

EXAMPLE 4. Find the length of the line segment joining the point $A(7, -2)$ to the midpoint of the line segment between the points $B(4, 1)$ and $C(3, -5)$.

Solution. The midpoint of the line segment between B and C is at

$$\bar{x} = \frac{4 + 3}{2} = \frac{7}{2}, \quad \bar{y} = \frac{1 - 5}{2} = -2.$$

From the distance formula applied to A and this midpoint we have

$$d = \sqrt{(7 - \tfrac{7}{2})^2 + (-2 - (-2))^2} = \tfrac{7}{2}.$$

EXAMPLE 5. A line segment AB has its midpoint at $C(5, -1)$. Point A has coordinates $(2, 3)$; find the coordinates of B.

Solution. In the midpoint formula, we know that $\bar{x} = 5, \bar{y} = -1, x_1 = 2, y_1 = 3$, and we have to find x_2, y_2. Substituting these values in the midpoint formula, we get

$$5 = \frac{2 + x_2}{2}, \quad -1 = \frac{3 + y_2}{2},$$

and

$$x_2 = 8, \quad y_2 = -5.$$

▶2. THE EQUATIONS OF CERTAIN LOCI

It frequently happens that a curve or a set of points in the plane is described by some geometric condition. For example, we may discuss the set of points each of which is 4 units away from some fixed point; or we may seek the set of points each of which is on a line parallel to some fixed line and a certain number of units from it. These curves or loci can sometimes be described in terms of an equation. Of course in doing so we assume that a specific Cartesian coordinate system has been introduced.

We now outline a step-by-step procedure for finding the equation of a locus, when a geometric description of the locus is given.

Step 1. Draw a figure and let $P(x, y)$ be a typical point on the locus; do not take P in a special position.

Step 2. Write the geometric condition.

Step 3. Write the equivalent equation involving the x and y coordinates of P.

Step 4. Simplify the resulting equation, using reversible steps where possible to make sure that the locus of the simplified equation coincides with that of the equation obtained in Step 3.

We illustrate with examples.

EXAMPLE 1. Find an equation of the locus of all points $P(x, y)$ which are equidistant from the points $A(3, 2)$ and $B(-1, 4)$.

Solution. We draw the points A, B, and P as shown in Fig. 3–3. The geometric condition (Step 2) is

$$|AP| = |BP|.$$

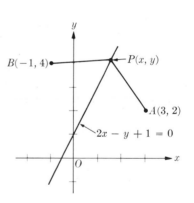

Using the distance formula and applying Step 3 above, we have

$$|AP| = |BP| \Leftrightarrow \sqrt{(x - 3)^2 + (y - 2)^2}$$
$$= \sqrt{(x + 1)^2 + (y - 4)^2}.$$

Figure 3–3

The right-hand equality holds if and only if

$$(x - 3)^2 + (y - 2)^2 = (x + 1)^2 + (y - 4)^2$$
$$\Leftrightarrow x^2 - 6x + 9 + y^2 - 4y + 4 = x^2 + 2x + 1 + y^2 - 8y + 16$$
$$\Leftrightarrow -8x + 4y - 4 = 0 \Leftrightarrow 2x - y + 1 = 0.$$

The graph of $2x - y + 1 = 0$ is sketched in Fig. 3–3.

EXAMPLE 2. Find the equation of the locus of all points $P(x, y)$ which are equidistant from the point $F(4, 0)$ and the y axis.

Solution. We label a point $P(x, y)$, draw a line parallel to the x axis through P, and denote by D the intersection of this line with the y axis (Fig. 3–4). The coordinates of D are $(0, y)$. The geometric condition of the locus (Step 2) is $|PF| = |PD|$. The corresponding condition as required in Step 3 is

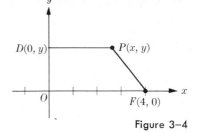

$$\sqrt{(x - 4)^2 + y^2} = |x|.$$

Squaring both sides, we see that the equation above holds

$$\Leftrightarrow x^2 - 8x + 16 + y^2 = x^2.$$

Hence the equation of the locus is $y^2 = 8x - 16$.

Figure 3–4

PROBLEMS

In problems 1 through 4, find the lengths of the sides of the triangles with the given points as vertices.

1. $A(1, 2)$, $B(4, 3)$, $C(2, 4)$ 2. $A(-3, -2)$, $B(5, 0)$, $C(0, 4)$

3. $A(2, -1)$, $B(4, 7)$, $C(5, 2)$ 4. $A(0, 2)$, $B(0, 3)$, $C(3, 4)$

In problems 5 through 8, locate the midpoint of the line segments joining the given points.

5. $P_1(2, 3)$, $P_2(4, 7)$ 6. $P_1(-2, -3)$, $P_2(4, -5)$

7. $P_1(-2, 4)$, $P_2(1, -1)$ 8. $P_1(0, 0)$, $P_2(4, 0)$

9. The midpoint of a line segment AB is at the point $(-1, 2)$. The point A has the coordinates $(2, -1)$. Find the coordinates of B.

10. The midpoint of a line segment AB is at the point $(-3, 2)$. The x coordinate of A is 5 and the y coordinate of B is -9. Find the points A and B.

11. Find the lengths of the medians of the triangle with vertices $A(2, -1)$, $B(4, 3)$, $C(-2, 5)$.

12. Same as problem 11, if $A(-1, -2)$, $B(2, 0)$, $C(4, 5)$.

13. Show that the triangle with vertices at $A(-3, 1)$, $B(1, 3)$, $C(-3, 6)$ is isosceles.

14. Same as 13, with $A(2, -3)$, $B(4, 1)$, $C(5, -2)$.

15. Show that the triangle with vertices at $A(-2, 1)$, $B(1, -1)$, $C(3, 2)$ is an isosceles right angle.

16. Same as 15, with $A(-3, -2)$, $B(-1, 1)$, $C(2, -1)$.

17. Show that the quadrilateral with vertices at $(2, 0)$, $(-1, 3)$, $(-4, 0)$, $(-1, -3)$ is a square.

18. Show that the diagonals of the following quadrilateral bisect each other: $(-3, 1)$, $(3, -7)$, $(9, -2)$, $(3, 6)$.

19. Same as 18, with $(6, 2)$, $(3, -4)$, $(-1, 0)$, $(2, 6)$.

20. The four points $A(2, -1)$, $B(4, 0)$, $C(8, 1)$, $D(1, 7)$ form a quadrilateral. Show that the midpoints of the sides are the vertices of a parallelogram.

21. The points $P(0, 0)$, $Q(a, 0)$, $R(b, c)$, $S(d, e)$ form the vertices of a quadrilateral. Let A, B, C, D denote the midpoints of the sides PQ, QR, RS, and SP, respectively. Show that A, B, C, D form the vertices of a parallelogram.

In problems 22 through 30, find the equation of the locus of all $P(x, y)$ which satisfy the given condition.

22. Equidistant from $(-1, 2)$ and $(3, 4)$.

23. At a distance of 2 from $(1, 1)$.

24. At a distance of 5 from $(3, 4)$.

25. Its distance from $(4, -1)$ equals its distance from the y axis.

26. Its distance from $(1, 2)$ equals its distance from the x axis.

27. It is twice as far from $(1, -1)$ as from $(4, -1)$.

28. The sum of the squares of its distances from the axes is 16.

29. It is half as far from $(2, -1)$ as it is from $(-4, 2)$.

30. Its distance from the y axis is half its distance from $(3, 0)$.

▶3. SLOPE OF A LINE

A line L, not parallel to the x axis, intersects it. Such a line and the x axis form two angles which are supplementary. To be definite, we denote by α the angle between the positive x axis and the upper ray of the line (see Fig. 3–5). The angle α will have a value between 0 and 180°. Two examples are shown in Fig. 3–5.

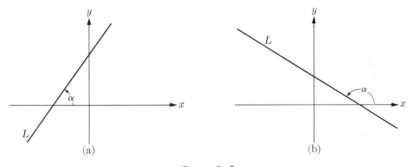

(a)   (b)

Figure 3–5

The angle α is called the **inclination** of the line L. All lines parallel to the x axis are said to make a zero angle with that axis and are said to have inclination zero. From plane geometry we recall the statement: "If, when two lines are cut by a transversal, corresponding angles are equal, the lines are parallel, and conversely." In Fig. 3–6, lines L_1 and L_2 each have inclination α. Applying the theorem of plane geometry with the x axis as transversal, we conclude that L_1 and L_2 are parallel. More generally, we say that *all lines with the same inclination are parallel* and, conversely, *all parallel lines have the same inclination.*

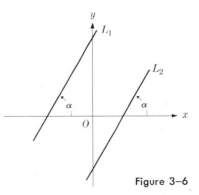

Figure 3–6

The notion of inclination is a simple one which is easy to understand. However, for purposes of analytic geometry, it is cumbersome and difficult to use. For this reason we introduce the notion of the **slope** of a line L. This is usually denoted by m and is defined in terms of the inclination of L by

$$m = \tan \alpha.$$

Before discussing slope, we recall some of the properties of the tangent function. It is positive for angles in the first quadrant. It starts at 0 for $\alpha = 0°$ and increases steadily, reaching 1 when $\alpha = 45°$. It continues to increase until, as the angle approaches 90°, the function increases without bound. The tangent of 90° is not defined. Loosely speaking, we sometimes say the tangent of 90° is infinite. In the second quadrant, the tangent function is negative, and its values are obtained with the help of the relation $\tan(180° - \alpha) = -\tan \alpha$. That is, the tangent of an obtuse angle is the negative of that of the corresponding acute supplementary angle.

From these facts about the tangent function, we see that any line parallel to the x axis has zero slope, while a line with an inclination between 0° and 90° has positive slope. A line parallel to the y axis has no slope, strictly speaking, although we sometimes say such a line has infinite slope. If the inclination is an obtuse angle, the slope is negative. The line in Fig. 3–5(a) has positive slope, and the one in Fig. 3–5(b) has negative slope.

Consider the line passing through the points $P(2, 1)$ and $Q(5, 3)$, as shown in Fig. 3–7. To find the slope of this line, draw a parallel to the x axis through P and a parallel to the y axis through Q, forming the right triangle PQR. The angle α at P is equal to the inclination, by corresponding angles of parallel lines. The definition of tangent function in a right triangle is "opposite over adjacent." The slope is therefore given by

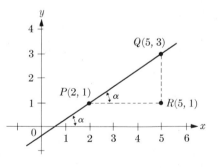

$$m = \tan \alpha = \frac{|RQ|}{|PR|}.$$

Since $|RQ| = 3 - 1 = 2$ and $|PR| = 5 - 2 = 3$, the slope m is $\frac{2}{3}$.

Figure 3–7

Suppose $P(x_1, y_1)$ and $Q(x_2, y_2)$ are any two points in the plane, and we wish to find the slope m of the line passing through these two points. The procedure is exactly the same as the one just described. In Fig. 3–8(a) we see that, if $y_2 > y_1$,

$$m = \tan \alpha = \frac{y_2 - y_1}{x_2 - x_1},$$

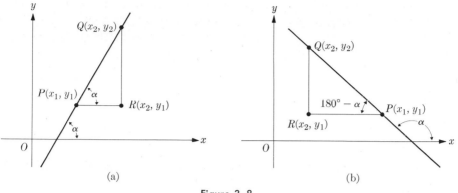

Figure 3-8

and in Fig. 3-8(b)

$$m = \tan \alpha = -\tan(180 - \alpha) = -\frac{y_2 - y_1}{x_1 - x_2} = \frac{y_2 - y_1}{x_2 - x_1}.$$

If $y_2 < y_1$, we get

$$m = \frac{y_1 - y_2}{x_1 - x_2} = \frac{y_2 - y_1}{x_2 - x_1} \tag{1}$$

so the formula holds in all cases if $x_1 \neq x_2$ and $y_1 \neq y_2$. A difficulty may arise if the points P and Q are on a vertical line, since then $x_1 = x_2$ and the denominator is zero. However, we know that a vertical line has no slope, and we state that the formula holds for all cases except when $x_1 = x_2$. Note that there is no difficulty if P and Q lie on a horizontal line. In that case, $y_1 = y_2$ and the slope is zero, as it should be. We conclude: the *slope of a line through the points* $P(x_1, y_1)$ *and* $Q(x_2, y_2)$ *with* $x_1 \neq x_2$ *is given by the formula*

$$m = \frac{y_2 - y_1}{x_2 - x_1}.$$

EXAMPLE 1. Find the slope of the line through the points $(4, -2)$ and $(7, 3)$.

Solution. In the formula for slope given by (1) it doesn't matter which point we label (x_1, y_1) [the other being labeled (x_2, y_2)]. We let $(4, -2)$ be (x_1, y_1) and $(7, 3)$ be (x_2, y_2). This gives

$$m = \frac{3 - (-2)}{7 - 4} = \frac{5}{3}.$$

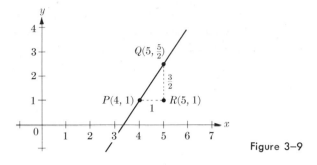

Figure 3–9

EXAMPLE 2. Through the point $P(4, 1)$ construct a line which has slope equal to $3/2$.

Solution. Starting at P, draw a parallel to the x axis extending to the point R one unit to the right (Fig. 3–9). Now draw a parallel to the y axis, stopping $3/2$ units above R. The coordinates of this point Q are $(5, \frac{5}{2})$. The line through P and Q has slope $3/2$.

▶4.   PARALLEL AND PERPENDICULAR LINES

We have seen that parallel lines always have the same inclination. Therefore, *two parallel lines will always have the same slope.* Conversely, we show that *lines with the same slope must be parallel.* This fact is a result of a simple property of the tangent function: if this function has a given value, say b, there is exactly one angle between $0°$ and $180°$, say β, such that $\tan \beta = b$. We conclude that if two lines have the same slope, their inclinations must also be the same, and the lines therefore are parallel. (This works even if the slopes are infinite, since then the lines are perpendicular to the x axis and so parallel.)

When are two lines perpendicular? If a line is parallel to the x axis, it has zero slope and is perpendicular to any line which has infinite slope. Suppose, however, that a line L_1 has slope m_1 which is not zero (Fig. 3–10). Its inclination will be α_1, also different from zero. Let L_2 be perpendicular to L_1 and have slope m_2 and inclination α_2, as shown. We recall from plane geometry: "An exterior angle of a triangle is equal to the sum of the remote interior angles." This means that

$$\alpha_2 = 90° + \alpha_1$$

and

$$\tan \alpha_2 = \tan (90° + \alpha_1).$$

We recall from trigonometry the formula

$$\tan (90 + A) = -\cot A,$$

from which we obtain

$$\tan \alpha_2 = -\cot \alpha_1 = -\frac{1}{\tan \alpha_1}.$$

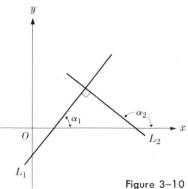

Figure 3–10

In this derivation we assumed that $\alpha_2 > \alpha_1$. But if this is not so, then $\alpha_2 = \alpha_1 - 90°$ and the same result holds. This last formula, in terms of slopes, states that

$$m_2 = -\frac{1}{m_1}$$

or, in words: *Two lines are perpendicular if their slopes are the negative reciprocals of each other, and conversely.*

EXAMPLE 1. Show that the line through $P_1(3, -4)$ and $Q_1(-2, 6)$ is parallel to the line through $P_2(-3, 6)$ and $Q_2(9, -18)$.

Solution. The slope of the line through P_1 and Q_1, according to the formula, is

$$m_1 = \frac{6 - (-4)}{-2 - 3} = \frac{10}{-5} = -2.$$

Similarly, the line through P_2 and Q_2 has slope

$$m_2 = \frac{-18 - 6}{9 + 3} = -2.$$

Since the slopes are the same, the lines must be parallel.

EXAMPLE 2. Determine whether or not the three points $P(-1, -5)$, $Q(1, 3)$, and $R(7, 12)$ lie on the same straight line.

Solution. The line through P and Q has slope

$$m_1 = \frac{3 - (-5)}{1 - (-1)} = 4.$$

If R were on this line, the line joining R and Q would have to be the very same line; therefore it would have the same slope. The slope of the line through R and Q is

$$m_2 = \frac{12 - 3}{7 - 1} = \frac{3}{2},$$

and this is different from m_1. Therefore, P, Q, and R do not lie on the same line.

EXAMPLE 3. Is the line through the points $P_1(5, -1)$ and $Q_1(-3, 2)$ perpendicular to the line through the points $P_2(-3, 1)$ and $Q_2(0, 9)$?

Solution. The line through P_1 and Q_1 has slope

$$m_1 = \frac{2 - (-1)}{-3 - 5} = -\frac{3}{8}.$$

The line through P_2 and Q_2 has slope

$$m_2 = \frac{9 - 1}{0 - (-3)} = \frac{8}{3}.$$

The slopes are the negative reciprocals of each other, and the lines are perpendicular.

EXAMPLE 4. Given the isosceles triangle with vertices at the points $P(-1, 4)$, $Q(0, 1)$, and $R(2, 5)$, show that the median drawn from P is perpendicular to the base QR (Fig. 3–11).

Solution. Let M be the point where the median from P intersects the base QR. From the definition of median, M must be the midpoint of the segment QR. The coordinates of M, from the midpoint formula, are

$$\bar{x} = \frac{0 + 2}{2} = 1, \qquad \bar{y} = \frac{5 + 1}{2} = 3.$$

Now we check the slopes of PM and QR. The slope of PM is

$$m_1 = \frac{3 - 4}{1 - (-1)} = -\frac{1}{2}.$$

The slope of QR is

$$m_2 = \frac{5 - 1}{2 - 0} = 2.$$

Since

$$m_1 = -\frac{1}{m_2},$$

the median is perpendicular to the base.

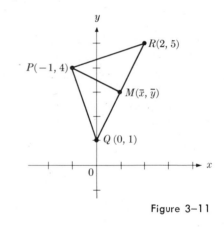

Figure 3–11

PROBLEMS

In problems 1 through 6, check to see whether the line through the pair of points P_1, Q_1 is parallel or perpendicular to the line through the pair of points P_2, Q_2.

1. $P_1(-3, 2)$, $Q_1(5, -1)$ and $P_2(2, 3)$, $Q_2(10, 0)$
2. $P_1(1, -3)$, $Q_1(4, 0)$ and $P_2(2, 2)$, $Q_2(3, 1)$
3. $P_1(9, 5)$, $Q_1(1, 5)$ and $P_2(5, 2)$, $Q_2(-4, 2)$
4. $P_1(5, 1)$, $Q_1(0, 7)$ and $P_2(4, 3)$, $Q_2(1, -2)$
5. $P_1(-6, 5)$, $Q_1(12, 2)$ and $P_2(5, 0)$, $Q_2(8, -4)$
6. $P_1(3, 3)$, $Q_1(6, 3)$ and $P_2(6, 5)$, $Q_2(6, -3)$

In problems 7 through 10, determine whether or not the three points all lie on the same straight line.

7. $P(4, -2)$, $Q(7, 2)$, $R(-5, 3)$ 8. $P(-5, -1)$, $Q(2, 0)$, $R(9, 1)$

9. $P(0, 2)$, $Q(5, 3)$, $R(10, 4)$ 10. $P(7, -1)$, $Q(-7, 0)$, $R(4, -7)$

11. Construct a line passing through the point $(3, 1)$ having slope $2/3$.

12. Construct a line passing through the point $(1, -2)$ and having slope $-3/4$.

In problems 13 through 23, the points P, Q, R, S are the vertices of a quadrilateral (in order). In each determine whether the figure is a trapezoid, parallelogram, rhombus, rectangle, square, or none of these.

13. $P(3, 5)$, $Q(5, 3)$, $R(1, -6)$, $S(-4, -1)$

14. $P(-2, -3)$, $Q(-1, -1)$, $R(3, 11)$, $S(2, 9)$

15. $P(0, 1)$, $Q(3, -3)$, $R(\frac{51}{13}, -\frac{21}{13})$, $S(-\frac{4}{13}, -\frac{5}{13})$

16. $P(6, 9)$, $Q(-7, -4)$, $R(-7, -7)$, $S(6, 6)$

17. $P(-1, -4)$, $Q(1, 0)$, $R(1, 4)$, $S(-1, 0)$

18. $P(-\frac{2}{3}, \frac{14}{3})$, $Q(2, 0)$, $R(3, -6)$, $S(3, 9)$

19. $P(2, 1)$, $Q(-1, -2)$, $R(3, 4)$, $S(4, -5)$

20. $P(2, 1)$, $Q(5, 1)$, $R(1, 2)$, $S(4, 2)$

21. $P(2, 0)$, $Q(1, -5)$, $R(0, 8)$, $S(-1, -7)$

22. $P(3, -5)$, $Q(8, 7)$, $R(-4, 12)$, $S(-9, 0)$

23. $P(1, -4)$, $Q(1, 1)$, $R(-2, 5)$, $S(-2, 0)$

24. The points $A(4, -3)$, $B(5, 0)$, and $C(-2, 4)$ are the vertices of a triangle. Show that the line through the midpoints of the sides AB and AC is parallel to the base BC of the triangle.

25. The points $A(0, 0)$, $B(a, 0)$, and $C(\frac{1}{2}a, b)$ are the vertices of a triangle. Show that the triangle is isosceles. Prove that the median from C is perpendicular to the base AB. How general is this proof?

26. The points $A(0, 0)$, $B(a, 0)$, $C(a + b, c)$, $D(b, c)$ form the vertices of a parallelogram. Prove that the diagonals bisect each other.

27. Let $A(-\frac{1}{2}a, 0)$ and $B(\frac{1}{2}a, 0)$ be two adjacent vertices of a regular hexagon situated above the side AB. Find the coordinates of the remaining vertices and the length of the diagonal joining two opposite vertices.

28. Let $A(-\frac{1}{2}a, 0)$ and $B(\frac{1}{2}a, 0)$ be two adjacent vertices of a regular octagon situated above the side AB. Find the coordinates of the remaining vertices and the length of the diagonal joining two opposite vertices.

▶5. THE STRAIGHT LINE

We saw earlier that the equation $y = 4$ represented all points on a line parallel to the x axis and four units above it. Similarly, the equation $x = -2$ represents all points on a line parallel to the y axis and two units to the left. More generally, any line parallel to the y axis has an equation of the form

$$x = a,$$

where a is the number denoting how far the line is to the right or left of the y axis.

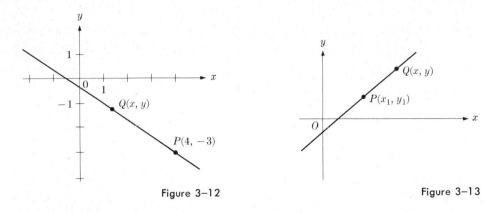

Figure 3–12   Figure 3–13

The equation

$$y = b$$

describes a line parallel to the x axis and b units from it. In this way we obtain the equations of all lines with zero or infinite slope. A line which is not parallel to either axis has a slope m which is different from zero. Suppose also that the line passes through a point $P(x_1, y_1)$. To be specific, we consider the slope m to be $-\frac{2}{3}$ and the point P to have coordinates $(4, -3)$. If a point $Q(x, y)$ is on this line, then the slope as calculated from P to Q must be $-\frac{2}{3}$, as shown in Fig. 3–12. That is (assuming $Q \neq P$)

$$\frac{y + 3}{x - 4} = -\tfrac{2}{3}, \quad \text{or} \quad y + 3 = -\tfrac{2}{3}(x - 4).$$

This is the equation of the line passing through the point $(4, -3)$ *with slope* $-\frac{2}{3}$. In the general case of a line with slope m passing through $P(x_1, y_1)$ the statement that $Q(x, y)$ is on the line is the same as the statement that the slope m as computed from P to Q (Fig. 3–13) is (if $Q \neq P$)

$$\frac{y - y_1}{x - x_1} = m, \quad \text{or} \quad \boxed{y - y_1 = m(x - x_1).}$$

This last equation is called **the point-slope form** *for the equation of a line.* The point P also satisfies the equation in the box. That is, if we are given the coordinates of a point and the numerical value of the slope, substitution in the above formula yields the equation of the line going through the point and having the given slope.

EXAMPLE 1. Find the equation of the line passing through the point $(-2, 5)$ and having slope $4/3$.

Solution. Substitution in the above formula gives

$$y - 5 = \tfrac{4}{3}[x - (-2)], \quad \text{or} \quad 3y = 4x + 23.$$

We know that two points determine a line. The problem of finding the equation of the line passing through the points $(3, -5)$ and $(-7, 2)$ can be solved in two steps. First we employ the formula for the slope of a line, as given in Section 3, to obtain the slope of the line through the given points. We get

$$m = \frac{2 - (-5)}{-7 - 3} = -\frac{7}{10}.$$

Then, knowing the slope, we use *either* point, together with the slope, in the point-slope formula. This gives [using the point $(3, -5)$]

$$y - (-5) = -\tfrac{7}{10}(x - 3), \quad \text{or} \quad 10y = -7x - 29.$$

We verify quite easily that the same equation is obtained if the point $(-7, 2)$ is used instead of $(3, -5)$.

The above process may be transformed into a general formula by applying it to two points, $P_1(x_1, y_1)$ and $P_2(x_2, y_2)$. The slope of the line through these points is

$$m = \frac{y_2 - y_1}{x_2 - x_1}.$$

Substituting this value for the slope into the point-slope form, we get the **two-point form** for the equation of a line:

$$y - y_1 = \frac{y_2 - y_1}{x_2 - x_1}(x - x_1).$$

Note that this is really not a new formula, but merely the point-slope form with an expression for the slope substituted into it.

Another variation of the point-slope form is obtained by introducing a number called the y intercept. Every line not parallel to the y axis must intersect it; if we denote by $(0, b)$ the point of intersection, *the number b is called* **the y intercept.*** Suppose a line has slope m and y intercept b. We substitute in the point-slope form to get

$$y - b = m(x - 0),$$

or

$$y = mx + b.$$

This is called the **slope-intercept form** for the equation of a straight line.

EXAMPLE 2. A line has slope 3 and y intercept -4. Find its equation.

Solution. Substitution in the slope-intercept formula gives

$$y = 3x - 4.$$

---

* The point $(0, b)$ is also called the y intercept.

If a line intersects the x axis at a point $(a, 0)$, *the number a is called* **the x intercept.**

Suppose a line has x intercept a and y intercept b (Fig. 3–14). This means the line must pass through the points $(a, 0)$ and $(0, b)$. Using the two-point form for the equation of a line, we have (assuming that $a \neq 0$)

$$y - 0 = \frac{b}{-a}(x - a), \quad \text{or} \quad bx + ay = ab.$$

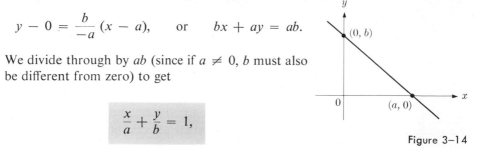

We divide through by ab (since if $a \neq 0$, b must also be different from zero) to get

$$\frac{x}{a} + \frac{y}{b} = 1,$$

Figure 3–14

which is called the **two-intercept form** for the equation of a line. It is valid whenever the x and y intercepts are different from zero.

EXAMPLE 3. Find the equation of a line with x intercept -3 and y intercept 4.

Solution. Substitution in the above formula yields

$$\frac{x}{-3} + \frac{y}{4} = 1,$$

which is the desired result.

The important thing to notice is that the point-slope form is the basic one for the equation of a straight line. All other formulas were derived as simple variations or particular cases.

All three examples led to equations of lines which could be put in the form

$$Ax + By + C = 0,$$

where A, B, and C are any numbers. This equation is the most general equation of the first degree in x and y. We shall establish the theorem:

Theorem 1. *Every equation of the form*

$$Ax + By + C = 0,$$

so long as A and B are not both zero, is the equation of a straight line.

Proof. We consider two cases, according as $B = 0$ or $B \neq 0$. If $B = 0$, then we must have $A \neq 0$, and the above equation becomes

$$x = -\frac{C}{A},$$

which we know is the equation of a straight line parallel to the y axis and $-C/A$ units from it.

If $B \neq 0$, we divide by B and solve for y, getting

$$y = -\frac{A}{B}x - \frac{C}{B}.$$

From the slope-intercept form for the equation of a line, we recognize this as the equation of a line with slope $-A/B$ and y intercept $-C/B$.

In the statement of the theorem it is necessary to make the requirement that A and B are not both zero. If both of them vanish and C is zero, the linear equation reduces to the triviality $0 = 0$, which is satisfied by every point $P(x, y)$ in the plane. If $C \neq 0$, then no point $P(x, y)$ satisfies the equation $0 = C$.

EXAMPLE 4. Given the linear equation

$$3x + 2y + 6 = 0,$$

find the slope and y intercept.

Solution. Solving for y, we have

$$y = -\tfrac{3}{2}x - 3.$$

From this we simply read off that $m = -\tfrac{3}{2}$ and $b = -3$.

EXAMPLE 5. Put the equation

$$2x - 5y + 7 = 0$$

in the two-intercept form.

Solution. We rewrite the equation in the form $2x - 5y = -7$, and divide by -7:

$$\frac{2x}{-7} - \frac{5y}{-7} = 1.$$

Now we again rewrite each term so that x and y appear alone in the numerator:

$$\frac{x}{-\frac{7}{2}} + \frac{y}{\frac{7}{5}} = 1.$$

The x and y intercepts are $-\tfrac{7}{2}$ and $\tfrac{7}{5}$, respectively. Note that if we merely wanted to find the intercepts we could set y equal to zero and solve for x to get the x intercept, and set x equal to zero and solve for y to get the y intercept. With this knowledge we could then write the answer to Example 5 by using the two-intercept formula.

An equation of the first degree in x and y is called a **linear equation**. When we solve for y in terms of x, as in the point-slope form, y becomes a function of x. Such a function is called a **linear function**. Any line not parallel to the y axis may be thought of as a function. Also, the equation of any line not parallel to the x axis may be solved for x in terms of y.

EXAMPLE 6. Find the equation of the line which is the perpendicular bisector of the line segment joining the points $P(-3, 2)$ and $Q(5, 6)$. (See Fig. 3–15.)

Solution. We give two methods.

Method 1. We first find the slope m of the line through P and Q. It is

$$m = \frac{6 - 2}{5 + 3} = \frac{1}{2}.$$

The slope of the perpendicular bisector must be -2, the negative reciprocal. Next we get the coordinates of the midpoint of the line segment PQ. They are

$$\bar{x} = \frac{5 - 3}{2} = 1,$$

$$\bar{y} = \frac{6 + 2}{2} = 4.$$

The equation of the line through $(1, 4)$ with slope -2 is

$$y - 4 = -2(x - 1),$$

or

$$2x + y - 6 = 0,$$

which is the desired equation.

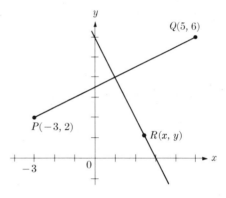

Figure 3–15

Method 2. We start by noting that any point on the perpendicular bisector must be equidistant from P and Q. Let $R(x, y)$ be any such point, and let d_1 denote the distance from R to P, and d_2 the distance from R to Q. From the formula for the distance between two points and the condition $d_1 = d_2$, we have

$$\sqrt{(x + 3)^2 + (y - 2)^2} = \sqrt{(x - 5)^2 + (y - 6)^2}.$$

Squaring both sides and multiplying out, we obtain

$$x^2 + 6x + 9 + y^2 - 4y + 4 = x^2 - 10x + 25 + y^2 - 12y + 36.$$

The terms of the second degree cancel and the remaining terms combine to give

$$6x - 4y + 13 = -10x - 12y + 61,$$

or

$$16x + 8y - 48 = 0.$$

Dividing by 8, we obtain the same answer as in the first method: $2x + y - 6 = 0$.

PROBLEMS

In problems 1 through 20, find the equation of the line with the given requirements.

1. Slope 2 and passing through (1, 3)
2. Slope -3 and passing through (3, -1)
3. Slope $-\frac{3}{2}$ and passing through (-1, 4)
4. Passing through the points (-4, -3) and (5, 6)
5. Passing through the points (-3, 4) and (2, -3)
6. Passing through the points (-2, 0) and (0, 4)
7. Slope 0 and passing through (2, 5)
8. Passing through the points (2, 3) and (2, 7)
9. Passing through the points (5, 1) and (-2, 1)
10. Slope $-\frac{3}{2}$ and y intercept -2
11. Slope 0 and y intercept 3
12. Slope $\frac{2}{3}$ and y intercept -4
13. Slope $\frac{1}{2}$ and x intercept 2
14. Slope -3 and x intercept 0
15. Slope -2 and y intercept 0
16. x intercept 2 and y intercept -3
17. x intercept -3 and y intercept -4
18. x intercept 3 and passing through the point (-4, 3)
19. y intercept -4 and passing through the point (2, 1)
20. Parallel to the y axis and passing through the point (2, -1)
21. Find the slope and y intercept of the line $3x + 2y = 6$.
22. Find the slope and x intercept of the line $2x + y + 4 = 0$.
23. Find both intercepts of the line $3x + 2y + 7 = 0$.
24. Find both intercepts of the line $x - 2y + 3 = 0$.
25. Find the equation of the line through the point (2, 3) and parallel to the line $2x + 3y - 6 = 0$.
26. Find the equation of the line through the point (-3, -1) and parallel to the line $2x + 3y + 4 = 0$.
27. Find the equation of the line through the point (2, 3) and perpendicular to the line $3x - 2y + 5 = 0$.
28. Find the equation of the line through the point (2, -1) and perpendicular to the line $4x - 3y + 4 = 0$.
29. Find the equation of the line through the point (2, 1) and parallel to the line through the points (1, 1) and (-4, 3).
30. Find the equation of the line passing through (1, 2) and parallel to the line through the points (1, -1) and (4, 4).

31. Find the equation of the line passing through $(-2, 2)$ and perpendicular to the line through the points $(4, 3)$ and $(-3, -2)$.

32. Find the equation of the line passing through $(1, 2)$ and perpendicular to the line through the points $(1, -1)$ and $(-3, 4)$.

33. Find the equation of the perpendicular bisector of the line segment joining $(5, -1)$ and $(2, 2)$.

34. Find the equation of the perpendicular bisector of the line segment joining $(2, 3)$ and $(-3, 1)$.

35. The points $P(0, 0)$, $Q(a, 0)$, $R(a, b)$, and $S(0, b)$ are the vertices of a rectangle. Show that if the diagonals meet at right angles, then $|a| = |b|$ and the rectangle is a square.

36. Show that the points $A(0, 0)$, $B(a, 0)$, $C(a + b, c)$, $D(b, c)$ are the vertices of a parallelogram and show that its area is ac. (Assume $a > 0$, $c > 0$.)

37. Show that the points $A(0, 0)$, $B(a, 0)$, $C(d, c)$, $D(b, c)$ are the vertices of a trapezoid and that its area is $\frac{1}{2}c(a + d - b)$. (Assume $c > 0$ and $d > b$.)

38. Let $A(0, 0)$ $B(a, 0)$ $C(b, c)$ be the vertices of a triangle. (Assume $a > 0$, $c > 0$.) Suppose that D and E are points on sides AC and BC, respectively, with line segment DE parallel to AB. Show that

$$\frac{|CD|}{|DE|} = \frac{|CA|}{|AB|}.$$

▶**6.   PARAMETRIC EQUATIONS OF A LINE.   DIRECTED LINES. POINT OF DIVISION FORMULA**

The correspondence between the real numbers, R_1, and the points on a line may be regarded as a **Cartesian coordinate system on the line.** Since the real numbers are ordered according to their size, each coordinate system on a line sets up an **ordering** of the points on this line. Suppose P_1 corresponds to the number t_1 and P_2 corresponds to t_2 (Fig. 3–16). Then we say that P_1 **precedes** P_2 if and only if $t_1 < t_2$. We write

$$P_1 \prec P_2 \Leftrightarrow t_1 < t_2,$$

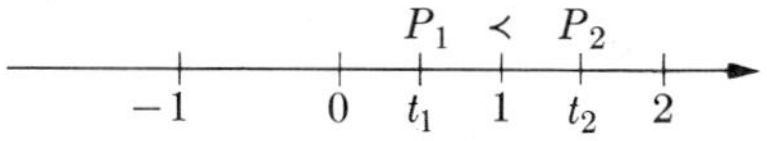

Figure 3–16

where "$\prec$" is the symbol for "precedes." If, for example, we have a horizontal line and we take a coordinate system with positive numbers to the right (as shown in Fig. 3–16), then "precedes" corresponds to "is to the left of." Of course, we could have taken positive numbers going to the left in which case "precedes" would correspond to "is to the right of." It is important to observe that for a line l, any two coordinate systems give either the same ordering to the points of l or they give the opposite ordering to the points of l. In other words, *for every line l there are exactly two possible orderings of the points on it.* We shall use a script letter such as α to stand for one of these two orderings.

DEFINITION. Let l be a line and α one of its two possible orderings. We call the pair (l, α) a **directed line** and we use the symbol $\vec{l}$ as an abbreviation for the pair.

Let $\vec{l}$ be a directed line so that all points are ordered according to the symbol $\prec$. We say that a coordinate system (with the letter t used for numbers) on l **agrees with the ordering** of $\vec{l}$ when

$$P_1 \prec P_2 \Leftrightarrow t_1 < t_2,$$

where t_1 and t_2 are the coordinates of P_1 and P_2, respectively.

Suppose A and B are two points on a line and α is the ordering which makes A precede B. That is, $A \prec B$ with the ordering α. Then we define the pair (AB, α) to be the **directed line segment from A to B** and we denote it by $\overrightarrow{AB}$.

We recall that the distance between any two distinct points A and B is the length of the line segment joining them and so is always a positive number. We use the symbol $|AB|$ for this length. Corresponding to directed segments, we now define the *directed distance* between two points, a number which may be positive or negative.

DEFINITION. If $\overrightarrow{AB}$ is a directed line segment on a directed line $\vec{l}$, we define the **directed distance from A to B along $\vec{l}$**, denoted by $\overline{AB}$, by the formulas:

$$\overline{AB} = |AB| \Leftrightarrow A \prec B \text{ according to the order on } \vec{l},$$
$$\overline{AB} = -|AB| \Leftrightarrow B \prec A \text{ according to the order on } \vec{l}.$$

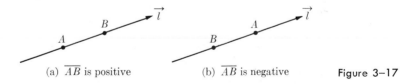

(a) $\overline{AB}$ is positive   (b) $\overline{AB}$ is negative   **Figure 3–17**

Figure 3–17 shows a directed line $\vec{l}$ with the arrow at the head indicating the ordering. In (a) of Fig. 3–17, the directed distance $\overline{AB}$ is positive, while in (b), the directed distance $\overline{AB}$ is negative.

Theorem 2. (i) *Let $\vec{l}$ be a directed line and suppose a Cartesian coordinate system is given on $\vec{l}$. If the Cartesian system agrees with the ordering of $\vec{l}$, then*

$$\overline{P_1P_2} = t_2 - t_1,$$

where t_1 and t_2 are the coordinates of P_1 and P_2, respectively. (ii) *If P_1, P_2,* *and P_3 are any points on a directed line $\vec{l}$, then*

$$\overline{P_1P_2} + \overline{P_2P_3} = \overline{P_1P_3}.$$

Proof. (i) We have $|P_1P_2| = |t_2 - t_1|$. From the definition of directed distance

$$\overline{P_1P_2} = t_2 - t_1 \qquad \text{if} \quad t_2 > t_1,$$
$$\overline{P_1P_2} = -|t_2 - t_1| = t_2 - t_1 \qquad \text{if} \quad t_2 < t_1.$$

Hence in all cases $\overline{P_1P_2} = t_2 - t_1$.

(ii) If t_1, t_2, and t_3 are the coordinates of P_1, P_2, and P_3, respectively, then we use the result of (i) to write

$$\overline{P_1P_2} + \overline{P_2P_3} = (t_2 - t_1) + (t_3 - t_2)$$
$$= t_3 - t_1 = \overline{P_1P_3}.$$

Suppose a Cartesian coordinate system has been set up on a geometric plane and $\vec{l}$ is a directed line in this plane. We now define the **inclination of the directed line** $\vec{l}$. Referring to Fig. 3–18, the inclination α is defined as that angle between $-180°$ and $+180°$ measured *from* the positive x axis *to* the positive ray of $\vec{l}$ as indicated by the arrow showing the direction of $\vec{l}$. If the line $\vec{l}$ is horizontal, we define the inclination α to be 0 if $\vec{l}$ is directed to the right and 180° if $\vec{l}$ is directed to the left. We note that while the inclination of a directed line may

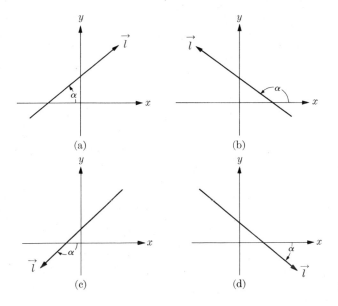

(a)   (b)

(c)   (d)   Figure 3–18

be a positive or negative angle, the inclination of an undirected line (as discussed on page 45) is always between $0°$ and $180°$, measured counterclockwise.

In Section 5 we learned a number of ways of finding the equation of a line. We now develop still another method, one which makes use of the notion of directed line. Suppose we have a Cartesian coordinate system in a plane in which $\vec{l}$ is a directed line. We introduce a coordinate system on $\vec{l}$ which agrees with the direction of $\vec{l}$ and has its origin at a point which we denote by P_0. See Fig. 3-19. In the customary way we use the letter t for units in the coordinate system along $\vec{l}$. Suppose P is a point on $\vec{l}$ with Cartesian coordinates (x, y). Denote the Cartesian coordinates of P_0 by (x_0, y_0) and the directed distance from P_0 to P by t. With α as the inclination of $\vec{l}$, we recall the formulas from trigonometry

$$\cos \alpha = \frac{x - x_0}{t}, \qquad \sin \alpha = \frac{y - y_0}{t}.$$

Then, solving for x and y,

$$x = x_0 + t \cos \alpha, \qquad y = y_0 + t \sin \alpha.$$

These are the **parametric equations** of a straight line through the point (x_0, y_0) with inclination α. Because of the properties of directed lines it is easy to see that these equations are valid for t negative as well as t positive.

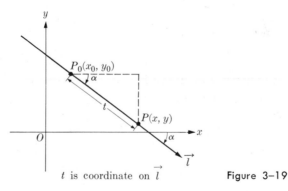

t is coordinate on $\vec{l}$ Figure 3-19

It is natural to inquire why *two* equations are used for the parametric description of a line when all previous formulas employed only *one* equation to represent a line. Two equations are required because the additional quantity t, known as a **parameter,** is present in the equations. By solving one of the equations for t and substituting in the other equation, we can *eliminate the parameter.* Performing this task (assuming $\alpha \neq \pm 90°$), we obtain

$$t = \frac{x - x_0}{\cos \alpha} \qquad \text{and} \qquad y = y_0 + \frac{(x - x_0)}{\cos \alpha} \sin \alpha.$$

Therefore $y - y_0 = (\tan \alpha)(x - x_0)$, which we recognize as the point-slope formula for the equation of a line.

EXAMPLE 1.   Find the parametric equations of the directed line through the point $(3, -2)$ having inclination $\alpha = -135°$.

Solution. We have

$$x = 3 + t\cos(-135°), \quad y = -2 + t\sin(-135°)$$
$$\Leftrightarrow x = 3 - \tfrac{1}{2}t\sqrt{2}, \quad y = -2 - \tfrac{1}{2}t\sqrt{2}.$$

EXAMPLE 2.   Find parametric equations of the line l which has equation $x - \sqrt{3}\,y = 1 + \sqrt{3}$.

Solution. We first find a point on the line. Letting $x = 1$, we get $y = -1$, and $(1, -1)$ is on the line. The slope of l is $1/\sqrt{3}$ and so α is $30°$ if l is directed one way and $\alpha = -150°$ if l is directed oppositely. Thus we find

$$x = 1 + t\cos(30°), \quad y = -1 + t\sin(30°)$$
$$\Leftrightarrow x = 1 + \tfrac{1}{2}t\sqrt{3}, \quad y = -1 + \tfrac{1}{2}t$$

or

$$x = 1 + t\cos(-150°), \quad y = -1 + t\sin(-150°)$$
$$\Leftrightarrow x = 1 - \tfrac{1}{2}t\sqrt{3}, \quad y = -1 - \tfrac{1}{2}t.$$

We observe that one pair of parametric equations is obtained from the other when t is replaced by $-t$.

By means of directed line segments we can extend the midpoint formula, given in Section 1, to obtain the coordinates of a point P which is any fraction of the distance along the segment from P_1 to P_2. For example, we may wish to find the point which is $\tfrac{1}{3}$ of the way from P_1 to P_2, or $\tfrac{2}{5}$ of the way, or even $1/\sqrt{3}$ of the way. In general, we say that **P is h of the way from P_1 to P_2** if and only if P is on the line $\overrightarrow{l}$ through P_1 and P_2 and $\overline{P_1P} = h\overline{P_1P_2}$.

We now derive formulas for the coordinates of P in terms of those of $P_1(x_1, y_1)$ and $P_2(x_2, y_2)$. Let P have coordinates (x, y) and be h of the way from P_1 to P_2.

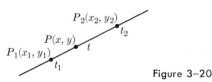

Figure 3–20

Suppose that a Cartesian coordinate system is set up which agrees with the ordering of $\overrightarrow{l}$, and that P_1, P_2, P have coordinates t_1, t_2, t, respectively (Fig. 3–20).

Then from Theorem 2, we know that

$$\overline{P_1P} = t - t_1, \qquad \overline{P_1P_2} = t_2 - t_1.$$

Since $\overline{P_1P} = h\overline{P_1P_2}$, we have

$$t - t_1 = h(t_2 - t_1). \tag{1}$$

Now we use the parametric equations of the line $\vec{l}$:

$$x = x_0 + t \cos \alpha, \qquad y = y_0 + t \sin \alpha. \tag{2}$$

Since (x_1, y_1) is on the line $\vec{l}$, we have

$$x_1 = x_0 + t_1 \cos \alpha, \qquad y_1 = y_0 + t_1 \sin \alpha.$$

Subtracting these equations from the equations of the line $\vec{l}$ as given by (2), we get

$$x - x_1 = (t - t_1) \cos \alpha, \qquad y - y_1 = (t - t_1) \sin \alpha, \tag{3}$$

which are also equations of the same line $\vec{l}$. Now we use the fact that (x_2, y_2) is on the line to obtain

$$x_2 - x_1 = (t_2 - t_1) \cos \alpha, \qquad y_2 - y_1 = (t_2 - t_1) \sin \alpha. \tag{4}$$

Since $t - t_1 = h(t_2 - t_1)$ [see Eq. (1)], it follows that

$$x - x_1 = h(x_2 - x_1), \qquad y - y_1 = h(y_2 - y_1),$$

from which we immediately obtain the **point of division formula**

$$x = x_1 + h(x_2 - x_1), \qquad y = y_1 + h(y_2 - y_1).$$

In developing this formula for the coordinates (x, y) of the point P, which is h of the way from P_1 to P_2, the reader probably assumed that h is some number between 0 and 1, such as $\frac{1}{2}$ for the midpoint, $\frac{1}{3}$ for the point which divides $\overrightarrow{P_1P_2}$ in the ratio 1:2, and so forth. However, everything which has been derived is valid for *any* value for h. Figure 3–21 shows the relative positions of P, P_1, and P_2 for various values of h.

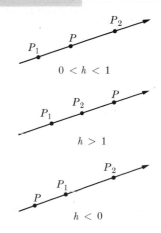

$0 < h < 1$

$h > 1$

$h < 0$

Figure 3–21

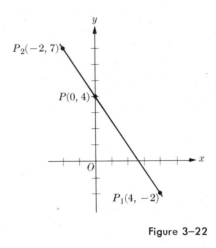

Figure 3–22

EXAMPLE 3. Find the coordinates of the point $P(x, y)$ which is $\frac{2}{3}$ of the way from $P_1(4, -2)$ to $P_2(-2, 7)$. Plot the line P_1P_2 and the point P.

Solution. Using the point of division formula with $h = \frac{2}{3}$, we obtain

$$x = 4 + \tfrac{2}{3}(-2 - 4) = 0, \qquad y = -2 + \tfrac{2}{3}(7 + 2) = 4.$$

The graph is drawn in Fig. 3–22.

PROBLEMS

In problems 1 through 10 find in each case parametric equations of the line with the given requirements.

1. Passing through $(2, -1)$ and $(3, 4)$
2. Passing through $(1, 2)$ and $(3, -1)$
3. Passing through $(1, -2)$, inclination $\alpha = 45°$
4. Passing through $(-1, 2)$, inclination $\alpha = -60°$
5. Passing through $(2, -3)$, slope 1
6. Passing through $(3, 1)$, parallel to $2x + y - 3 = 0$
7. Passing through $(1, -1)$, perpendicular to $3x - 2y = 6$
8. y intercept -2, inclination $\alpha = 120°$
9. x intercept 1, inclination $\alpha = -135°$
10. x intercept 2, perpendicular to the line $y = -2$

In problems 11 through 18, find the coordinates of the point P which is h of the way from P_1 to P_2. Plot the line and the points.

11. $P_1(2, 1)$, $P_2(8, 4)$, $h = \frac{2}{3}$
12. $P_1(1, 2)$, $P_2(6, -3)$, $h = \frac{1}{5}$
13. $P_1(3, 2)$, $P_2(1, -1)$, $h = 3$
14. $P_1(-1, 3)$, $P_2(1, 0)$, $h = -3$

15. $P_1(2, -1)$, $P_2(0, 1)$, $h = -\frac{1}{3}$ 16. $P_1(1, 1)$, $P_2(-2, -5)$, $h = -\frac{3}{5}$

17. $P_1(0, -2)$, $P_2(3, 1)$, $h = -2$ 18. $P_1(2, -1)$, $P_2(0, 3)$, $h = \frac{3}{2}$

In problems 19 through 22, find h so that P is h of the way from P_1 to P_2. Illustrate.

19. $P_1(2, 1)$, $P_2(2, -2)$, $P(4, -1)$ 20. $P_1(3, -5)$, $P_2(-2, 5)$, $P(1, -1)$

21. $P_1(-1, -1)$, $P_2(1, 4)$, $P(-\frac{1}{7}, \frac{8}{7})$ 22. $P_1(-2, 5)$, $P_2(0, 2)$, $P(4, -4)$

23. The point $(1, 3)$ is $\frac{1}{3}$ of the way from $P_1(3, 1)$ to P_2. Find the coordinates of P_2. Draw a figure.

24. Given $P_1(3, 1)$ and $P_2(0, 2)$; find the coordinates of the point or points on the line P_1P_2 such that $|PP_2| = 2|P_1P_2|$.

25. Show that the medians of any triangle $P_1(x_1, y_1)$, $P_2(x_2, y_2)$, $P_3(x_3, y_3)$ intersect at a point $\frac{2}{3}$ of the way from each vertex to the midpoint of the opposite side.

▶7. DISTANCE FROM A POINT TO A LINE. PROJECTIONS

The **distance from a point to a line** is defined as the length of the perpendicular segment dropped from the point to the line. The number representing this distance is always positive except that it is taken to be zero if the point happens to be on the line. The following theorem gives a simple formula for this distance.

Theorem 3. *The distance d from the point $P_0(x_0, y_0)$ to the line L whose equation is $Ax + By + C = 0$ is given by the formula*

$$d = \frac{|Ax_0 + By_0 + C|}{\sqrt{A^2 + B^2}}. \tag{1}$$

Proof. L has slope $-A/B$, so that the slope of the line L_0 through P_0 and perpendicular to L has slope B/A.

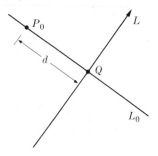

Figure 3-23

Accordingly, L_0 has the parametric equations

$$x = x_0 + t \cos \alpha_0, \qquad y = y_0 + t \sin \alpha_0 \tag{2}$$

where

$$\cos \alpha_0 = \frac{A}{\sqrt{A^2 + B^2}}, \qquad \sin \alpha_0 = \frac{B}{\sqrt{A^2 + B^2}}. \tag{3}$$

For (x, y) any point on L_0, the distance from (x_0, y_0) to (x, y) is just $|t|$. To see this we note from (2) that $(x - x_0) = t \cos \alpha_0$, $(y - y_0) = t \sin \alpha_0$ and, squaring these equations, we get $t^2 = (x - x_0)^2 + (y - y_0)^2 = d^2$. Now let $(\bar{x}, \bar{y})$ be the coordinates of Q (see Fig. 3–23) and suppose $\bar{t}$ is the value of t in (2) corresponding to Q. We have

$$A\bar{x} + B\bar{y} + C = 0 = Ax_0 + By_0 + C + \bar{t}(A \cos \alpha_0 + B \sin \alpha_0),$$

and, from (3),

$$0 = Ax_0 + By_0 + C + \bar{t}\sqrt{A^2 + B^2}.$$

Solving for $\bar{t}$, we obtain formula (1).

EXAMPLE 1. Find the distance d from the point $(2, -1)$ to the line

$$3x + 4y - 5 = 0.$$

Solution. We see that $A = 3$, $B = 4$, $C = -5$, $x_0 = 2$, $y_0 = -1$. Substituting in the formula, we get

$$d = \frac{|3 \cdot 2 + 4(-1) - 5|}{\sqrt{3^2 + 4^2}} = \frac{|-3|}{5} = \frac{3}{5}.$$

DEFINITION. The **distance between parallel lines** is the shortest distance from any point on one of the lines to the other line.

EXAMPLE 2. Find the distance between the parallel lines

$$L_1: 2x - 3y + 7 = 0$$

and

$$L_2: 2x - 3y - 6 = 0.$$

Solution. First find any point on one of the lines, say L_1. To do this, let x be any value and solve for y in the equation for L_1. If $x = 1$ in L_1, then $y = 3$ and the point $(1, 3)$ is on L_1. The distance from $(1, 3)$ to L_2 is

$$d = \frac{|2(1) - 3(3) - 6|}{\sqrt{2^2 + (-3)^2}} = \frac{|-13|}{\sqrt{13}}$$

$$= \frac{13}{\sqrt{13}} = \sqrt{13}.$$

The distance between the parallel lines is $\sqrt{13}$. The answer could be checked by taking any point on L_2 and finding its distance to L_1.

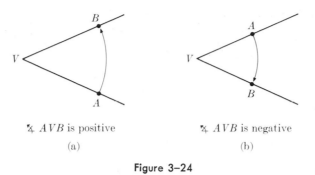

$\measuredangle AVB$ is positive

(a)

$\measuredangle AVB$ is negative

(b)

Figure 3–24

A **directed angle** is an angle in which one side is designated as the **initial side** and the other is designated as the **terminal side**. If, in an angle with vertex V, the point A is on the initial side and B is on the terminal side, then the directed angle from VA to VB is denoted by the symbol $\measuredangle AVB$. We continue to use the symbol $\angle AVB$ for an undirected angle. The measure, in degrees or radians, of $\measuredangle AVB =$ meas $\angle AVB$ if the angle is generated by revolving VA in a counterclockwise sense into VB. See Fig. 3–24(a). If the rotation is clockwise the measure of the directed angle is negative, i.e., $\measuredangle AVB = -$meas $\angle AVB$. See Fig. 3–24(b). If $\vec{L}_1$ and $\vec{L}_2$ are two directed lines (see Fig. 3–25a), then the **directed angle from** $\vec{L}_1$ **to** $\vec{L}_2$ is the directed angle formed by the positive ray of $\vec{L}_1$ as initial side and that of $\vec{L}_2$ as terminal side. If, as seen in Fig. 3–25(a), the inclination of $\vec{L}_1$ is α_1 and that of $\vec{L}_2$ is α_2, then the directed angle φ from $\vec{L}_1$ to $\vec{L}_2$ is given by

$$\varphi = \alpha_2 - \alpha_1.$$

In some cases φ may differ from $\alpha_2 - \alpha_1$ by either $360°$ or $-360°$ as shown in Fig. 3–25(b).

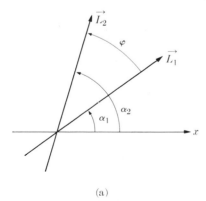

(a)

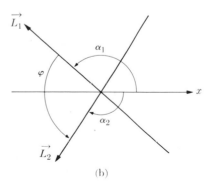

(b)

Figure 3–25

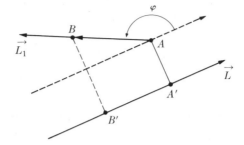

Figure 3–26

DEFINITION. Let A and B be points in R_2 and $\vec{L}$ a directed line. We define the **projection of $\overrightarrow{AB}$ on $\vec{L}$** as the number $A'B'$ where A' and B' are the feet of the perpendiculars dropped from A and B onto $\vec{L}$ (see Fig. 3–26). We denote this number by

$$\text{Proj}_{\vec{L}}\ \overrightarrow{AB}$$

and we also call it the **component of $\overrightarrow{AB}$ along $\vec{L}$.**

Theorem 4. *Suppose that A and B are given with coordinates (x_A, y_A) and (x_B, y_B), respectively, and that $\vec{L}$ is a directed line with inclination α. Then we have the formula*

$$\text{Proj}_{\vec{L}}\ AB = (x_B - x_A) \cos \alpha + (y_B - y_A) \sin \alpha. \tag{4}$$

Proof. Let $\vec{L}_1$ be the directed line through A and B directed so that A precedes B. (See Fig. 3–27.) Let α_1 be the inclination of $\vec{L}_1$ and suppose that φ is the angle from $\vec{L}$ to $\vec{L}_1$. Then we have

$$\cos \alpha_1 = \frac{x_B - x_A}{|AB|}, \qquad \sin \alpha_1 = \frac{y_B - y_A}{|AB|}.$$

From trigonometry, we know that

$$\begin{aligned}
\text{Proj}_{\vec{L}}\ \overrightarrow{AB} &= |AB| \cos \varphi = |AB| \cos (\alpha_1 - \alpha) \\
&= |AB| (\cos \alpha_1 \cos \alpha + \sin \alpha_1 \sin \alpha) \\
&= (|AB| \cos \alpha_1) \cos \alpha + (|AB| \sin \alpha_1) \sin \alpha,
\end{aligned}$$

from which we obtain (4).

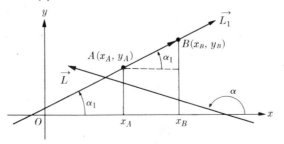

Figure 3–27

PROBLEMS

In problems 1 through 8, find the distance from the given points to the given lines. Draw figures.

1. $(1, -2)$, $4x + 3y + 12 = 0$ 2. $(-1, 2)$, $x + 2y - 2 = 0$
3. $(2, -1)$, $5x + 12y + 13 = 0$ 4. $(2, 1)$, $3x + 2y + 6 = 0$
5. $(3, 2)$, $x + 2y = 4$ 6. $(1, -3)$, $x = -4$
7. $(-2, 2)$, $y + 2 = 0$ 8. $(3, 2)$, $2x - y = 5$

In problems 9 through 12, find the distance between the given parallel lines. Draw a figure.

9. $4x - 3y - 12 = 0$ and $4x - 3y - 2 = 0$
10. $2x - y - 2 = 0$ and $4x - 2y + 7 = 0$
11. $4x + 5y = 0$ and $4x + 5y = 8$
12. $Ax + By + C = 0$ and $Ax + By + C' = 0$

In problems 13 through 16, the vertices of triangles are given. Find the lengths of the altitudes from the first vertex to the side joining the other two. Then find the areas of the triangles.

13. $(5, 1)$, $(0, 0)$, $(4, 5)$ 14. $(1, 1)$, $(2, 6)$, $(6, 4)$
15. $(-1, 3)$, $(3, -3)$, $(5, 3)$ 16. $(2, 1)$, $(0, 0)$, $(-1, 2)$

17. Find two points on the y axis which are at a distance 3 from the line $4x + 3y - 6 = 0$. Draw a figure.

18. Find two points on the x axis which are at a distance 5 from the line $2x + 3y = 6$. Draw a figure.

19. Find two points on the line $3x - 4y + 12 = 0$ which are at a distance 3 from the line $2x + 3y - 6 = 0$.

20. Find the equations of the two lines parallel to the line $4x + 3y - 12 = 0$ and at a distance 5 from it.

21. Find the equations of the two lines parallel to the line $x + 2y - 3 = 0$ and at a distance 2 from it.

22. Find the distance of the midpoint of the segment joining $(2, 4)$ and $(3, 2)$ from the line $x + 2y - 4 = 0$.

23. By means of the formula for the distance from a point to a line, find the two lines which bisect the angles made by L_1: $x + 2y - 3 = 0$ and L_2: $2x - y + 2 = 0$.

In problems 24 through 31, find the projection of $\overrightarrow{AB}$ on the directed line $\vec{L}$ satisfying the given conditions. Draw a figure.

24. $A(2, 3)$, $B(-1, 1)$, $\cos \alpha = \frac{3}{5}$, $\sin \alpha = \frac{4}{5}$.
25. $A(-3, 1)$, $B(0, 5)$, $\cos \alpha = \frac{4}{5}$, $\sin \alpha = -\frac{3}{5}$.
26. $A(-2, 1)$, $B(2, 4)$, $\alpha = 150°$
27. $A(1, -1)$, $B(3, 1)$, $\alpha = -120°$

28. $A(2, 0)$, $B(4, -2)$, $\vec{L}$ through $C(1, -3)$, $D(4, 1)$, directed from C to D.

29. $A(-2, -1)$, $B(4, -3)$, $\vec{L}$ through $C(1, 2)$ and $D(-1, -4)$ and directed from C to D.

30. $A(1, -2)$, $B(3, 4)$, $\vec{L}$ has slope 2 and is directed upwards.

31. $A(3, 1)$, $B(4, 4)$, $\vec{L}$ has slope $\frac{1}{2}$ and is directed to the left.

▶8.   FAMILIES OF LINES

The equation

$$y = 2x + k$$

represents a line of slope 2 for each value of k. The totality of lines obtained by letting k have any value—positive, negative, or zero—is called the **family of lines** of slope 2. In this instance, the family consists of a collection of parallel lines. More generally, a family of lines is a collection of lines which have a particular geometric property. The equation

$$y = kx + 3$$

represents for each value of k a line which has y intercept 3. The totality of such lines obtained by letting k take on all possible values is another example of a family of lines. Note that there is one line which passes through the point $(0, 3)$ which is not in this family: the y axis itself. Its equation is $x = 0$ and it is not included in the above family. It occasionally happens that families of lines have exceptions, and these must be taken into account. The following examples illustrate other families.

EXAMPLE 1.   Write the equation for the family of lines parallel to $3x + 2y + 4 = 0$.

Solution. The slope of every member of the family must be $-\frac{3}{2}$. Therefore $3x + 2y + k = 0$, where k can have any value, represents the family.

EXAMPLE 2.   Write the equation for the family of all lines passing through $(3, -2)$.

Solution. The point-slope form for the equation of a line through $(3, -2)$ is

$$y + 2 = m(x - 3),$$

and this represents every line through $(3, -2)$, with one exception. The line $x = 3$ must be added to the above family to get *all* lines through $(3, -2)$.

EXAMPLE 3.   Write the equation for the family of all lines having x intercept -2.

Solution. The two-intercept form for the equation of a line is

$$\frac{x}{a} + \frac{y}{b} = 1.$$

The equation $(x/-2) + (y/b) = 1$, where b can have any value, represents all lines with x intercept -2 except for two: the x axis ($y = 0$) and the line $x = -2$.

The two lines

$$L_1: 3x - 2y + 7 = 0 \quad \text{and} \quad L_2: 2x + y - 6 = 0$$

intersect at a point (call it P) which may be found by solving the equations simultaneously. The first degree equation

$$(3x - 2y + 7) + k(2x + y - 6) = 0$$

represents a family of lines passing through the intersection of L_1 and L_2. To see this we first note that no matter what k is, the above line must pass through P because $3x - 2y + 7 = 0$ at P and $2x + y - 6 = 0$ at P. As k varies, the equation changes, so the above system represents a family of lines through P. The family consists of *all* lines through P, with one exception. No value of k gives L_2 itself. Recombining the terms, we conclude that the equation

$$(3 + 2k)x + (k - 2)y + 7 - 6k = 0$$

represents the family of all lines through P except for L_2.

EXAMPLE 4. Find the equation of the line passing through $(2, -3)$ and the intersection of the lines

$$L_1: 3x - y + 8 = 0$$

and

$$L_2: 2x + 5y + 7 = 0.$$

Solution. We could find the intersection point of L_1 and L_2 and then use the two-point form of the equation of a line to get the answer. Instead we shall use the idea of a family of lines. The equation

$$3x - y + 8 + k(2x + 5y + 7) = 0 \tag{1}$$

represents the family through the intersection of L_1 and L_2. The particular member passing through $(2, -3)$ must satisfy

$$3(2) - (-3) + 8 + k[2(2) + 5(-3) + 7] = 0.$$

This equation yields

$$k = \tfrac{17}{4}$$

and, substituting this value of k in the equation of the family given by (1), we obtain

$$3x - y + 8 + \tfrac{17}{4}(2x + 5y + 7) = 0,$$

$$\Leftrightarrow 46x + 81y + 151 = 0.$$

PROBLEMS

In problems 1 through 12, write the equations of the families of lines satisfying the given conditions.

1. Parallel to the line $2x + y - 4 = 0$

2. Having y intercept -2

3. Having x intercept 3

4. Passing through the point $(1, -3)$

5. Passing through the point $(-1, 2)$

6. Perpendicular to the line $2x + y - 3 = 0$

7. Having equal x and y intercepts

8. Having x intercept equal to the negative of the y intercept

9. Having y intercept equal to twice the x intercept

10. Having the product of the intercepts equal to 6

11. Having the sum of the intercepts $= 5$

12. The area of the right triangle formed by the line and the axes being 6

13. Find the equation of the line passing through $(5, -1)$ and through the intersection of the lines $2x - y + 3 = 0$ and $x + 2y - 7 = 0$.

14. Find the equation of the line passing through $(2, -4)$ and through the intersection of the lines $3x + 4y + 1 = 0$ and $2x + 6y + 3 = 0$.

15. Find the equation of the line passing through the intersection of the lines $x + 2y - 3 = 0$ and $2x + y - 3 = 0$ and having slope 2.

16. Find the equation of the line passing through the intersection of the lines $2x - y + 4 = 0$ and $x + 2y + 6 = 0$ and having slope -3.

17. Find the equation of the line passing through the intersection of the lines $x - 2y + 3 = 0$ and $x - y + 1 = 0$ and having y intercept -3.

18. Find the equations of the lines which bisect the angles formed by the lines $L_1: x - y + 4 = 0$ and $L_2: 4x + 3y - 5 = 0$.

19. Find the equations of the lines which pass through the intersection of the lines $\overrightarrow{L_1}: x - 2y - 3 = 0$ and $\overrightarrow{L_2}: x + y - 4 = 0$ and which are two units from the origin.

20. Find the equation of the line which belongs to both the families $2x + 3y = k$ and $3x + my = 6$.

21. Find the member or members of the family $y = 2x + k$ which intersect the curve $x^2 + y^2 = \frac{4}{5}$ in exactly one point.

▶9. ANGLE BETWEEN TWO LINES. BISECTORS OF ANGLES

We recall that the inclination α of a line L is the angle that the upward pointing ray of L makes with the positive x axis. If the line is parallel to the x axis, its inclination is zero. The slope m of the line L is $m = \tan \alpha (\alpha \neq 90°)$. Two nonparallel lines L_1 and L_2 make four angles at their intersection: two equal obtuse angles and two equal acute angles (unless the lines are perpendicular). The obtuse angle is the supplement of the acute angle.

The angle swept out when the line L_1 is rotated counterclockwise to L_2 about the point of intersection is called the **angle from L_1 to L_2**. Let φ be the angle (measured in radians) from L_1 to L_2 (see Fig. 3–28). Then, clearly, the angle from L_2 to L_1 is $\pi - \varphi$.

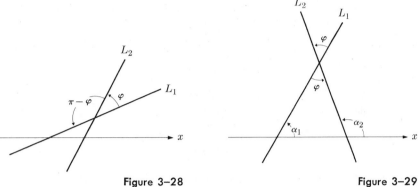

Figure 3–28   Figure 3–29

Let L_1 and L_2 be two intersecting lines (neither of which is vertical) with inclinations α_1 and α_2 and slopes $m_1 = \tan \alpha_1$ and $m_2 = \tan \alpha_2$, respectively. Let φ be the angle from L_1 to L_2 and suppose $\alpha_1 < \alpha_2$. Then, since the exterior angle of a triangle is the sum of the remote interior angles (see Fig. 3–29), we have

$$\alpha_2 = \alpha_1 + \varphi, \qquad \varphi = \alpha_2 - \alpha_1, \qquad \tan \varphi = \tan (\alpha_2 - \alpha_1).$$

Using the formula for the tangent of the difference of two angles, we get

$$\tan \varphi = \frac{\tan \alpha_2 - \tan \alpha_1}{1 + \tan \alpha_1 \tan \alpha_2}.$$

In terms of slopes, we obtain

$$\tan \varphi = \frac{m_2 - m_1}{1 + m_1 m_2}, \tag{1}$$

which is the formula for *the tangent of the angle from L_1 to L_2*. The derivation above assumed that $\alpha_2 > \alpha_1$. In case $\alpha_1 > \alpha_2$, we see that

$$\pi - \varphi = \alpha_1 - \alpha_2, \qquad \tan (\pi - \varphi) = \frac{m_1 - m_2}{1 + m_1 m_2}.$$

Since $\tan (\pi - \varphi) = -\tan \varphi$, the formula (1) holds in this case also.

EXAMPLE 1. Find the tangent of the angle from the line $L_1 = \{(x, y): 2x + 3y = 5\}$ to the line $L_2 = \{(x, y): 4x - 3y = 2\}$.

Solution. Here $m_1 = -\frac{2}{3}$ and $m_2 = \frac{4}{3}$. The formula yields

$$\tan \varphi = \frac{\frac{4}{3} - (-\frac{2}{3})}{1 + (-\frac{2}{3})(\frac{4}{3})} = 18.$$

Formula (1) fails if one of the lines is vertical. We divide both the numerator and the denominator by m_2 in the formula for $\tan \varphi$, getting

$$\tan \varphi = \frac{1 - (m_1/m_2)}{(1/m_2) + m_1}.$$

Consider now that L_2 becomes vertical; that is, $m_2 \to \infty$ as L_2 approaches a vertical line. Then $m_1/m_2 \to 0$, $1/m_2 \to 0$. Therefore the formula for the angle from L_1 to L_2 when L_2 is vertical is simply

$$\tan \varphi = \frac{1}{m_1}.$$

From plane geometry we recall that the angle bisectors of two given intersecting lines represent the locus of all points which are equidistant from the given lines. The equations of these angle bisectors are found by using the formula for the distance from a point to a line and the methods of solving locus problems. The following example illustrates the technique.

EXAMPLE 2. Find the equations of the bisectors of the angles formed by the lines $3x + 4y - 7 = 0$ and $4x + 3y + 2 = 0$.

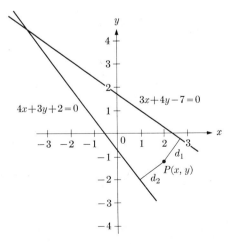

Figure 3–30

Solution. We draw a figure, as shown (Fig. 3–30), and let $P(x, y)$ be a point on the locus (angle bisector, in this case). Let d_1 be the distance from P to the line $3x + 4y - 7 = 0$ and d_2 the distance from P to the line $4x + 3y + 2 = 0$. The conditions of the locus are $d_1 = d_2$. Then, from the formula for the distance from a point to a line, we have

$$d_1 = \frac{|3x + 4y - 7|}{\sqrt{3^2 + 4^2}} \quad \text{and} \quad d_2 = \frac{|4x + 3y + 2|}{\sqrt{4^2 + 3^2}}.$$

The conditions of the locus state that

$$|3x + 4y - 7| = |4x + 3y + 2|.$$

At this time we make use of our knowledge of absolute values. If $|a| = |b|$, then either $a = b$ or $a = -b$. This tells us that either

$$3x + 4y - 7 = 4x + 3y + 2 \quad \text{or} \quad 3x + 4y - 7 = -(4x + 3y + 2).$$

Simplifying, we obtain

$$x - y + 9 = 0 \quad \text{and} \quad 7x + 7y - 5 = 0$$

as the equations of the two angle bisectors. Note that the angle bisectors have slopes 1 and -1 and are therefore perpendicular, as we know they should be.

To distinguish the bisectors from each other, we have merely to find the tangent of the angle from one of the given lines to the other, and then the tangent of the angle from one of the given lines to the angle bisectors. A comparison of the sizes indicates where the angle bisectors fall.

PROBLEMS

In problems 1 through 9, find $\tan \varphi$, where φ is the angle from L_1 to L_2.

1. $L_1: x + 2y + 1 = 0$; $L_2: x - y + 2 = 0$
2. $L_1: x + y - 3 = 0$; $L_2: 2x + y + 3 = 0$
3. $L_1: 2x - y + 3 = 0$; $L_2: x + y - 2 = 0$
4. $L_1: 2x + y - 2 = 0$; $L_2: x + 2y = 4$
5. $L_1: x + 2y = 3$; $L_2: 2x - y = 4$
6. $L_1: 3x + y = 0$; $L_2: 3x - 2y = 5$
7. $L_1: x - 2y = 3$; $L_2: y = 3$
8. $L_1: 2x + 2y = 3$; $L_2: x = 2$
9. $L_1: 3x + 2y = 6$; $L_2: x = -2$

In problems 10 through 13, find the tangents of the angles of the triangles ABC with the vertices given.

10. $A(-2, 2)$, $B(4, -1)$, $C(4, 4)$
11. $A(1, -2)$, $B(-1, 0)$, $C(3, 4)$
12. $A(0, -2)$, $B(3, 0)$, $C(-2, 5)$
13. $A(-2, 0)$, $B(3, -1)$, $C(-1, 3)$

In problems 14 through 18, find the equations of the lines through the given point P_0 making the given angle θ with the given line L.

14. $P_0 = (2, -1)$, $\theta = \pi/4$, $L: 2x - y = 4$

15. $P_0 = (1, 1)$, $\tan \theta = -\frac{1}{2}$, $L: x + y - 2 = 0$

16. $P_0 = (2, 3)$, $\tan \theta = 2$, $L: 2x + 3y = 6$

17. $P_0 = (-2, -3)$, $\theta = \pi/6$, $L: x - y - 2 = 0$

18. $P_0 = (1, 2)$, $\theta = \pi/2$, $L: x - 2y = 1$

In problems 19 through 23, find the equations of the two bisectors of the angles formed by the given lines. Draw figures.

19. $4x - 3y = 6$, $3x + 4y = 12$ 20. $2x + y = 3$, $x - 2y = 2$

21. $2x + y = 3$, $2y - 11x = 25$ 22. $y = 0$, $3x - 2y = 0$

23. $x = 3$, $3x - 2y + 2 = 0$

24. Given a triangle with vertices $A(2, 1)$, $B(7, 1)$, $C(5, 5)$; show that the three angle bisectors meet in a point.

25. Show that the bisectors of the angles of any triangle meet in a point.

In problems 26 through 28, find the equation of the bisector of $\angle AVB$. Draw a figure.

26. $A(1, 0)$, $V(-1, 2)$, $B(6, 3)$ 27. $A(0, 2)$, $V(-3, 1)$, $B(5, 1)$

28. $A(3, -3)$, $V(-1, 0)$, $B(2, 4)$

29. Find the equation of the family or families of lines which make an angle of $\pi/4$ with the line $2x + y = 5$.

30. The points $A(1, 0)$, $B(5, 0)$, $C(7, 4)$, $D(3, 8)$ are the vertices of a quadrilateral. Find the tangent of the angle the diagonals make with each other.

▶10.   GEOMETRIC THEOREMS

Many theorems in plane geometry can be proved most easily if we introduce a Cartesian coordinate system and use the methods of analytic geometry. Since we shall be concerned with lines, triangles, quadrilaterals, and so forth, the location of the particular figures with respect to the coordinate axes is of the utmost importance. The work in solving a problem is often made simpler by a strategic selection of the coordinate axes. We illustrate the idea with two examples.

EXAMPLE 1.   Suppose that $ABCD$ is a parallelogram with $CD \parallel AB$ and $AD \parallel BC$. See Fig. 3–31. We choose a coordinate system with the origin at A and with the positive x axis along the line segment going from A to B. The positive direction of the y axis is chosen so that C and D have positive y coordinates. Denote the coordinates of B by $(a, 0)$, where a is a positive number; denote the coordinates of D by (b, c). Show that the coordinates of C are $(a + b, c)$.

Solution.   Let (x_C, y_C) denote the coordinates of C. Since $CD \parallel AB$, the line CD is parallel to the x axis and so the y coordinate of C is equal to the y coordinate of D;

that is, $y_C = c$. As for the x coordinate of C, we observe that since $BC \parallel AD$, these lines have the same slope. Therefore

$$\text{slope } AD = \frac{c}{b} = \text{slope } BC = \frac{c}{x_C - a}, \qquad \text{if} \quad x_C \neq a.$$

Solving for x_C, we get $x_C = a + b$.

If $x_C = a$, we have $b = 0$ and the result follows in all cases.

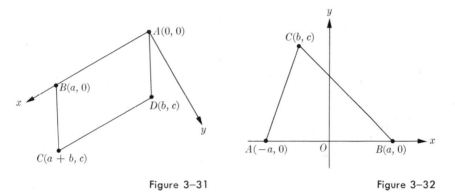

Figure 3–31   Figure 3–32

EXAMPLE 2. Prove that the perpendicular bisectors of the sides of a triangle meet in a point.

Solution. Let $\triangle ABC$ be given. We choose the x axis along AB (see Fig. 3–32), and we select the y axis along the perpendicular bisector of the line segment AB. The positive y axis is taken in the direction of C. With this selection of a coordinate system A has coordinates $(-a, 0)$, with a some positive number, and B has coordinates $(+a, 0)$. The point C may be anywhere (not on the x axis) and we label its coordinates (b, c).

We now obtain the equations of the perpendicular bisectors of AC and BC. The slope of AC is $c/(b + a)$ and the slope of the perpendicular bisector is $-(b + a)/c$. The midpoint of the segment AC has coordinates $((b - a)/2, c/2)$. From the point-slope form for the equation of a line, we obtain for the perpendicular bisector of AC

$$y - \frac{c}{2} = -\frac{(b + a)}{c}\left(x - \frac{b - a}{2}\right). \tag{1}$$

Similarly, the perpendicular bisector of the line segment BC has the equation

$$y - \frac{c}{2} = -\frac{(b - a)}{c}\left(x - \frac{a + b}{2}\right). \tag{2}$$

By solving simultaneously, we see that the point of intersection of lines (1) and (2) is on the y axis. (The student will find it instructive to carry out the necessary steps.) Since we selected the y axis as the perpendicular bisector of AB, the result follows.

In the examples above, we assumed that the given figure was in any position in the plane, and then chose axes to simplify the coordinates of the points. Since the sides of the objects considered are arbitrary in length, we are required to use letters instead of numbers for the coordinates of the various points in the figures. The student should be careful to use letters to denote such unknown magnitudes, for otherwise he would be verifying the theorem only in a special case. Since figures may usually be translated and rotated without loss of generality, it is always possible to choose the x axis horizontally. This was done in Example 2, whereas in Example 1 the axes were left in an arbitrary position for emphasis.

Actually, it is possible to develop all of Euclidean plane geometry by the methods of analytic geometry. However, it is also true that synthetic methods (that is, the methods using traditional geometric proofs of theorems) are sometimes easier than analytic methods. For this reason both techniques continue to develop, and it is desirable that the student become familiar with the methods of both synthetic and analytic geometry.

PROBLEMS

Prove the following theorems by the methods of this section.

1. The sum of the squares of the distances of any point in the plane to two opposite vertices of a rectangle equals the sum of the squares of its distances from the other two vertices.

2. If D is the midpoint of side BC in $\triangle ABC$, then
$$|AB|^2 + |AC|^2 = 2|AD|^2 + 2|BD|^2.$$

3. The lines joining the midpoints of the opposite sides of a quadrilateral bisect each other.

4. The lines joining the midpoints of adjacent sides of a quadrilateral form a parallelogram.

5. If the diagonals of a parallelogram are equal, the parallelogram is a rectangle.

6. The diagonals of a parallelogram bisect each other.

7. If two medians of a triangle are equal, the triangle is isosceles.

8. The altitudes on the legs of an isosceles triangle are equal.

9. The medians of a triangle meet at a point.

10. The altitudes of a triangle meet at a point.

11. In a trapezoid, the diagonals and the line joining the midpoints of the parallel sides meet in a point.

12. In any parallelogram $ABCD$ (vertices in order around it), the vertex D, the midpoint of AB, and the point E on AC which is $\frac{1}{3}$ of the way from A to C lie on a line.

13. In a trapezoid, the nonparallel sides and the line joining the midpoints of the parallel sides meet in a point.

14. If two altitudes of a triangle are equal, the triangle is isosceles.

15. In $\triangle ABC$, $|AB| + |BC| > |AC|$.

16. In a regular hexagon, the diagonals joining opposite vertices meet in a point.

▶**11.   LINEAR INEQUALITIES IN TWO UNKNOWNS**

We are interested in finding the solution set of a linear inequality such as

$$2x - 3y + 6 > 0.$$

First, we observe that the locus

$$L = \{(x, y): 2x - 3y + 6 = 0\}$$

is a straight line as shown in Fig. 3–29. Next we note that

$$2x - 3y + 6 > 0 \Leftrightarrow y < \tfrac{2}{3}x + 2.$$

Suppose $P_0(x_0, y_0)$ is a point on L. Then we have

$$2x_0 - 3y_0 + 6 = 0 \quad \text{or} \quad y_0 = \tfrac{2}{3}x_0 + 2.$$

If we keep x_0 fixed, then we see that any point (x_0, y) with y less than y_0 satisfies the inequality

$$y < \tfrac{2}{3}x_0 + 2.$$

Geometrically, we draw the line $x = x_0$ as shown in Fig. 3–33 and conclude that every point on this line *below* P_0 satisfies the inequality $2x - 3y + 6 > 0$. Denoting by S the set

$$S = \{(x, y): 2x - 3y + 6 > 0\},$$

we find, since P_0 may be *any* point on L, that every point in the plane below L is in S. The set S is shown as the shaded portion in Fig. 3–34.

Instead of employing the inequality $y < \tfrac{2}{3}x + 2$, we could use the equivalence

$$2x - 3y + 6 > 0 \Leftrightarrow x > \tfrac{3}{2}y - 3.$$

Since a point P_0 on L satisfies $x_0 = \tfrac{3}{2}y_0 - 3$, we keep y_0 fixed and conclude that any point $P(x, y_0)$ with $x > x_0$ is in S. Geometrically, this statement says that every point on the line $y = y_0$ which lies to the right of P_0 is in S (Fig. 3–33).

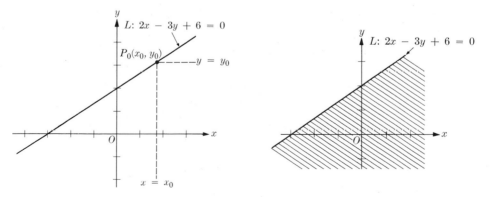

Figure 3–33   Figure 3–34

A set consisting of all points on one side of a line is called a **half-plane.** The solution set of the opposite inequality $2x - 3y + 6 < 0$ consists of the half-plane shown as the unshaded portion of Fig. 3–34.

Suppose we wish to find the solution set of a pair of simultaneous linear inequalities. Let S_1 be the solution set of the first inequality and S_2 the solution set of the second. Then a point (x, y) in the number plane, R_2, satisfies both inequalities if and only if it lies in both solution sets. By definition, the totality of such points is precisely the **intersection** of the two sets and is denoted by $S_1 \cap S_2$. (This terminology and notation was introduced in Section 2 of Chapter 1.) The same idea holds for the solution set S of several simultaneous inequalities, where S is then composed of the intersection of the solution sets of the individual inequalities.

EXAMPLE 1. Draw a graph of the solution set of the pair of inequalities

$$2x - y - 4 > 0 \quad \text{and} \quad 3x + y - 11 > 0.$$

Solution. We introduce the notation

$$L_1: 2x - y - 4 = 0,$$
$$L_2: 3x + y - 11 = 0,$$
$$S_1 = \{(x, y): 2x - y - 4 > 0\},$$
$$S_2 = \{(x, y): 3x + y - 11 > 0\}.$$

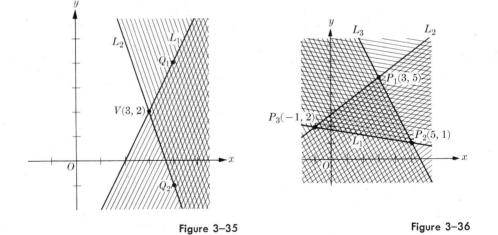

Figure 3–35   Figure 3–36

By solving simultaneously the equations for L_1 and L_2, we obtain their point of intersection. The result is $V(3, 2)$ as shown in Fig. 3–35. Then we have

$$2x - y - 4 > 0 \Leftrightarrow y < 2x - 4$$
$$\Leftrightarrow x > \tfrac{1}{2}y + 2.$$

Using either of the inequalities on the right above, we see that S_1 consists of the shaded half-plane to the right of L_1 (Fig. 3–35). Similarly,

$$3x + y - 11 > 0 \Leftrightarrow y > -3x + 11$$
$$\Leftrightarrow x > -\tfrac{1}{3}y + \tfrac{11}{3},$$

so that S_2 is the shaded half-plane to the right of L_2, as shown. The solution to the pair of inequalities is the *doubly shaded* portion consisting of those points in the interior of the angle Q_1VQ_2 in Fig. 3-35.

EXAMPLE 2. Describe and draw a graph of the solution set of the three inequalities

$$x + 6y - 11 > 0, \qquad 3x - 4y + 11 > 0, \qquad -2x - y + 11 > 0.$$

Solution. Let L_1, L_2, and L_3 be the straight lines which are the solution sets of the corresponding *equations*. Denote by S_1, S_2, and S_3 the solution sets

$$S_1 = \{(x, y): x + 6y - 11 > 0\}, \qquad S_2 = \{(x, y): 3x - 4y + 11 > 0\},$$
$$S_3 = \{(x, y): -2x - y + 11 > 0\}.$$

Then S_1 is the shaded half-plane above the line L_1 as shown in Fig. 3–36. The set S_2 is the shaded half-plane below L_2, and S_3 is the shaded half-plane below L_3. By solving simultaneously the equations for L_2 and L_3 we get $P_1(3, 5)$ as the point of intersection. Similarly $P_2(5, 1)$ is the point of intersection of L_1 and L_3; finally, $P_3(-1, 2)$ is the point of intersection of L_1 and L_2. The desired solution set is $S_1 \cap S_2 \cap S_3$ and this set is the triply shaded triangle, $\triangle P_1P_2P_3$, shown in Fig. 3–36.

PROBLEMS

In problems 1 through 11, describe and plot a graph of the solution set (which might be empty) of the given system of inequalities.

1. $2x + y - 3 > 0$ and $x - y + 2 < 0$
2. $-x + y - 2 < 0$ and $2x + y - 4 > 0$
3. $x + 2y - 2 \geq 0$ and $2x + y - 2 > 0$
4. $2x - 6y + 12 > 0$ and $6x + 4y - 8 > 0$
5. $2x + 3y - 6 \geq 0$ and $3x - 2y + 5 \geq 0$
6. $3x + 4y + 5 \geq 0$ and $2x - y + 3 \leq 0$
7. $2x - y - 1 > 0$, $-6x + 8y + 13 > 0$, $-8x - y + 14 > 0$
8. $-2x + 2y + 7 > 0$, $12x + y - 3 > 0$, $-8x - 5y + 15 > 0$
9. $4x + y - 2 > 0$, $-4x - y + 8 > 0$
10. $2x + 1 > 0$, $y + 2 > 0$, $-4x - y + 6 > 0$, $-y + 2 > 0$
11. $2x - 1 \geq 0$, $y + 7 \geq 0$, $2x - y - 1 \geq 0$, $-4x - y + 16 > 0$
12. Can every trapezoid be described by a system of four linear inequalities? Prove your result.

13. What types of quadrilaterals can be described by four linear inequalities?

14. State a condition which guarantees that a polygon with n sides can be described by n linear inequalities.

15. Give an example of a system of three linear inequalities in which the solution set is not empty and is not a triangle. Describe all possible types of regions a system of three linear inequalities might depict.

▶12. SYSTEMS OF LINEAR INEQUALITIES; RELATION TO LINEAR PROGRAMMING

In Section 11 we discussed regions in the plane bounded by straight lines which are defined by two or three linear inequalities. Suppose we have 10 such inequalities, or 1000. We may ask two questions: (1) Are there any points in the plane which satisfy all these inequalities? (2) If so, how would we go about finding them? To simplify the discussion, we shall consider a particular example and give some hints about the more general case.

Suppose we have the following five inequalities:

$$L_1: \quad x + y - 4 \geq 0; \quad L_2: \quad x + 3y - 8 \geq 0;$$
$$L_3: \quad 8x + 7y - 27 \geq 0; \quad L_4: \quad -4x - y + 21 \geq 0;$$
$$L_5: \quad 2x - 3y + 7 \geq 0.$$

To learn if there are any points which satisfy all the inequalities, we start by replacing the first two inequalities by the equations

$$x + y - 4 = 0 \quad \text{and} \quad x + 3y - 8 = 0$$

and solving them simultaneously. The solution of this system is the point $P_1(2, 2)$. We know that this point satisfies L_1 and L_2. We substitute in the remaining three inequalities, to obtain

$$8(2) + 7(2) - 27 \geq 0, \text{ yes};$$
$$-4(2) - 2 + 21 \geq 0, \text{ yes};$$
$$2(2) - 3(2) + 7 \geq 0, \text{ yes}.$$

We conclude that P_1 is a point which satisfies all the inequalities. Next we try two other inequalities, say L_1 and L_3. Solving the equations

$$x + y - 4 = 0 \quad \text{and} \quad 8x + 7y - 27 = 0$$

simultaneously, we get $P_2(-1, 5)$. We substitute this in the remaining three inequalities:

$$-1 + 3(5) - 8 \geq 0, \text{ yes};$$
$$-4(-1) - 5 + 21 \geq 0, \text{ yes};$$
$$2(-1) - 3(5) + 7 \geq 0, \text{ no!}$$

So P_2 is not a point which satisfies all the inequalities. Continuing in this way by solving all possible pairs of equations corresponding to L_1 through L_5, we find that of the ten points of intersection the four points

$$P_1(2, 2), \quad P_3(5, 1), \quad P_4(4, 5), \quad P_5(1, 3)$$

satisfy all the inequalities, while the other six do not. Figure 3–37 shows the four points. It can be shown that the shaded quadrilateral contains all the points which satisfy the system L_1 through L_5. The points P_1, P_3, P_4, P_5 are the vertices of this quadrilateral.

We now give a general procedure for determining the solution set of an arbitrary number, say n, of linear inequalities. Let l_i be the bounding line of the solution set of the ith inequality. First, we find all possible points of intersection of the l_i by pairs. To do so, we must solve $n(n-1)/2$ pairs of simultaneous linear equations. Second, each point of intersection is substituted into the remaining $n-2$ inequalities to see if all of them are satisfied at the point. Third, if no point of intersection satisfies all the inequalities, it can be shown that the solution set S is empty. Fourth, if the points of intersection $P_1, \ldots, P_k$ belong to S, then the boundary of S consists of segments or rays of certain of the l_i, or may contain an entire l_i (if $n = 2$ and $l_1 \parallel l_2$). If each of these particular l_i contains two of the P_j, then S is the interior and boundary of the convex* polygon having the P_j as vertices (as in Fig. 3–37). But it can happen that exactly two of these l_i contain only one P_j; in this case S is unbounded and its boundary consists of a ray of each of these two l_i and segments of the remaining particular l_i (as in Fig. 3–38). An angle is the special case where $n = 2$ and l_1 and l_2 intersect at P_1.

Clearly, if $n = 100$ the task is almost hopeless, and if $n = 1000$ or $10,000$, the problem staggers the imagination. However, with modern electronic computers

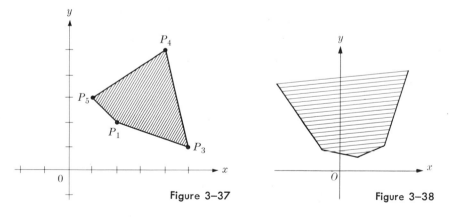

Figure 3–37   Figure 3–38

---

* A set is called **convex** if and only if the line segment joining each two points in the set is also in the set.

capable of performing thousands of additions, subtractions, multiplications, and divisions each second, problems of this type become feasible.

Systems of linear inequalities arise in many branches of engineering, physics, economics, statistics, and so forth. In fact, the problem usually doesn't end when the region satisfying the inequalities is obtained. Further questions may be asked. For example, what point in the region is closest to the x axis? Does the region intersect a specific line? What point in the region is farthest from the origin? Computers can answer these and many other questions by modern programming methods.

In actual practice, the problems are much more complicated (usually involving many unknowns), and as a result the techniques developed for determining the region in question are much more sophisticated than the one outlined above.

PROBLEMS

In problems 1 through 4, find the region, if any, satisfied by each of the systems of inequalities.

1. $x + y - 1 \geq 0, x + y - 3 \geq 0, x - 2 \geq 0, x + 2y - 1 \geq 0, -y + 2 \geq 0,$
   $-x + 6 \geq 0$

2. $x \geq 0, y \geq 0, x - y + 1 \geq 0, -x - y + 5 \geq 0, -3x - y + 8 \geq 0$

3. $-x \geq 0, y \geq 1, x - 2y + 4 \geq 0, x - 3y + 4 \geq 0, -x - y + 7 \geq 0$

4. $x + y + 3 \geq 0, \quad 2x - y + 6 \geq 0, \quad -x + 7 \geq 0, \quad -x - y + 12 \geq 0,$
   $-y + 8 \geq 0, x - 2y \geq 0, -x - 3y \geq 0$

5. Does the solution set in problem 1 intersect the x axis?

6. In the solution set of problem 2, find the point which is closest to the line $x = -10$.

7. In the solution set of problem 3, find the point which is closest to the line $x - y = 30$.

8. Does the line $x - 2y = 7$ intersect the solution set of the inequalities of problem 4?

Vectors in a Plane 4

▶1. DIRECTED LINE SEGMENTS AND VECTORS

Let A and B be two points in a plane. We recall that the directed segment from A to B is defined as the segment AB which is ordered so that A precedes B. (see Section 6 of Chapter 3). We shall use the symbol $\overrightarrow{AB}$ to denote such a directed segment, and we call A its **base** and B its **head**. The directed segment $\overrightarrow{BA}$ is the same line segment AB with the opposite ordering; in this case B is the base and A the head. To distinguish base and head we usually draw an arrow at the head as shown in Fig. 4–1. The **magnitude** of a directed line segment $\overrightarrow{AB}$ is its length $|AB|$.

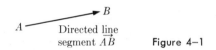

Directed line
segment $\overrightarrow{AB}$ Figure 4–1

Two directed line segments $\overrightarrow{AB}$ and $\overrightarrow{CD}$ are said to **have the same magnitude and direction** if and only if

$$\text{Proj}_{\vec{L}}\ \overrightarrow{AB} = \text{Proj}_{\vec{L}}\ \overrightarrow{CD}$$

(see Section 8 of Chapter 3) for every directed line $\vec{L}$ (see Fig. 4.2).

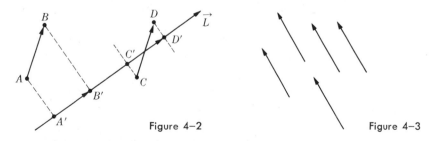

Figure 4–2   Figure 4–3

Figure 4–3 shows several line segments having the same magnitude and direction. Whenever two directed line segments $\overrightarrow{AB}$ and $\overrightarrow{CD}$ have the same magnitude and direction, we say they are **equivalent** and write

$$\overrightarrow{AB} \approx \overrightarrow{CD}.$$

We have seen in Chapter 2, Section 1, and Chapter 3, Section 1, that there are many ways of introducing a Cartesian coordinate system in a Euclidean plane. It is usually convenient to denote such coordinate systems by ordered pairs of numbers such as (x, y) or (r, s). If A is a point in the plane we frequently denote its (x, y) coordinate by (x_A, y_A) and its (r, s) coordinates by (r_A, s_A).

We recall the definition of the projection of a directed line segment $\overrightarrow{AB}$ on a directed line $\vec{L}$ given in Chapter 3, Section 7. From the very definition it is clear that this quantity is a *geometric* one, that is, its value does not depend on the particular coordinate system. Hence the relation $\overrightarrow{AB} \approx \overrightarrow{CD}$ is also geometric in character. We now establish a relationship for equivalent directed line segments in terms of a coordinate system.

Theorem 1. *Suppose that A, B, C, and D are points in the plane with coordinates (x_A, y_A), (x_B, y_B), (x_C, y_C), and (x_D, y_D), respectively. If the equations*

$$x_B - x_A = x_D - x_C \qquad \text{and} \qquad y_B - y_A = y_D - y_C \tag{1}$$

hold, then $\overrightarrow{AB} \approx \overrightarrow{CD}$. Conversely, if $\overrightarrow{AB} \approx \overrightarrow{CD}$, then equations (1) hold for any Cartesian coordinate system.

Proof. Assume equations (1) hold. For any directed line $\vec{L}$,

$$\text{Proj}_{\vec{L}} \overrightarrow{AB} = (x_B - x_A) \cos \alpha + (y_B - y_A) \sin \alpha \tag{2}$$

where α is the inclination of $\vec{L}$. Now using (1), we have

$$\text{Proj}_{\vec{L}} \overrightarrow{AB} = (x_D - x_C) \cos \alpha + (y_D - y_C) \sin \alpha = \text{Proj}_{\vec{L}} \overrightarrow{CD}. \tag{3}$$

Since $\vec{L}$ is any directed line, we conclude that $\overrightarrow{AB} \approx \overrightarrow{CD}$.

Now suppose that $\overrightarrow{AB} \approx \overrightarrow{CD}$. Then, since $\text{Proj}_{\vec{L}} \overrightarrow{AB} = \text{Proj}_{\vec{L}} \overrightarrow{CD}$, we find from (2) and (3) that

$$(x_B - x_A) \cos \alpha + (y_B - y_A) \sin \alpha = (x_D - x_C) \cos \alpha + (y_D - y_C) \sin \alpha.$$

This relation is valid for every line $\vec{L}$ and hence for all values of α. If $\alpha = 0$, we get $x_B - x_A = x_D - x_C$. If $\alpha = \pi/2$, we get $y_B - y_A = y_D - y_C$. That is, equations (1) hold and the proof is complete.

Remark. Equations (1) in the above Theorem may be interpreted geometrically. They state that two directed line segments are equivalent if either one of the following two conditions holds:

(i) $\overrightarrow{AB}$ and $\overrightarrow{CD}$ are both on the same directed line $\vec{L}$ and their directed lengths are equal; or

(ii) the points A, B, C, and D are the vertices of a parallelogram with $\overrightarrow{AB}$ and $\overrightarrow{CD}$ as two parallel sides.

The proof of the above statement follows immediately from equations (1) in Theorem 1.

If we are given a directed line segment $\overrightarrow{AB}$, we see at once that there is an unlimited number of equivalent ones. In fact, if C is any given point in the plane, we can use Eqs. (1) of Theorem 1 to find the coordinates of the unique point D such that $\overrightarrow{CD} \approx \overrightarrow{AB}$. Theorem 1 also yields various simple properties of the relation $\approx$. For example, if $\overrightarrow{AB} \approx \overrightarrow{CD}$, then $\overrightarrow{CD} \approx \overrightarrow{AB}$; also if $\overrightarrow{AB} \approx \overrightarrow{CD}$ and $\overrightarrow{CD} \approx \overrightarrow{EF}$, then $\overrightarrow{AB} \approx \overrightarrow{EF}$ (see problem 23 in Section 2 below).

The definition of a vector involves an abstract concept—that of a collection of directed line segments.

DEFINITION. A **vector** is a collection of all directed line segments having a given magnitude and a given direction. We shall use boldface letters to denote vectors; thus when we write **v** for a vector, it stands for an entire collection of directed line segments. A particular directed line segment in the collection is called a **representative** of the vector **v**. Any member of the collection may be used as a representative.

Figure 4–3 shows five representatives of the same vector. Since any two representative directed line segments of the same vector are equivalent, the collection used to define a vector is called an **equivalence class.**

[The vector as we have defined it is sometimes called a **free vector.** There are other ways of introducing vectors; one is to call a directed line segment a vector. We then would make the convention that directed line segments with the same magnitude and direction (i.e., equivalent) are equal vectors. This leads to certain logical difficulties which we wish to avoid.]

[Vectors occur with great frequency in various branches of physics and engineering. Problems in mechanics, especially those involving forces, are concerned with "lines of action," i.e., the direction in which forces act. In such problems it is convenient to define a vector as the equivalence class of all directed line segments which lie along a given straight line and have a given magnitude.]

DEFINITIONS. The **length** of a vector is the common length of all its representative segments. A **unit vector** is a vector of length one. Two vectors are said to be **orthogonal** (or **perpendicular**) if any representative of one vector is perpendicular to any representative of the other (i.e., the representatives lie along perpendicular lines).

For convenience, we consider directed line segments of zero length; these are simply points. The **zero vector,** denoted by **0**, is the class of directed line segments of zero length.* We make the convention that the zero vector is orthogonal to all vectors.

---

* We could develop vectors by using ordered pairs of points instead of directed line segments. Then the zero vector would be the collection of all ordered pairs of points (A, B) in which $B = A$.

▶2.   OPERATIONS WITH VECTORS

Vectors may be added to yield other vectors. Suppose **u** and **v** are vectors, i.e., each is a collection of directed line segments. To add **u** and **v**, we first select a representative of **u**, say $\overrightarrow{AB}$, as shown in Fig. 4–4(a). Next we take the particular representative of **v** which has its base at the point B, and label it $\overrightarrow{BC}$. We then draw the directed line segment $\overrightarrow{AC}$. The sum **w** of **u** and **v** is the equivalence class of directed line segments of which $\overrightarrow{AC}$ is a representative. We write

$$\mathbf{u} + \mathbf{v} = \mathbf{w}.$$

It is important to note that we could have started with any representative of **u**, say $\overrightarrow{A'B'}$ in Fig. 4–4(b). Then we could have selected the representative of **v** with base at B'. The directed line segments $\overrightarrow{A'C'}$ and $\overrightarrow{AC}$ are representatives of the same vector, as is easily seen from Theorem 1. (See problem 24 at the end of this section.)

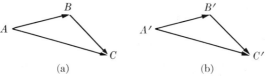

(a)   (b)   Figure 4–4

Vectors may be multiplied by numbers to yield new vectors. Suppose **v** is a vector and c is a number. Let $\overrightarrow{AB}$ be a representative of **v** and let C be the point which is c of the way from A to B. Then $\overrightarrow{AC}$ is a representative of the vector $c\mathbf{v}$. It follows easily from Theorem 1 that if $\overrightarrow{A'B'}$ is another representative of **v** and C' is c of the way from A' to B', then $\overrightarrow{A'C'}$ is also a representative of $c\mathbf{v}$; that is, $\overrightarrow{A'C'} \approx \overrightarrow{AC}$. (See problem 25 at the end of this section.) If c is positive, the representatives of $c\mathbf{v}$ have the same direction as those of **v** but are c times as long. If c is negative, the representatives of $c\mathbf{v}$ are oppositely directed from those of **v** and are $|c|$ times as long. If $c = 0$, then $c\mathbf{v} = \mathbf{0}$. Figure 4–5 shows representatives of various multiples of the vector **v** having the representative $\overrightarrow{AB}$. We write $-\mathbf{v}$ for the vector $(-1)\mathbf{v}$.

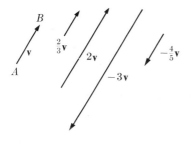

Figure 4–5

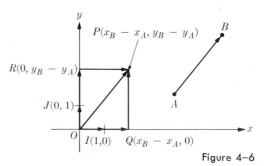

Figure 4–6

DEFINITIONS. Suppose we are given a Cartesian coordinate system in the plane. We call I the point with coordinates $(1, 0)$ and J the point with coordinates $(0, 1)$ as shown in Fig. 4–6. The **unit vector i** is defined as the vector which has $\overrightarrow{OI}$ as one of its representatives. The **unit vector j** is defined as the vector which has $\overrightarrow{OJ}$ as a representative.

Theorem 2. *Suppose a vector* **w** *has* $\overrightarrow{AB}$ *as a representative. Denote the coordinates of A and B by* (x_A, y_A) *and* (x_B, y_B)*, respectively. Then* **w** *may be expressed in the form*

$$\mathbf{w} = (x_B - x_A)\mathbf{i} + (y_B - y_A)\mathbf{j}.$$

Proof. From Eqs. (1) in Theorem 1, we know that **w** has the representative $\overrightarrow{OP}$ where P has coordinates $(x_B - x_A, y_B - y_A)$. (See Fig. 4–6.) Let $Q(x_B - x_A, 0)$ and $R(0, y_B - y_A)$ be the points on the coordinate axes as shown in Fig. 4–6. It is clear geometrically that $\overrightarrow{OQ}$ is $(x_B - x_A)$ times as long as $\overrightarrow{OI}$ [i.e., Q is $(x_B - x_A)$ of the way from O to I] and $\overrightarrow{OR}$ is $(y_B - y_A)$ times as long as $\overrightarrow{OJ}$. It is clear from the point of division formula that Q is $(x_B - x_A)$ of the way from O to I, R is $(y_B - y_A)$ of the way from O to J, and $\overrightarrow{QP} \approx \overrightarrow{OR}$. Letting **u** and **v** denote the vectors which have $\overrightarrow{OQ}$ and $\overrightarrow{OR}$ as representatives, respectively, we use the rule for addition of vectors to obtain

$$\mathbf{w} = \mathbf{u} + \mathbf{v}$$

since $\overrightarrow{QP}$ is also a representative of **v**. Since $\mathbf{u} = (x_B - x_A)\mathbf{i}$ and $\mathbf{v} = (y_B - y_A)\mathbf{j}$, the result of the theorem is established.

EXAMPLE 1. A vector **v** has $\overrightarrow{AB}$ as a representative. If A has coordinates $(3, -2)$ and B has coordinates $(1, 1)$, express **v** in terms of **i** and **j**. Draw a figure.

Solution. Using the formula in Theorem 2, we have (see Fig. 4–7)

$$\mathbf{v} = (1 - 3)\mathbf{i} + (1 + 2)\mathbf{j} = -2\mathbf{i} + 3\mathbf{j}.$$

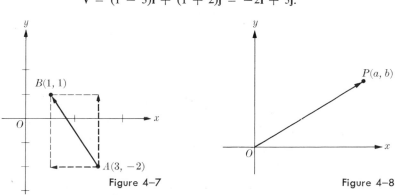

Figure 4–7   Figure 4–8

In the next theorem and in general, the *length of a vector* **v** *will be denoted by* $|\mathbf{v}|$.

Theorem 3. *If* $\mathbf{v} = a\mathbf{i} + b\mathbf{j}$, *then*

$$|\mathbf{v}| = \sqrt{a^2 + b^2}.$$

Therefore $\mathbf{v} = \mathbf{0}$ *if and only if* $a = b = 0$.

Proof. By means of Theorem 2, we know that **v** has as one of its representatives the directed segment $\overrightarrow{OP}$ where P has coordinates (a, b). (See Fig. 4–8.) Then we have $|\overrightarrow{OP}| = \sqrt{a^2 + b^2}$; since, by definition, the length of a vector is the length of any of its representatives, the result follows.

The next theorem is useful for problems concerned with the addition of vectors and the multiplication of vectors by numbers.

Theorem 4. *If* $\mathbf{v} = a\mathbf{i} + b\mathbf{j}$ *and* $\mathbf{w} = c\mathbf{i} + d\mathbf{j}$, *then*

$$\mathbf{v} + \mathbf{w} = (a + c)\mathbf{i} + (b + d)\mathbf{j}.$$

Further, if h is any number, then

$$h\mathbf{v} = (ha)\mathbf{i} + (hb)\mathbf{j}.$$

Proof. Let P, Q, R, and S have coordinates as shown in Fig. 4–9. Then $\overrightarrow{OP}$ and $\overrightarrow{OQ}$ are representatives of **v** and **w**, respectively. Since $\overrightarrow{PS} \approx \overrightarrow{OQ}$, we use the rule for addition of vectors to find that $\overrightarrow{OS}$ is a representative of $\mathbf{v} + \mathbf{w}$. The point R is h of the way from O to P. Hence $\overrightarrow{OR}$ is a representative of $h\mathbf{v}$.

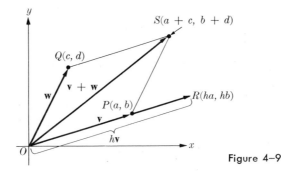

Figure 4–9

We conclude from the theorems above that the addition of vectors and their multiplication by numbers satisfy the following laws:

$$\left.\begin{array}{c} \mathbf{u} + (\mathbf{v} + \mathbf{w}) = (\mathbf{u} + \mathbf{v}) + \mathbf{w} \\ c(d\mathbf{v}) = (cd)\mathbf{v} \end{array}\right\} \text{Associative laws}$$

$$\mathbf{u} + \mathbf{v} = \mathbf{v} + \mathbf{u} \qquad\quad \text{Commutative laws}$$

$$\left.\begin{array}{c} (c + d)\mathbf{v} = c\mathbf{v} + d\mathbf{v} \\ c(\mathbf{u} + \mathbf{v}) = c\mathbf{u} + c\mathbf{v} \end{array}\right\} \text{Distributive laws}$$

$$1 \cdot \mathbf{u} = \mathbf{u}, \qquad 0 \cdot \mathbf{u} = \mathbf{0}, \qquad (-1)\mathbf{u} = -\mathbf{u}$$

where $-\mathbf{u}$ denotes that vector such that $\mathbf{u} + (-\mathbf{u}) = \mathbf{0}$. These laws hold for all $\mathbf{u}$, $\mathbf{v}$, $\mathbf{w}$ and all numbers c and d. It is important to note that multiplication and division of vectors is not (and will not be) defined. However, subtraction is defined. If $\overrightarrow{AB}$ is a representative of $\mathbf{v}$ and $\overrightarrow{AC}$ is one of $\mathbf{w}$, then $\overrightarrow{CB}$ is a representative of $\mathbf{v} - \mathbf{w}$ (see Fig. 4–10).

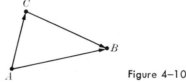

Figure 4–10

EXAMPLE 2.   Given the vectors $\mathbf{u} = 2\mathbf{i} - 3\mathbf{j}$, $\mathbf{v} = -4\mathbf{i} + \mathbf{j}$.   Express the vector $2\mathbf{u} - 3\mathbf{v}$ in terms of $\mathbf{i}$ and $\mathbf{j}$.

Solution. $2\mathbf{u} = 4\mathbf{i} - 6\mathbf{j}$ and $-3\mathbf{v} = 12\mathbf{i} - 3\mathbf{j}$. Adding these vectors, we get $2\mathbf{u} - 3\mathbf{v} = 16\mathbf{i} - 9\mathbf{j}$.

DEFINITION. Let $\mathbf{v}$ be any vector except $\mathbf{0}$. The unit vector $\mathbf{u}$ in the direction of $\mathbf{v}$ is defined by

$$\mathbf{u} = \left(\frac{1}{|\mathbf{v}|}\right) \mathbf{v}.$$

EXAMPLE 3. Given the vector $\mathbf{v} = -2\mathbf{i} + 3\mathbf{j}$, find a unit vector in the direction of $\mathbf{v}$.

Solution. We have $|\mathbf{v}| = \sqrt{4 + 9} = \sqrt{13}$. The desired vector $\mathbf{u}$ is

$$\mathbf{u} = \frac{1}{\sqrt{13}} \mathbf{v} = -\frac{2}{\sqrt{13}} \mathbf{i} + \frac{3}{\sqrt{13}} \mathbf{j}.$$

EXAMPLE 4. Given the vector $\mathbf{v} = 2\mathbf{i} - 4\mathbf{j}$. Find a representative $\overrightarrow{AB}$ of $\mathbf{v}$, given that A has coordinates $(3, -5)$.

Solution. Denote the coordinates of B by x_B, y_B. Then we have (by Theorem 2)

$$x_B - 3 = 2 \quad \text{and} \quad y_B + 5 = -4.$$

Therefore $x_B = 5$, $y_B = -9$.

PROBLEMS

In problems 1 through 7, express $\mathbf{v}$ in terms of $\mathbf{i}$ and $\mathbf{j}$, given that the endpoints A and B of the representative $\overrightarrow{AB}$ of $\mathbf{v}$ have the given coordinates. Draw a figure for each.

1. $A(1, 2)$, $B(-2, -1)$ 2. $A(2, -1)$, $B(-1, 2)$ 3. $A(-1, -2)$, $B(2, 2)$
4. $A(2, -2)$, $B(-2, 3)$ 5. $A(3, 0)$, $B(0, -4)$ 6. $A(2, -1)$, $B(6, -1)$
7. $A(-1, 2)$, $B(-5, 2)$

In problems 8 through 11, in each case find a unit vector $\mathbf{u}$ in the direction of $\mathbf{v}$. Express $\mathbf{u}$ in terms of $\mathbf{i}$ and $\mathbf{j}$.

8. $\mathbf{v} = 3\mathbf{i} - 4\mathbf{j}$ 9. $\mathbf{v} = 12\mathbf{i} - 5\mathbf{j}$

10. $\mathbf{v} = 2\sqrt{3}\,\mathbf{i} - 2\mathbf{j}$ 11. $\mathbf{v} = 2\mathbf{i} + 3\mathbf{j}$

In problems 12 through 17, find the representative $\overrightarrow{AB}$ of the vector $\mathbf{v}$ from the information given. Draw a figure for each.

12. $\mathbf{v} = 4\mathbf{i} - 2\mathbf{j}$, $A(1, 2)$ 13. $\mathbf{v} = -3\mathbf{i} + 5\mathbf{j}$, $A(4, 3)$
14. $\mathbf{v} = 2\mathbf{i} - 3\mathbf{j}$, $B(1, 3)$ 15. $\mathbf{v} = -2\mathbf{i} - 3\mathbf{j}$, $B(1, 4)$
16. $\mathbf{v} = 2\mathbf{i} - 3\mathbf{j}$, midpoint of segment AB has coordinates $(2, -1)$
17. $\mathbf{v} = 3\mathbf{i} - 2\mathbf{j}$, midpoint of segment AB has coordinates $(-3, 2)$
18. Find a representative of the vector $\mathbf{v}$ of unit length making an angle of $150°$ with the positive x axis. Express $\mathbf{v}$ in terms of $\mathbf{i}$ and $\mathbf{j}$ (two solutions).
19. Find the vector $\mathbf{v}$ (in terms of $\mathbf{i}$ and $\mathbf{j}$) which has length $2\sqrt{2}$ and makes an angle of $45°$ with the positive y axis (two solutions).
20. Given that $\mathbf{u} = 2\mathbf{i} - \mathbf{j}$, $\mathbf{v} = 3\mathbf{i} + 2\mathbf{j}$. Find $\mathbf{u} + \mathbf{v}$ in terms of $\mathbf{i}$ and $\mathbf{j}$. Draw a figure.
21. Given that $\mathbf{u} = -\mathbf{i} + 4\mathbf{j}$, $\mathbf{v} = 2\mathbf{i} - 3\mathbf{j}$. Find $\mathbf{u} + \mathbf{v}$ in terms of $\mathbf{i}$ and $\mathbf{j}$. Draw a figure.
22. Given that $\mathbf{u} = -2\mathbf{i} - \mathbf{j}$, $\mathbf{v} = \mathbf{i} + 2\mathbf{j}$. Find $2\mathbf{u} + 3\mathbf{v}$ in terms of $\mathbf{i}$ and $\mathbf{j}$. Draw a figure.
23. Show that if $\overrightarrow{AB} \approx \overrightarrow{CD}$ and $\overrightarrow{CD} \approx \overrightarrow{EF}$, then $\overrightarrow{AB} \approx \overrightarrow{EF}$.
24. Show that if $\overrightarrow{AB} \approx \overrightarrow{DE}$ and $\overrightarrow{BC} \approx \overrightarrow{EF}$, then $\overrightarrow{AC} \approx \overrightarrow{DF}$. Draw a figure.
25. Show that if $\overrightarrow{AB} \approx \overrightarrow{DE}$, c is any real number, C is the point c of the way from A to B, and F is the point c of the way from D to E, then $\overrightarrow{AC} \approx \overrightarrow{DF}$. Draw a figure.
26. Show that the vectors $\mathbf{v} = 2\mathbf{i} + 4\mathbf{j}$ and $\mathbf{w} = 10\mathbf{i} - 5\mathbf{j}$ are orthogonal.
27. Show that the vectors $\mathbf{v} = -3\mathbf{i} + \sqrt{2}\,\mathbf{j}$ and $\mathbf{w} = 4\sqrt{2}\,\mathbf{i} + 12\mathbf{j}$ are orthogonal.

28. Let **u** and **v** be two nonzero vectors. Show that they are orthogonal if and only if the following equation holds:

$$|\mathbf{u} + \mathbf{v}|^2 = |\mathbf{u}|^2 + |\mathbf{v}|^2.$$

29. Write out a proof establishing the associative, commutative, and distributive laws for vectors (see page 91).

▶ 3. OPERATIONS WITH VECTORS, CONTINUED

Two vectors **v** and **w** are said to be **parallel** or **proportional** when each is a scalar multiple of the other (and neither is zero). The representatives of parallel vectors are all parallel directed line segments.

By the **angle between two vectors v** and **w** (neither $=$ **0**), we mean the measure of the angle between any representatives of **v** and **w** having the same base (see Fig. 4–11). Two parallel vectors make an angle either of 0 or of π, depending on whether they are pointing in the same or opposite directions (i.e., whether the scalar multiple is positive or negative).

Suppose that **i** and **j** are the usual unit vectors pointing in the direction of the x and y axes, respectively. Then we have the following theorem.

Theorem 5. *If θ is the angle between the vectors*

$$\mathbf{v} = a\mathbf{i} + b\mathbf{j} \quad and \quad \mathbf{w} = c\mathbf{i} + d\mathbf{j},$$

then

$$\cos \theta = \frac{ac + bd}{|\mathbf{v}|\,|\mathbf{w}|}.$$

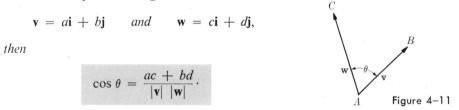

Figure 4–11

Proof. We draw the representatives of **v** and **w** with base at the origin of the coordinate system, as shown in Fig. 4–12. Using Theorem 2 of Section 2, we see that the coordinates of P are (a, b) and those of Q are (c, d). The length of **v** is $|OP|$ and the length of **w** is $|OQ|$. We apply the law of cosines to $\triangle OPQ$, obtaining

$$\cos \theta = \frac{|OP|^2 + |OQ|^2 - |QP|^2}{2|OP|\,|OQ|}.$$

Therefore

$$\cos \theta = \frac{a^2 + b^2 + c^2 + d^2 - (a - c)^2 - (b - d)^2}{2|OP|\,|OQ|}$$

$$= \frac{ac + bd}{|\mathbf{v}|\,|\mathbf{w}|}.$$

EXAMPLE 1. Given the vectors $\mathbf{v} = 2\mathbf{i} - 3\mathbf{j}$ and $\mathbf{w} = \mathbf{i} - 4\mathbf{j}$, compute the cosine of the angle between $\mathbf{v}$ and $\mathbf{w}$.

Solution. We have

$$|\mathbf{v}| = \sqrt{4 + 9} = \sqrt{13}, \qquad |\mathbf{w}| = \sqrt{1 + 16} = \sqrt{17}.$$

Therefore

$$\cos \theta = \frac{2 \cdot 1 + (-3)(-4)}{\sqrt{17}\sqrt{13}} = \frac{14}{\sqrt{221}}.$$

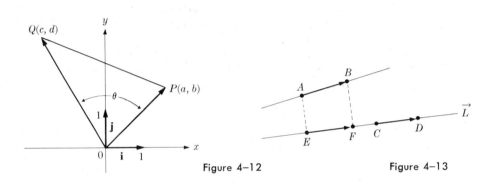

Figure 4–12   Figure 4–13

Suppose that $\overrightarrow{AB}$ and $\overrightarrow{CD}$ are directed line segments as shown in Fig. 4–13. We denote by $\overrightarrow{L}$ the directed line through $\overrightarrow{CD}$, directed so that C precedes D. Let E and F be the feet of the perpendiculars from A and B, respectively, onto the line $\overrightarrow{L}$ (Fig. 4–13). We define the projection of $\overrightarrow{AB}$ in the direction of $\overrightarrow{CD}$ as the directed distance $\overline{EF}$ along $\overrightarrow{L}$. We observe that this projection is the same as the projection of $\overrightarrow{AB}$ along $\overrightarrow{L}$.

Let $\mathbf{v}$ and $\mathbf{w}$ be nonzero vectors with $\overrightarrow{AB}$ a representative of $\mathbf{v}$ and $\overrightarrow{CD}$ a representative of $\mathbf{w}$. We define the projection of $\mathbf{v}$ along $\mathbf{w}$ to be the projection of $\overrightarrow{AB}$ in the direction of $\overrightarrow{CD}$. We write $\text{Proj}_{\mathbf{w}}\,\mathbf{v}$ for this quantity, and if θ is the angle between the vectors $\mathbf{v}$ and $\mathbf{w}$, it follows that

$$\text{Proj}_{\mathbf{w}}\,\mathbf{v} = |\mathbf{v}| \cos \theta.$$

If θ is an acute angle the projection is positive, while if θ is obtuse the projection is negative. Finally, if the vectors $\mathbf{v}$ and $\mathbf{w}$ have the representations $\mathbf{v} = a\mathbf{i} + b\mathbf{j}$, $\mathbf{w} = c\mathbf{i} + d\mathbf{j}$, then

$$\text{Proj}_{\mathbf{w}}\,\mathbf{v} = \frac{ac + bd}{|\mathbf{w}|}.$$

EXAMPLE 2. Find the projection of $\mathbf{v} = 3\mathbf{i} - 2\mathbf{j}$ on $\mathbf{w} = -2\mathbf{i} - 4\mathbf{j}$.

Solution. We obtain the desired projection, by using the formula

$$|\mathbf{v}| \cos \theta = \frac{-6 + 8}{\sqrt{20}} = \frac{1}{\sqrt{5}} \cdot$$

The **scalar product** of two nonzero vectors $\mathbf{v}$ and $\mathbf{w}$, written $\mathbf{v} \cdot \mathbf{w}$, is defined by the formula

$$\mathbf{v} \cdot \mathbf{w} = |\mathbf{v}|\,|\mathbf{w}| \cos \theta,$$

where θ is the angle between $\mathbf{v}$ and $\mathbf{w}$. If one of the vectors is $\mathbf{0}$, the scalar product is defined to be 0. The terms **dot product** and **inner product** are also used to designate scalar product. It is evident from the definition that scalar product satisfies the relations

$$\mathbf{v} \cdot \mathbf{w} = \mathbf{w} \cdot \mathbf{v}, \qquad \mathbf{v} \cdot \mathbf{v} = |\mathbf{v}|^2.$$

Furthermore, if $\mathbf{v}$ and $\mathbf{w}$ are orthogonal, then

$$\mathbf{v} \cdot \mathbf{w} = 0,$$

and conversely. If $\mathbf{v}$ and $\mathbf{w}$ are parallel, we have $\mathbf{v} \cdot \mathbf{w} = \pm|\mathbf{v}|\,|\mathbf{w}|$, and conversely. In terms of the orthogonal unit vectors $\mathbf{i}$ and $\mathbf{j}$, the vectors $\mathbf{v} = a\mathbf{i} + b\mathbf{j}$ and $\mathbf{w} = c\mathbf{i} + d\mathbf{j}$ have as their scalar product (see Theorem 5)

$$\mathbf{v} \cdot \mathbf{w} = ac + bd.$$

In addition, it can be verified that the **distributive law**

$$\mathbf{u} \cdot (\mathbf{v} + \mathbf{w}) = \mathbf{u} \cdot \mathbf{v} + \mathbf{u} \cdot \mathbf{w}$$

holds for any three vectors.

EXAMPLE 3. Given the vectors $\mathbf{u} = 3\mathbf{i} + 2\mathbf{j}$ and $\mathbf{v} = 2\mathbf{i} + a\mathbf{j}$. Determine the number a so that $\mathbf{u}$ and $\mathbf{v}$ are orthogonal. Determine a so that $\mathbf{u}$ and $\mathbf{v}$ are parallel. For what value of a will $\mathbf{u}$ and $\mathbf{v}$ make an angle of $\pi/4$?

Solution. If $\mathbf{u}$ and $\mathbf{v}$ are orthogonal, we have

$$3 \cdot 2 + 2 \cdot a = 0 \qquad \text{and} \qquad a = -3.$$

For $\mathbf{u}$ and $\mathbf{v}$ to be parallel, we must have

$$\mathbf{u} \cdot \mathbf{v} = \pm|\mathbf{u}|\,|\mathbf{v}|,$$

or

$$6 + 2a = \pm\sqrt{13} \cdot \sqrt{4 + a^2}.$$

Solving, we obtain $a = \frac{4}{3}$. From the formula

$$\tfrac{1}{2}\sqrt{2} = \cos\frac{\pi}{4} = \frac{6 + 2a}{\sqrt{13}\cdot\sqrt{4 + a^2}},$$

we see that $\mathbf{u}$ and $\mathbf{v}$ make an angle of $\pi/4$ when

$$a = 10, \ -\tfrac{2}{5}.$$

Because they are geometric quantities which are independent of the coordinate system, vectors are well suited for establishing certain types of theorems in plane geometry. We give two examples to exhibit the technique.

EXAMPLE 4. Let $\overrightarrow{OA}$ be a representative of $\mathbf{u}$ and $\overrightarrow{OB}$ a representative of $\mathbf{v}$. Let C be the point on the line AB which is $\frac{2}{3}$ of the way from A to B. Express in terms of $\mathbf{u}$ and $\mathbf{v}$ the vector $\mathbf{w}$ which has $\overrightarrow{OC}$ as representative (Fig. 4–14).

Solution. Let $\mathbf{z}$ be the vector with $\overrightarrow{AC}$ as representative, and $\mathbf{t}$ the vector with $\overrightarrow{AB}$ as representative. We have

$$\mathbf{w} = \mathbf{u} + \mathbf{z} = \mathbf{u} + \tfrac{2}{3}\mathbf{t}.$$

Also, we know that

$$\mathbf{t} = \mathbf{v} - \mathbf{u},$$

and so

$$\mathbf{w} = \mathbf{u} + \tfrac{2}{3}(\mathbf{v} - \mathbf{u}) = \tfrac{1}{3}\mathbf{u} + \tfrac{2}{3}\mathbf{v}.$$

Let $\overrightarrow{AB}$ be a directed line segment. We introduce a convenient symbol for the vector $\mathbf{v}$ which has $\overrightarrow{AB}$ as one of its representatives.

Notation. The symbol $\mathbf{v}[\overrightarrow{AB}]$ denotes the vector which has $\overrightarrow{AB}$ as a representative.

EXAMPLE 5. Let $ABDC$ be a parallelogram, as shown in Fig. 4–15. Suppose that E is the midpoint of CD and F is $\frac{2}{3}$ of the way from A to E on AE. Show that F is $\frac{2}{3}$ of the way from B to C.

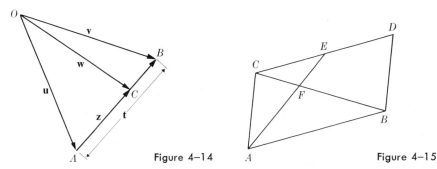

Figure 4–14   Figure 4–15

Solution. Let $\overrightarrow{AB}$, $\overrightarrow{AC}$, $\overrightarrow{AF}$, $\overrightarrow{AE}$, $\overrightarrow{BF}$, $\overrightarrow{BC}$, and $\overrightarrow{CE}$ all be representatives of vectors. Then $\mathbf{v}[\overrightarrow{AB}]$, $\mathbf{v}[\overrightarrow{AC}]$, $\mathbf{v}[\overrightarrow{AF}]$, etc., are the vectors which have the directed line segments shown in brackets as representatives. By hypothesis, we have

$$\mathbf{v}[\overrightarrow{AB}] = \mathbf{v}[\overrightarrow{CD}] \quad \text{and} \quad \mathbf{v}[\overrightarrow{CE}] = \tfrac{1}{2}\mathbf{v}[\overrightarrow{CD}].$$

The rule for addition of vectors gives us

$$\mathbf{v}[\overrightarrow{AE}] = \mathbf{v}[\overrightarrow{AC}] + \mathbf{v}[\overrightarrow{CE}] = \mathbf{v}[\overrightarrow{AC}] + \tfrac{1}{2}\mathbf{v}[\overrightarrow{AB}].$$

Also, since $\mathbf{v}[\overrightarrow{AF}] = \tfrac{2}{3}\mathbf{v}[\overrightarrow{AE}]$, we obtain

$$\mathbf{v}[\overrightarrow{AF}] = \tfrac{2}{3}\mathbf{v}[\overrightarrow{AC}] + \tfrac{1}{3}\mathbf{v}[\overrightarrow{AB}].$$

The rule for subtraction of vectors gives us

$$\mathbf{v}[\overrightarrow{BF}] = \mathbf{v}[\overrightarrow{AF}] - \mathbf{v}[\overrightarrow{AB}] = \tfrac{2}{3}\mathbf{v}[\overrightarrow{AC}] - \tfrac{2}{3}\mathbf{v}[\overrightarrow{AB}] = \tfrac{2}{3}(\mathbf{v}[\overrightarrow{AC}] - \mathbf{v}[\overrightarrow{AB}]).$$

Since $\mathbf{v}[\overrightarrow{BC}] = \mathbf{v}[\overrightarrow{AC}] - \mathbf{v}[\overrightarrow{AB}]$, we conclude that

$$\mathbf{v}[\overrightarrow{BF}] = \tfrac{2}{3}\mathbf{v}[\overrightarrow{BC}],$$

which is the desired result.

PROBLEMS

In problems 1 through 8, given that θ is the angle between $\mathbf{v}$ and $\mathbf{w}$, find $|\mathbf{v}|$, $|\mathbf{w}|$, $\cos\theta$, and the projection of $\mathbf{v}$ on $\mathbf{w}$.

1. $\mathbf{v} = 3\mathbf{i} + 4\mathbf{j}$, $\mathbf{w} = 4\mathbf{i} + 3\mathbf{j}$
2. $\mathbf{v} = 4\mathbf{i} + 3\mathbf{j}$, $\mathbf{w} = 5\mathbf{i} + 12\mathbf{j}$
3. $\mathbf{v} = 7\mathbf{i} - 24\mathbf{j}$, $\mathbf{w} = 12\mathbf{i} + 5\mathbf{j}$
4. $\mathbf{v} = 24\mathbf{i} + 7\mathbf{j}$, $\mathbf{w} = 4\mathbf{i} - 3\mathbf{j}$
5. $\mathbf{v} = 3\mathbf{i} - 2\mathbf{j}$, $\mathbf{w} = 4\mathbf{i} + 3\mathbf{j}$
6. $\mathbf{v} = \mathbf{i} + \mathbf{j}$, $\mathbf{w} = 3\mathbf{i} + 4\mathbf{j}$
7. $\mathbf{v} = 2\mathbf{i} + 3\mathbf{j}$, $\mathbf{w} = 3\mathbf{i} - \mathbf{j}$
8. $\mathbf{v} = 7\mathbf{i} - \mathbf{j}$, $\mathbf{w} = 4\mathbf{i} + 3\mathbf{j}$

In problems 9 through 13, find $\cos\theta$ and $\cos\alpha$, given that $\theta = \angle ABC$ and $\alpha = \angle BAC$. Use vector methods and draw figures.

9. $A(-2, 2)$, $B(3, 1)$, $C(3, 5)$
10. $A(3, 1)$, $B(-2, 1)$, $C(1, 4)$
11. $A(2, 3)$, $B(6, -1)$, $C(2, -1)$
12. $A(2, 2)$, $B(0, -1)$, $C(4, 3)$
13. $A(0, 0)$, $B(2, -4)$, $C(5, 7)$

In problems 14 through 19, determine the a (if possible) such that the given condition on $\mathbf{v}$ and $\mathbf{w}$ is satisfied.

14. $\mathbf{v} = 3\mathbf{i} + a\mathbf{j}$, $\mathbf{w} = 2\mathbf{i} + \mathbf{j}$, $\mathbf{v}$ and $\mathbf{w}$ orthogonal (i.e., $\perp$)
15. $\mathbf{v} = 2\mathbf{i} - \mathbf{j}$, $\mathbf{w} = a\mathbf{i} + 2\mathbf{j}$, $\mathbf{v}$ and $\mathbf{w}$ orthogonal
16. $\mathbf{v} = 2\mathbf{i} + \mathbf{j}$, $\mathbf{w} = \mathbf{i} + 2a\mathbf{j}$, $\mathbf{v}$ and $\mathbf{w}$ parallel
17. $\mathbf{v} = a\mathbf{i} - \mathbf{j}$, $\mathbf{w} = \mathbf{i} - 2a\mathbf{j}$, $\mathbf{v}$ and $\mathbf{w}$ parallel
18. $\mathbf{v} = a\mathbf{i}$, $\mathbf{w} = 3\mathbf{i} + 2\mathbf{j}$, $\mathbf{v}$ and $\mathbf{w}$ parallel
19. $\mathbf{v} = 3\mathbf{i} + 4\mathbf{j}$, $\mathbf{w} = \mathbf{i} + a\mathbf{j}$, $\mathbf{v}$ and $\mathbf{w}$ make an angle of $\pi/3$

20. Prove the distributive law for the scalar product.

21. Let $\mathbf{i}$ and $\mathbf{j}$ be the usual unit vectors of one coordinate system, and let $\mathbf{i}_1$ and $\mathbf{j}_1$ be the unit orthogonal vectors corresponding to another Cartesian system of coordinates. Given that

$$\mathbf{v} = a\mathbf{i} + b\mathbf{j}, \qquad \mathbf{w} = c\mathbf{i} + d\mathbf{j},$$
$$\mathbf{v} = a_1\mathbf{i}_1 + b_1\mathbf{j}_1, \qquad \mathbf{w} = c_1\mathbf{i}_1 + d_1\mathbf{j}_1,$$

show that

$$ac + bd = a_1c_1 + b_1d_1.$$

In problems 22 through 25, the quantity $|AB|$ denotes (as is customary) the length of the line segment AB, the quantity $|AC|$, the length of AC, etc.

22. Given $\triangle ABC$, in which $\angle A = 120°$, $|AB| = 4$, and $|AC| = 7$. Find $|BC|$ and the directed lengths of the projections of $\overrightarrow{AB}$ and $\overrightarrow{AC}$ on $\overrightarrow{BC}$. Draw a figure.

23. Given $\triangle ABC$, with $\angle A = 45°$, $|AB| = 8$, $|AC| = 6\sqrt{2}$. Find $|BC|$ and the directed lengths of the projections of $\overrightarrow{AB}$ and $\overrightarrow{AC}$ on $\overrightarrow{BC}$. Draw a figure.

24. Given $\triangle ABC$, with $|AB| = 10$, $|AC| = 9$, $|BC| = 7$. Find the directed lengths of the projections of $\overrightarrow{AC}$ and $\overrightarrow{BC}$ on $\overrightarrow{AB}$. Draw a figure.

25. Given $\triangle ABC$, with $|AB| = 5$, $|AC| = 7$, $|BC| = 9$. Find the directed lengths of the projections of $\overrightarrow{AB}$ and $\overrightarrow{AC}$ on $\overrightarrow{CB}$. Draw a figure.

In problems 26 through 33, the notation used in Example 5 will be employed.

26. Given the line segments AB and AC, with D on AB $\frac{2}{3}$ of the way from A to B. Let E be the midpoint of AC. Express $\mathbf{v}[\overrightarrow{DE}]$ in terms of $\mathbf{v}[\overrightarrow{AB}]$ and $\mathbf{v}[\overrightarrow{AC}]$. Draw a figure.

27. Suppose that $\mathbf{v}[\overrightarrow{AD}] = \frac{1}{4}\mathbf{v}[\overrightarrow{AB}]$ and $\mathbf{v}[\overrightarrow{BE}] = \frac{1}{2}\mathbf{v}[\overrightarrow{BC}]$. Find $\mathbf{v}[\overrightarrow{DE}]$ in terms of $\mathbf{v}[\overrightarrow{AB}]$ and $\mathbf{v}[\overrightarrow{BC}]$. Draw a figure.

28. Given $\square ABDC$, a parallelogram, with E $\frac{2}{3}$ of the way from B to D, and F as the midpoint of segment CD. Find $\mathbf{v}[\overrightarrow{EF}]$ in terms of $\mathbf{v}[\overrightarrow{AB}]$ and $\mathbf{v}[\overrightarrow{AC}]$.

29. Given parallelogram $ABDC$, with E $\frac{1}{4}$ of the way from B to C, and F $\frac{1}{4}$ of the way from A to D. Find $\mathbf{v}[\overrightarrow{EF}]$ in terms of $\mathbf{v}[\overrightarrow{AB}]$ and $\mathbf{v}[\overrightarrow{AC}]$.

30. Given parallelogram $ABDC$, with E $\frac{1}{3}$ of the way from B to D, and F $\frac{1}{4}$ of the way from B to C. Show that F is $\frac{3}{4}$ of the way from A to E.

31. Suppose that on the sides of $\triangle ABC$, $\mathbf{v}[\overrightarrow{BD}] = \frac{2}{3}\mathbf{v}[\overrightarrow{BC}]$, $\mathbf{v}[\overrightarrow{CE}] = \frac{2}{3}\mathbf{v}[\overrightarrow{CA}]$, and $\mathbf{v}[\overrightarrow{AF}] = \frac{2}{3}\mathbf{v}[\overrightarrow{AB}]$. Draw a figure and show that

$$\mathbf{v}[\overrightarrow{AD}] + \mathbf{v}[\overrightarrow{BE}] + \mathbf{v}[\overrightarrow{CF}] = \mathbf{0}.$$

32. Show that the conclusion of problem 31 holds when the fraction $\frac{2}{3}$ is replaced by any real number h.

33. Let $\mathbf{a} = \mathbf{v}[\overrightarrow{OA}]$, $\mathbf{b} = \mathbf{v}[\overrightarrow{OB}]$, and $\mathbf{c} = \mathbf{v}[\overrightarrow{OC}]$. Show that the medians of $\triangle ABC$ meet at a point P, and express $\mathbf{v}[\overrightarrow{OP}]$ in terms of $\mathbf{a}$, $\mathbf{b}$, and $\mathbf{c}$. Draw a figure.

The Circle 5

▶ 1. THE EQUATION OF A CIRCLE

Straight lines correspond to equations of the first degree in x and y. A systematic approach to the study of equations in x and y would proceed to equations of the second degree, third degree, and so on. Curves of the second degree are sufficiently simple that we can study them thoroughly in this course. Curves corresponding to equations of the third degree and higher will not be treated systematically, since the complications rapidly become too great. However, the study of equations of any degree—the theory of algebraic curves—is an interesting subject on its own and may be studied in advanced courses.

The set of all points (x, y) which are at a given distance r from a fixed point (h, k) represents a **circle.** We also use the phrase, *the locus of all such points is a circle.* From the formula for the distance between two points we know that

$$r = \sqrt{(x - h)^2 + (y - k)^2},$$

or, upon squaring,

$$(x - h)^2 + (y - k)^2 = r^2.$$

This is the equation of a circle with center at (h, k) and radius r.

EXAMPLE 1. Find the equation of the circle with center at $(-3, 4)$ and radius 6.

Solution. We have $h = -3, k = 4, r = 6$. Substituting in the equation of the circle, we obtain

$$(x + 3)^2 + (y - 4)^2 = 36.$$

We can multiply out to get

$$x^2 + y^2 + 6x - 8y - 11 = 0$$

as another form for the answer.

99

We now start the other way around. Suppose that we are given an equation such as

$$x^2 + y^2 - 4x + 7y - 8 = 0;$$

we ask whether or not this represents a circle and, if so, what the center and radius are. The process for determining these facts consists of "completing the square." We write

$$(x^2 - 4x \qquad) + (y^2 + 7y \qquad) = 8,$$

in which we leave the appropriate spaces, as shown. We add the square of half the coefficient of x and the square of half the coefficient of y to both sides and so obtain

$$(x^2 - 4x + 4) + (y^2 + 7y + \tfrac{49}{4}) = 8 + 4 + \tfrac{49}{4},$$

or

$$(x - 2)^2 + (y + \tfrac{7}{2})^2 = \tfrac{97}{4}.$$

This represents a circle with center at $(2, -\tfrac{7}{2})$ and radius $\tfrac{1}{2}\sqrt{97}$.

The process of completing the square may be performed for a *general* equation of the form

$$x^2 + y^2 + Dx + Ey + F = 0,$$

where D, E, F are any numbers. We get

$$x^2 + Dx + \frac{D^2}{4} + y^2 + Ey + \frac{E^2}{4} = -F + \frac{D^2}{4} + \frac{E^2}{4},$$

or

$$\left(x + \frac{D}{2}\right)^2 + \left(y + \frac{E}{2}\right)^2 = \tfrac{1}{4}(D^2 + E^2 - 4F).$$

We conclude that such an equation represents a circle with center at $(-D/2, -E/2)$ and radius $\tfrac{1}{2}\sqrt{D^2 + E^2 - 4F}$. But there is a problem: If the quantity under the radical is negative, there is no circle and consequently no locus. If it is zero, the circle reduces to a point. *The equation does represent a circle of radius $\tfrac{1}{2}\sqrt{D^2 + E^2 - 4F}$ if $D^2 + E^2 > 4F$, and only then.*

A circle is determined if the numbers $D, E,$ and F are given. This means that in general three conditions determine a circle. We all know that there is exactly one circle through three points (not all on a straight line). The following example shows how the equation is found when three points are prescribed.

EXAMPLE 2.   Find the equation of the circle passing through the points $(-3, 4)$, $(4, 5)$, and $(1, -4)$. Locate the center and determine the radius.

Solution. If the circle has the equation

$$x^2 + y^2 + Dx + Ey + F = 0,$$

then each point on the circle must satisfy this equation. By substituting, we get

$$(-3, 4): \quad -3D + 4E + F = -25;$$
$$(4, 5): \quad 4D + 5E + F = -41;$$
$$(1, -4): \quad D - 4E + F = -17.$$

We subtract the first equation from the second and the third from the second and eliminate F, to obtain

$$7D + E = -16, \quad 3D + 9E = -24.$$

We solve these simultaneously and get $D = -2$, $E = -2$. Substituting back into any one of the original three equations gives $F = -23$. The equation of the circle is

$$x^2 + y^2 - 2x - 2y - 23 = 0.$$

The center and radius are obtained by completing the square. We write

$$(x^2 - 2x \quad) + (y^2 - 2y \quad) = 23, \quad \text{and so} \quad (x - 1)^2 + (y - 1)^2 = 25.$$

The center is at $(1, 1)$; the radius is 5.

The problem of Example 2 could be solved in another way. (1) Find the perpendicular bisector of the line segment joining the first two points. (2) Do the same with the first and third (or second and third) points. (3) The point of intersection of the two perpendicular bisectors is the center of the circle. (4) The formula for the distance from the center of the circle to any of the original three points gives the radius. (5) Knowing the center and radius, we write the equation of the circle. (See Fig. 5–1.)

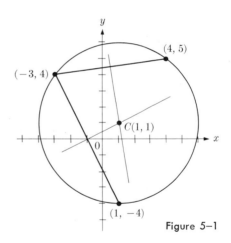

Figure 5–1

PROBLEMS

In problems 1 through 14, find the equations of the circles determined by the given conditions, where C denotes the center and r the radius.

1. $C = (0, 0)$, $r = 4$
2. $C = (4, 3)$, $r = 5$
3. $C = (2, 3)$, $r = 3$
4. $C = (-1, 2)$, $r = 2$
5. A diameter is the segment from $(3, 1)$ to $(-1, 5)$.
6. A diameter is the segment from $(-1, 2)$ to $(4, 4)$.
7. $C = (1, 2)$, passing through $(-2, -2)$
8. $C = (-2, -1)$, passing through $(3, -1)$

9. $C = (2, 3)$, tangent to the y axis

10. $C = (3, -1)$, tangent to the x axis

11. $C = (3, 2)$, tangent to $3x + 4y + 2 = 0$

12. $C = (2, -3)$ tangent to $12x - 5y = 0$

13. Tangent to both axes, radius 4, in third quadrant

14. Tangent to the x axis and to the line $y = x$ and having radius 5 (4 solutions)

In problems 15 through 22, determine the loci of the given equations by completing the squares.

15. $x^2 + y^2 - 4x + 2y = 0$ 16. $x^2 + y^2 - 6x + 4y + 5 = 0$

17. $x^2 + y^2 + 6x - 4y + 13 = 0$ 18. $x^2 + y^2 - 4x + 6y + 17 = 0$

19. $x^2 + y^2 + 3x - 5y + \frac{1}{2} = 0$ 20. $x^2 + y^2 - 3x + 4y + 4 = 0$

21. $2x^2 + 2y^2 - 5x + 3y + 2 = 0$ 22. $2x^2 + 2y^2 - 3x + 5y + 14 = 0$

23. Show that when $D = E$ the locus of $x^2 + y^2 + Dx + Ey + F = 0$ is always a circle if $F < D^2/2$. Note that if $F < 0$, the locus is always a circle, regardless of the sizes of D and E.

In problems 24 through 27, find the equations of the circles through the given points by the method used in Example 2. Find the centers and radii.

24. $(-2, -5), (-4, -3), (-2, 1)$ 25. $(-1, -3), (-2, -2), (2, 6)$

26. $(2, 7), (4, 3), (3, 0)$ 27. $(1, 1), (-1, 0), (3, -3)$

In problems 28 and 29, find the equations of the circles by the method described after Example 2.

28. $(4, -2), (6, 2), (-2, 4)$ 29. $(1, -2), (-2, 0), (-1, 4)$

30. Determine whether or not the points $(2, 0)$, $(0, 4)$, $(2, 2)$, and $(1, 1)$ lie on a circle.

31. Find the equation of the circle or circles with center on the y axis and passing through $(-2, 3)$ and $(4, 1)$.

32. Find the equation of the circle with center on the line $x + y - 1 = 0$ and passing through the points $(0, 5)$ and $(2, 1)$.

In problems 33 through 35, find the point or points of intersection, if any, of the given circle and the given line.

33. $x^2 + y^2 - 2x - 4y - 3 = 0$ 34. $x^2 + y^2 + 2x - 6y - 19 = 0$

    $2x - y - 2 = 0$ $2x - 5y - 12 = 0$

35. $x^2 + y^2 - 6x = 0$

    $3x - y - 23 = 0$

▶2. TANGENTS. FAMILIES OF CIRCLES

Let P be a point on a circle. Then a line L is said to be **tangent to the circle at the point** P if and only if L is perpendicular to the radius of the circle passing through P. (See Fig. 5–2.) We now show how to find the equation of such a

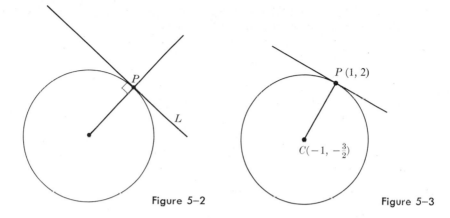

Figure 5-2   Figure 5-3

line L when the circle and the point P are given. We illustrate the method by working an example. The point $P(1, 2)$ is on the circle

$$x^2 + y^2 + 2x + 3y - 13 = 0.$$

By the methods of Section 1 we determine that the circle has center at $C(-1, -\frac{3}{2})$.
Referring to Fig. 5-3, we can calculate the slope of the radius through the points C and P. It is

$$\frac{2 - (-\frac{3}{2})}{1 - (-1)} = \frac{7}{4}.$$

Therefore the slope of the line L is $-\frac{4}{7}$, the negative reciprocal. Using the point-slope formula for the equation of a line, we obtain

$$y - 2 = -\tfrac{4}{7}(x - 1) \quad \text{or} \quad 4x + 7y - 18 = 0$$

as the equation of the desired line L.

   Let $P(x_1, y_1)$ be a point outside the circle K which has its center at $C(h, k)$ and has radius r. In set notation, we may write

$$K = \{(x, y): (x - h)^2 + (y - k)^2 - r^2 = 0\}.$$

A tangent from P to the circle has contact at a point Q (Fig. 5-4).

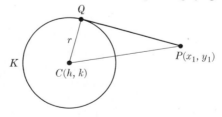

Figure 5-4

Since triangle PQC is a right triangle, we see that $|PQ| = \sqrt{|PC|^2 - r^2}$ and, using the distance formula for the length $|PC|$, we have

$$|PQ| = \sqrt{(x_1 - h)^2 + (y_1 - k)^2 - r^2}. \tag{1}$$

From the procedure for completing the square explained in Section 1, it follows that if the circle having equation

$$x^2 + y^2 + Dx + Ey + F = 0 \tag{2}$$

has center at (h, k) and radius $r > 0$, then

$$(x - h)^2 + (y - k)^2 - r^2 \equiv x^2 + y^2 + Dx + Ey + F.$$

Hence if the circle in Fig. 5–4 has the equation (2), then

$$|PQ| = \sqrt{x_1^2 + y_1^2 + Dx_1 + Ey_1 + F}.$$

EXAMPLE 1. A tangent is drawn from $P(8, 4)$ to the circle K given by

$$K = \{(x, y): x^2 + y^2 + 2x + y - 3 = 0\}.$$

Find the distance from P to the point of tangency Q.

Solution. Using the second formula above for $|PQ|$, we obtain

$$|PQ| = \sqrt{8^2 + 4^2 + 2 \cdot 8 + 4 - 3} = \sqrt{97}.$$

The equation

$$x^2 + (y - k)^2 = 25$$

represents a **family of circles.** All circles of this family have radius 5 and, furthermore, the centers are always on the y axis. Since a line is determined by two conditions while a circle is fixed by three conditions, families of circles are more extensive than families of lines, from some points of view. For example, all circles of radius 3 comprise a "larger" class than all lines of slope 7. This has meaning in the following sense: consider any point in the plane; only one member of the above family of lines passes through this point; however, an infinite number of circles of radius 3 passes through this same point. The family of lines of slope 7 is called a *one-parameter family.* The equation $y = 7x + k$ has *one* constant k which can take on any real value. The equation

$$(x - h)^2 + (y - k)^2 = 9$$

is a *two-parameter family* of circles. The numbers h and k may independently take on any numerical value.

EXAMPLE 2. Find the equation of the family of circles with centers on the line $y = x$ and radius 7.

Solution. The center $C(h, k)$ must have $h = k$ if the center is on the line $y = x$. The required equation is
$$(x - h)^2 + (y - h)^2 = 49.$$
This is a one-parameter family of circles.

EXAMPLE 3. Find the equation of the family of circles with center at $(3, -2)$.

Solution. The equation is
$$(x - 3)^2 + (y + 2)^2 = r^2.$$
This is a one-parameter family of concentric circles.

The two circles
$$x^2 + y^2 + Dx + Ey + F = 0,$$
$$x^2 + y^2 + D'x + E'y + F' = 0,$$

may or may not intersect. If we subtract one of these equations from the other, we obtain
$$(D - D')x + (E - E')y + (F - F') = 0,$$

which is the equation of a straight line. This straight line has slope $-(D - D')/(E - E')$ (if $E \neq E'$). The circles have centers at $(-D/2, -E/2)$ and $(-D'/2, -E'/2)$, respectively. The line joining the centers has slope
$$\frac{(-E'/2) + (E/2)}{(-D'/2) + (D/2)} = \frac{E - E'}{D - D'},$$

which is the negative reciprocal of $-(D - D')/(E - E')$. *The equation of the line obtained by subtraction of the equations of two circles is called the* **radical axis,** *and it is perpendicular to the line joining the centers of the circles.* Furthermore, if the circles intersect, the points of intersection satisfy *both* equations of circles. These points must therefore be on the radical axis.

EXAMPLE 4. Find the points of intersection of the circles.
$$x^2 + y^2 - 2x - 4y - 4 = 0,$$
$$x^2 + y^2 - 6x - 2y - 8 = 0.$$

Solution. By subtraction we get

$$4x - 2y + 4 = 0,$$

and we solve this equation simultaneously with the equation of either circle. The substitution of the equation of the line into the equation for the first circle gives

$$5x^2 - 2x - 8 = 0$$

or

$$x = \tfrac{1}{5} \pm \tfrac{1}{5}\sqrt{41}.$$

The points of intersection are $(\tfrac{1}{5} + \tfrac{1}{5}\sqrt{41}, \tfrac{12}{5} + \tfrac{2}{5}\sqrt{41})$ and $(\tfrac{1}{5} - \tfrac{1}{5}\sqrt{41}, \tfrac{12}{5} - \tfrac{2}{5}\sqrt{41})$.

PROBLEMS

In problems 1 through 4, find the equations of the lines tangent to the given circles at the points P on the circles.

1. $x^2 + y^2 + 4x + 2y = 0,\ P(-3, 1)$
2. $x^2 + y^2 - 2x - 3y - 7 = 0,\ P(-1, -1)$
3. $x^2 + y^2 + 6x + 5y - 21 = 0,\ P(1, 2)$
4. $x^2 + y^2 + 2y - 19 = 0,\ P(-4, -3)$

In problems 5 through 8, find the equations of the lines normal to the given circles at the given points P on the circles.

5. $x^2 + y^2 - 4x - 1 = 0,\ P(3, 2)$
6. $x^2 + y^2 + 8x - 3y + 18 = 0,\ P(-4, 1)$
7. $x^2 + y^2 + 7x - 4y - 6 = 0,\ P(-1, 6)$
8. $x^2 + y^2 - 2x + 8y + 16 = 0,\ P(2, -4)$

In problems 9 through 12, lines are drawn tangent to the given circles through the given points P, not on the circles. Find the equations of these lines.

9. $x^2 + y^2 - 6x + 4y - 21 = 0,\ P(-5, -4)$
10. $x^2 + y^2 - 5x - 2y + 7 = 0,\ P(2, -1)$
11. $x^2 + y^2 + 4x - y - 7 = 0,\ P(0, 3)$
12. $x^2 + y^2 - 6x + 2y + 2 = 0,\ P(-3, -3)$

In problems 13 through 19, write the equations of the families of circles with the given properties.

13. Center on the x axis, radius 3
14. Center on the y axis, radius 4
15. Center on the line $x + y - 6 = 0$, radius 1
16. Center on the x axis, tangent to the y axis
17. Center on the line $y = x$, tangent to both coordinate axes

18. Center on the line $y = 3x$, tangent to the line $x + 3y + 4 = 0$

19. Center on the line $x + 2y - 7 = 0$ and tangent to the line $2x + 3y - 1 = 0$

20. Give a reasonable definition for the statement that two circles intersect at right angles. Do the circles $x^2 + y^2 - 6y = 0$ and $x^2 + y^2 - 6x = 0$ satisfy your definition at their intersection points?

In problems 21 through 23, find the points of intersection of the two circles, if they intersect. If they do not intersect, draw a graph showing the location of the radical axis.

21. $x^2 + y^2 + 4x - 4y + 4 = 0$, $x^2 + y^2 - 2x + 2y - 8 = 0$

22. $x^2 + y^2 + 2x - 4y + 4 = 0$, $x^2 + y^2 - 2x + 4y + 1 = 0$

23. $2x^2 + 2y^2 - 4x - y - 1 = 0$, $3x^2 + 3y^2 - 12x + 3y + 11 = 0$

24. Given the circles $x^2 + y^2 + 2y - 4 = 0$ and $x^2 + y^2 + x - y - 2 = 0$. What is the geometric property of the family of circles

$$(x^2 + y^2 + x - y - 2) + k(x^2 + y^2 + 2y - 4) = 0?$$

Sketch a few members of this family.

25. Find the locus of all points $P(x, y)$ such that the distance from $(-2, 3)$ is always twice the distance from $(-1, 4)$. Describe the locus.

26. Find the locus of all points $P(x, y)$ such that the distance from $(4, 1)$ is always $\frac{2}{3}$ the distance from $(-1, 2)$. Describe the locus.

27. Given the three circles

$$x^2 + y^2 + x - y = 0$$
$$x^2 + y^2 + x + 2y - 3 = 0$$
$$x^2 + y^2 + 2x + y - 4 = 0,$$

show that the three radical axes obtained from each pair of circles are concurrent (i.e., the three lines meet in one point).

28. Work problem 27 again for the circles

$$x^2 + y^2 + Dx + Ey + F = 0,$$
$$x^2 + y^2 + D'x + E'y + F' = 0,$$
$$x^2 + y^2 + D''x + E''y + F'' = 0.$$

What conditions are necessary for the result? [*Hint:* Can you think of a pair of circles which do not have a radical axis?]

29. Show that the circles

$$x^2 + y^2 - 2x - 4y - 13 = 0$$
$$x^2 + y^2 + 4x + 2y - 67 = 0$$

are tangent at some point. Find the point at which they are tangent.

6 The Conics

▶1. THE PARABOLA

In Chapter 5 we began a systematic study of equations of the second degree in x and y. We now continue our analysis of second degree curves.

> DEFINITIONS. **A parabola** is the locus of all points whose distances from a fixed point equal their distances from a fixed line. The fixed point is called the **focus** and the fixed line the **directrix.**

The definition of the parabola is purely geometric in the sense that it says nothing about coordinates—that is, the values of x and y. Recall that the definition of the circle as the locus of points at a given distance from a fixed point is also a geometric one. It will turn out that parabolas are always given by second-degree equations, as is true in the case of the circle.

Figure 6–1 shows a typical situation. The line L is the directrix, the point F is the focus, and P, Q, R, S are points which satisfy the conditions of the locus. That is,

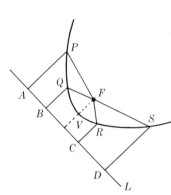

$$|AP| = |PF|,$$
$$|BQ| = |QF|,$$
$$|CR| = |RF|,$$

and

$$|DS| = |SF|.$$

Figure 6–1

The perpendicular to L through the focus intersects the parabola at a point V. This point is called the **vertex** of the parabola.

To find an equation for a parabola, we set up a coordinate system and place the directrix and the focus at convenient places. Suppose that the distance from the focus F to the directrix L is p units. We place the focus at $(p/2, 0)$ and let the line L be $x = -p/2$, as shown in Fig. 6–2. If $P(x, y)$ is a typical point on the locus, the conditions are such that the distance $|PF| = \sqrt{[x - (p/2)]^2 + (y - 0)^2}$

108

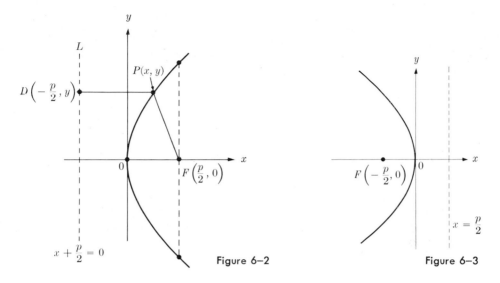

Figure 6–2   Figure 6–3

is equal to the distance from P to the line L. From an examination of the figure (or from the formula for the distance from a point to a line) we deduce that the distance from P to the line L is $|x + p/2|$. Therefore we have

$$\sqrt{\left(x - \frac{p}{2}\right)^2 + y^2} = \left|x + \frac{p}{2}\right|,$$

and, since both sides are positive, we may square to obtain

$$\left(x - \frac{p}{2}\right)^2 + y^2 = \left(x + \frac{p}{2}\right)^2,$$

and

$$x^2 - px + \frac{p^2}{4} + y^2 = x^2 + px + \frac{p^2}{4}.$$

This equation yields

$$y^2 = 2px, \qquad p > 0.$$

Each point on the locus satisfies this equation. Conversely, by reversing the steps, it can be shown that every point which satisfies this equation fulfills the conditions of the locus. We call this *the equation of the parabola with focus at* $(p/2, 0)$ *and with the line* $x = -p/2$ *as directrix*. The vertex of the parabola is at the origin. It is apparent from the equation that the x axis is an axis of symmetry, for if y is replaced by $-y$, the equation is unchanged. The line of symmetry of a parabola is called the **axis of the parabola**.

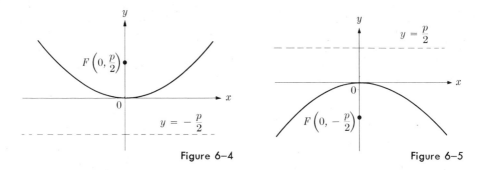

Figure 6–4   Figure 6–5

If the focus and directrix are placed in other positions, the equation will of course be different. There are four *standard positions* for the focus-directrix combination, of which the first has just been described. The second standard position has its focus at $(-p/2, 0)$ and the line $x = p/2$ as its directrix. Using the same method as before, we solve the locus problem which defines the parabola. We obtain the equation

$$y^2 = -2px, \qquad p > 0,$$

and a parabola in the position shown in Fig. 6–3. The third position has its focus at $(0, p/2)$ and the line $y = -p/2$ as the directrix. The resulting equation is

$$x^2 = 2py, \qquad p > 0,$$

and the parabola is in the position shown in Fig. 6–4. In the fourth position the focus is at $(0, -p/2)$ and the directrix is the line $y = p/2$. The equation then becomes

$$x^2 = -2py, \qquad p > 0,$$

and the parabola is in the position shown in Fig. 6–5. In all four positions the vertex is at the origin.

EXAMPLE 1. A parabola is given by

$$S = \{(x, y): y^2 = -12x\}.$$

Find the focus, directrix, and axis. Sketch the graph.

Solution. Since $2p = 12, p = 6$ and, since the equation is in the second of the standard forms, the focus is at $(-3, 0)$. The directrix is the line $x = 3$, and the x axis is the axis of the parabola. The curve is sketched in Fig. 6–6.

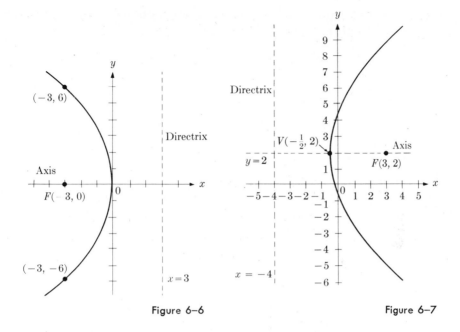

Figure 6–6   Figure 6–7

When the parabola is not in one of the standard positions, the equation is more complicated. If the directrix is horizontal or vertical, the equation is only slightly different; however, if the directrix is parallel to neither axis, the equation is altered considerably. The following example shows how the equation of the parabola may be obtained directly from the definition.

EXAMPLE 2. Find the equation of the parabola with focus at $F(3, 2)$ and with the line $x = -4$ as directrix. Locate the vertex and the axis of symmetry.

Solution. From the definition of parabola, $P(x, y)$ is a point on the curve if and only if $|PF| = \sqrt{(x - 3)^2 + (y - 2)^2}$ is equal to the distance from P to the directrix, which is $|x + 4|$. We have

$$\sqrt{(x - 3)^2 + (y - 2)^2} = |x + 4|.$$

Since both sides are positive, we square both sides and find

$$x^2 - 6x + 9 + y^2 - 4y + 4 = x^2 + 8x + 16.$$

This yields

$$(y - 2)^2 = 14x + 7 \Leftrightarrow (y - 2)^2 = 14(x + \tfrac{1}{2}).$$

A sketch of the equation (Fig. 6–7) shows that the axis is the line $y = 2$ and the vertex is at the point $(-\tfrac{1}{2}, 2)$.

The four forms for the parabola when the axis is vertical or horizontal and the vertex is at the point (a, b) instead of the origin are:*

$$(y - b)^2 = 2p(x - a), \quad p > 0, \tag{1}$$

with focus at $(a + p/2, b)$; directrix, $x = a - p/2$;

$$(y - b)^2 = -2p(x - a), \quad p > 0, \tag{2}$$

with focus at $(a - p/2, b)$; directrix, $x = a + p/2$;

$$(x - a)^2 = 2p(y - b), \quad p > 0, \tag{3}$$

with focus at $(a, b + p/2)$; directrix, $y = b - p/2$;

$$(x - a)^2 = -2p(y - b), \quad p > 0, \tag{4}$$

with focus at $(a, b - p/2)$; directrix, $y = b + p/2$.

EXAMPLE 3.   Given the parabola $x^2 = -4y - 3x + 2$, find the vertex, focus, directrix, and axis.   Sketch the curve.

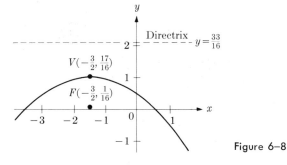

Figure 6–8

Solution. We complete the square in x, writing

$$x^2 + 3x = -4y + 2,$$

and then obtain

$$(x + \tfrac{3}{2})^2 = -4y + 2 + \tfrac{9}{4} = -4y + \tfrac{17}{4}.$$

This yields

$$(x + \tfrac{3}{2})^2 = -4(y - \tfrac{17}{16}),$$

which is in the fourth form above. We read off that the vertex is at $(-\tfrac{3}{2}, \tfrac{17}{16})$. Since $2p = 4$ we know that $p = 2$ and the focus is at $(-\tfrac{3}{2}, \tfrac{1}{16})$; the directrix is the line $y = \tfrac{33}{16}$, and the axis is the line $x = -\tfrac{3}{2}$. The curve is sketched in Fig. 6–8.

---

* These equations will be derived in Section 7 below. The student may also derive them directly from the definition.

PROBLEMS

In problems 1 through 6, find the coordinates of the foci and the equations of the directrices. Sketch the curves.

1. $y^2 = 4x$
2. $x^2 = -8y$
3. $y^2 = 12x$
4. $x^2 = -12y$
5. $2y^2 = 9x$
6. $4x^2 = -25y$

In problems 7 through 14, find the equation of the parabola with vertex at the origin which satisfies the given additional condition.

7. Focus at $(-4, 0)$
8. Focus at $(0, 4)$
9. Directrix: $y = -2$
10. Directrix: $x = -\frac{10}{3}$
11. Focus at $(0, \frac{5}{2})$
12. Directrix: $y = \frac{7}{3}$
13. Passing through $(-2, 5)$ and axis along the x axis
14. Passing through $(3, 2)$ and axis along the y axis

In problems 15 through 22, find the focus, vertex, directrix, and axis of the parabola. Sketch the curve.

15. $y = x^2 + 2x + 5$
16. $x = y^2 + 4y + 7$
17. $y = -2x^2 - 4x + 7$
18. $y = -4x^2 + 5x - 5$
19. $x = -y^2 + 2y - 7$
20. $x^2 + 2y - 3x + 5 = 0$
21. $y^2 + 4x + 2y - 5 = 0$
22. $3y^2 = 2x - 3y$

In problems 23 through 28, find the equations of the parabolas from the definition.

23. Directrix $x = 0$, focus at $(10, 0)$
24. Directrix $y = 0$, focus at $(0, 3)$
25. Vertex $(0, 3)$, focus $(0, -1)$
26. Vertex $(-3, 0)$, directrix $x = 2$
27. Focus $(2, 1)$, directrix $3x + 4y + 5 = 0$
28. Focus $(-3, 2)$, directrix $x + y + 4 = 0$

29. An arch has the shape of a parabola with vertex at the top and axis vertical. If it is 36 ft wide at the base and 24 ft high in the center, how wide is it 15 ft above the base?

30. The line through the focus of a parabola and parallel to the directrix is called the *line of the latus rectum*. The portion of this line cut off by the parabola is called the **latus rectum**. Show that for the parabola $y^2 = 2px$ the length of the latus rectum is $2p$. Is this true for any parabola where the distance from the focus to the directrix is p units?

31. Find the locus of all points whose distances from $(4, 0)$ are 1 unit less than their distances from the line $x = -6$. Describe the curve.

32. Find the equation of the parabola the axis of which is parallel to the x axis and passes through the points $(-1, 3)$, $(-4, 0)$, and $(8, -6)$.

33. A projectile hurled at an angle α with the horizontal follows (approximately) a parabolic path given by

$$y = x (\tan \alpha) - \frac{16}{v_0^2 \cos^2 \alpha} x^2,$$

where v_0 is the initial velocity. If a projectile initially has a velocity of 30 ft/sec and it makes an angle of $30°$ with the horizontal, find the distance it travels and the highest point of the trajectory.

34. Let P_1 and P_2 be any two points on the parabola $y^2 = 2px$ and suppose that the slope of the line through P_1 and P_2 is m. Let A be the midpoint of P_1P_2. Show that the y coordinate of A is p/m.

35. Find the locus of all points whose distances from $(4, 1)$ are equal to their distances from the line $y = x$. Describe the curve.

▶2. TANGENTS. GEOMETRICAL PROPERTIES OF A PARABOLA

In Chapter 5 we defined the tangent line to a circle and showed how to find its equation in any particular case. Since no other curve has the complete symmetry of a circle, the corresponding definition for more general curves is much more complicated. In this section we give a definition of a tangent line to a parabola.

Consider the parabola

$$x^2 = 2py$$

and let $P_0(x_0, y_0)$ be a point on it. The equation

$$y - y_0 = m(x - x_0)$$

represents a family of lines through P_0. In Fig. 6–9 the lines L_1, L_2, and L_3 are three members of this family. It appears that the members of the family intersect the parabola in two points: the point P_0 and another point Q. The second inter-section point Q is shown in Fig. 6–9 for the lines L_1 and L_2. To obtain the coordi-nates of the points P_0 and Q, we solve simultaneously the equations for the parabola and the straight line. By sub-stitution we find

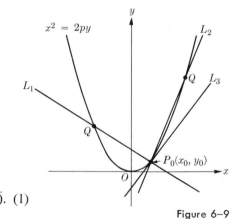

$$x^2 = 2p(y_0 + mx - mx_0)$$

$$\Leftrightarrow x^2 - 2pmx + 2p(mx_0 - y_0) = 0$$

$$\Leftrightarrow x = pm \pm \sqrt{p^2m^2 - 2p(mx_0 - y_0)}. \quad (1)$$

Figure 6–9

Now the points of intersection P_0 and Q *coincide* if and only if the two values of x coincide. The solution (1) for x can have one value if and only if the quantity under the radical is zero. In other words, when $p^2m^2 = 2p(mx_0 - y_0)$ we have the single solution $x = x_0 = pm$. Since m is the slope, we see that the line through P_0 with slope x_0/p touches the parabola at exactly one point. The line L_3 in Fig. 6–9 has this property. When $P_0(x_0, y_0)$ is on the parabola, we define the line with equation

$$y - y_0 = \frac{x_0}{p} (x - x_0)$$

as **the tangent line to the parabola** $x^2 = 2py$.

We observe that for any parabola each line parallel to the axis of the parabola intersects the parabola in one point. By definition, every other line in the plane having only one point of contact with the parabola is tangent to the parabola at the point of contact.

If we carry through the above analysis for the parabola $y^2 = 2px$, we may define the line l given by

$$l = \left\{ (x, y): y - y_0 = \frac{p}{y_0} (x - x_0) \right\}$$

as **the tangent line to the parabola** $y^2 = 2px$ at $P_0(x_0, y_0)$ provided P_0 is a point on the parabola and $y_0 \neq 0$.

EXAMPLE 1. Given the parabola

$$S = \{(x, y): x^2 = 12y\},$$

find the equation of the tangent line at the point on the parabola whose x coordinate is 4.

Solution. The y coordinate of the point on the curve is $\frac{16}{12} = \frac{4}{3}$. Since $p = 6$, the slope of the tangent line is $m = \frac{4}{6} = \frac{2}{3}$. The desired line has equation

$$y - \tfrac{4}{3} = \tfrac{2}{3}(x - 4) \Leftrightarrow 2x - 3y - 4 = 0.$$

Except for position in the plane, every parabola has the general character of $x^2 = 2py$ or $y^2 = 2px$. Therefore, the definition of tangent line to any parabola is identical with the one we gave above. The next example makes use of this definition.

EXAMPLE 2. Given the family of lines

$$l(k) = \{(x, y): 3x + 2y - k = 0\},$$

find the particular member of this family which is tangent to the parabola

$$S = \{(x, y): y = 3x^2 - 2x + 1\}.$$

Solution. If we solve simultaneously the equations of the parabola and the line, the solution set consists of the points of intersection of the two curves. Thus, depending on the value of k, the solution set $l(k) \cap S$ consists of two points, one point, or the empty set. When $l(k) \cap S$ consists of one point, the line is tangent to the parabola or is parallel to the axis of the parabola. By substitution, we find

$$\frac{k - 3x}{2} = 3x^2 - 2x + 1$$

$$\Leftrightarrow 6x^2 - x + 2 - k = 0$$

$$\Leftrightarrow x = \frac{1 \pm \sqrt{1 - 24(2 - k)}}{12}.$$

There is a single value of x if and only if

$$1 - 24(2 - k) = 0 \Leftrightarrow k = \tfrac{47}{24}.$$

The required member of the family is

$$l(\tfrac{47}{24}) = \{(x, y): 3x + 2y - \tfrac{47}{24} = 0\}.$$

Since the axis of the parabola is parallel to the y axis, the line $l(\tfrac{47}{24})$ is tangent to the parabola. The parabola and a few members of the family $l(k)$ are shown in Fig. 6–10.

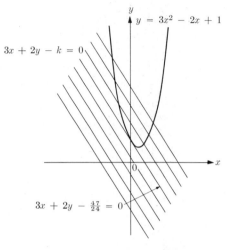

Figure 6–10

Suppose we are given a parabola and a point $P_1(x_1, y_1)$ which is *not* on the parabola. Then through the point P_1 there will either be two lines tangent to the parabola or no lines tangent to the parabola. A typical situation is shown in Fig. 6–11. To find the equations of the tangent lines when they exist, we perform the following steps:

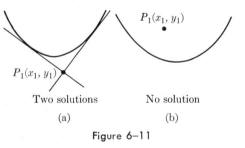

Two solutions   No solution
(a)   (b)

Figure 6–11

Step 1. Write the equation of the family of lines through P_1: $(y - y_1) = m(x - x_1)$.

Step 2. Solve simultaneously the equations of the line and the parabola. Find the values of m which make the solution set consist of one point.

Step 3. If the values of m obtained in Step 2 are complex, there are no tangent lines through P_1. If the values are real, inserting them in the equation for the family yields the tangent lines (provided neither line so obtained is parallel to the axis of the parabola).

EXAMPLE 3. Find the equations of the lines which are tangent to the parabola $S = \{(x, y): y^2 - y - 2x + 4 = 0\}$ and which pass through $P_1(-5, 2)$.

Solution. The family of lines through P_1 is $l(m) = \{(x, y): (y - 2) = m(x + 5)\}$. Solving this equation simultaneously with that of the parabola, we get

$$y^2 - y - \frac{2(y - 2 - 5m)}{m} + 4 = 0$$

$$\Leftrightarrow my^2 - (2 + m)y + 2(7m + 2) = 0$$

$$\Leftrightarrow y = \frac{(2 + m) \pm \sqrt{(2 + m)^2 - 8m(7m + 2)}}{2m}.$$

There is a single solution for y if and only if the quantity under the radical in the above expression is zero. That is,

$$l(m) \cap S \text{ consists of one point} \Leftrightarrow (2 + m)^2 - 8m(7m + 2) = 0$$
$$\Leftrightarrow 55m^2 + 12m - 4 = 0$$
$$\Leftrightarrow m = \tfrac{2}{11}, -\tfrac{2}{5}.*$$

We conclude that $l(\tfrac{2}{11})$ and $l(-\tfrac{2}{5})$ are tangent to S. The desired equations are

$$y - 2 = \tfrac{2}{11}(x + 5) \qquad \text{and} \qquad y - 2 = -\tfrac{2}{5}(x + 5).$$

The parabola has many interesting geometrical properties that make it suitable for practical applications. Probably the most familiar application is the use of parabolic shapes in headlights of cars, flashlights, and so forth. If a parabola is revolved about its axis, the surface generated is called a **paraboloid.** Such surfaces are used in headlights, optical and radio telescopes, radar, etc., because of the following geometric property. If a source of light (or sound or other type of wave) is placed at the focus of a parabola, and if the parabola is a reflecting surface, then the wave will bounce back in a line parallel to the axis of the parabola (Fig. 6–12). This creates a parallel beam without dispersion. (In actual practice there will be some dispersion, since the source of light must occupy more than one point.) Clearly the converse is also true. If a series of incoming waves is parallel to the axis of the reflecting paraboloid, the resulting signal will be concentrated at the focus (where the receiving equipment is therefore located).

We shall establish the reflecting property for the parabola $y^2 = 2px$. Let $P(x_1, y_1)$ be a point (not the vertex) on the parabola, and draw the line segment from P to the focus (Fig. 6–13). Now construct a line L through P parallel to the x axis and the line M tangent to the parabola at P. The law for the reflecting property of the parabola states that the angle formed by M and PF (α in the figure) must be equal to the angle formed by L and M (β in the figure). The slope

---

* This solution of this problem obviously assumed that $m \neq 0$. The axis of the parabola is seen by completing the square to be the line $y = \tfrac{1}{2}$. The line $y - 2 = 0$, corresponding to $m = 0$ in our family, is parallel to the axis of the parabola.

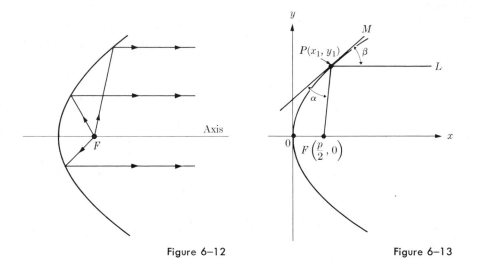

Figure 6–12   Figure 6–13

of the line M is $\tan \beta$, which is also the slope of the parabola at (x_1, y_1). The definition of the slope of the tangent line given in the middle of page 115 yields the relation

$$\tan \beta = \frac{p}{y_1}.$$

We now find $\tan \alpha$ from the formula for the angle between two lines. The slope of PF is

$$\frac{y_1 - 0}{x_1 - \frac{p}{2}}.$$

Therefore

$$\tan \alpha = \frac{\dfrac{y_1}{x_1 - p/2} - \dfrac{p}{y_1}}{1 + \dfrac{y_1}{x_1 - p/2}\dfrac{p}{y_1}},$$

and

$$\tan \alpha = \frac{y_1^2 - px_1 + \frac{1}{2}p^2}{x_1 y_1 - \frac{1}{2}py_1 + py_1}.$$

Since $y_1^2 = 2px_1$, this simplifies to

$$\tan \alpha = \frac{p}{y_1},$$

and the result is established.

The main cables of suspension bridges are parabolic in shape, since it can be shown that if the total weight of a bridge is distributed uniformly along its length, a cable in the shape of a parabola bears the load evenly. It is a remarkable fact that no other shape will perform this task.

PROBLEMS

In problems 1 through 8, find the equation of the line tangent to the parabola at the given point P on the parabola.

1. $x^2 = 9y$, $P(-3, 1)$
2. $x^2 = -3y$, $P(3, -3)$
3. $y^2 = -12x + 7$, $P(\frac{1}{2}, 1)$
4. $y^2 + 2x + 3y - 4 = 0$, $P(-3, -2)$
5. $x^2 - 3y - 2x + 4 = 0$, $P(0, \frac{4}{3})$
6. $(x + 2)^2 = -4(y + 3)$, $P(-4, -4)$
7. $y^2 - 2y - x = 0$, $P(15, -3)$
8. $-y^2 + 2x + 2y + 4 = 0$, $P(2, 4)$
9. Find the equations of the lines tangent to the parabola

$$x^2 - 3x + 4y - 8 = 0$$

at the points on the parabola where $y = 1$.

In each of problems 10 through 20 find, when possible, the equations of the lines tangent to the given parabola through the given point. When no tangent lines can be drawn, verify this fact.

10. $x^2 = 8y$, $P(-1, -3)$
11. $x^2 = 4y$, $P(-5, 1)$
12. $x^2 = 8y$, $P(-1, 3)$
13. $x^2 = 4y$, $P(5, 1)$
14. $y^2 = 4x$, $P(-1, 4)$
15. $y^2 = -12x$, $P(3, -1)$
16. $y^2 - 4y - 2x - 1 = 0$, $P(-3, -1)$
17. $2x^2 + 3x + y - 4 = 0$, $P(1, 1)$
18. $y^2 - 4y - 2x - 1 = 0$, $P(8, 1)$
19. $y - 2x - 2y^2 + 1 = 0$, $P(-3, 1)$
20. $y - 2x - 2y^2 + 1 = 0$, $P(3, 1)$
21. Find the member of the family $4y + x - k = 0$ which is tangent to the parabola $-x = 2y^2 - y + 1$.
22. Find a member of the family $ky + x - 5 = 0$ which is tangent to the parabola $-x = 3y^2 + 2y + 4$.
23. Find the member of the family of parabolas $-x = ky^2 + 4y - 3$ which is tangent to the line $2x - y + 1 = 0$.
24. For the parabola $x^2 = 2py$, establish the "optical property." That is, if (x_1, y_1) is a point on the parabola and M is the line tangent at (x_1, y_1), show that the angle formed between M and a line from (x_1, y_1) through the focus is equal to the angle M forms with a line through (x_1, y_1) parallel to the y axis.
25. Let $P(x_1, y_1)$ be a point (not the vertex) on the parabola $y^2 = 2px$. Find the equation of the line normal to the parabola at this point. Show that this normal line intersects the x axis at the point $Q(x_1 + p, 0)$.
26. A line through the focus of the parabola $x^2 = -12y$ intersects the parabola at the point $(5, -\frac{25}{12})$. Find the other point of intersection of this line with the parabola.
27. The two towers of a suspension bridge are 300 ft apart and extend 80 ft above the road surface. If the cable is tangent to the road at the center of the bridge, find the height of the cable above the road at 50 ft and also at 100 ft from the center of the bridge. (Assume that the road is horizontal.)

28. At a point P on a parabola a tangent line is drawn. This line intersects the axis of the parabola at a point A. If F denotes the focus, show that $\triangle APF$ is isosceles.

29. A stone wall borders one side of a field. There is 200 ft of fencing available to form the three sides of a rectangular enclosure, the wall forming the fourth side. Find the dimensions which provide the largest possible area.

30. (a) The points P and Q are on the parabola $y^2 = 8x$, and P has coordinates (2, 4). If triangle POQ is a right triangle with right angle at O (the origin), find the coordinates of Q. (b) Repeat the procedure with the equation $y^2 = 2px$, and P with coordinates $(x_1, \sqrt{2px_1})$.

▶ 3. **THE ELLIPSE**

Continuing our discussion of curves of the second degree, we now take up the ellipse.

DEFINITIONS. An **ellipse** is the locus of all points the sum of whose distances from two fixed points is constant. The two fixed points are called the **foci**.

For the definition to make sense, the constant (the sum of the distances from the foci) must be larger than the distance between the foci. Note that the definition is purely geometric and says nothing about coordinates or equations. It will turn out, that the equation of an ellipse is of the second degree.

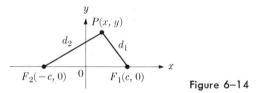

Figure 6–14

Suppose that the distance between the foci is $2c$ ($c > 0$). We place the foci—label them F_1 and F_2—at convenient points, say F_1 at $(c, 0)$ and F_2 at $(-c, 0)$. Let $P(x, y)$ be a point on the ellipse, d_1 the distance from P to F_1, and d_2 the distance from P to F_2 (Fig. 6–14). The conditions of the locus state that $d_1 + d_2$ is always constant. As we said, this constant (which we shall denote by $2a$) must be larger than $2c$. Thus we have

$$d_1 + d_2 = 2a,$$

or, from the distance formula,

$$\sqrt{(x - c)^2 + (y - 0)^2} + \sqrt{(x + c)^2 + (y - 0)^2} = 2a.$$

We simplify by transferring one radical to the right and squaring both sides:

$$(x - c)^2 + y^2 = 4a^2 - 4a\sqrt{(x + c)^2 + y^2} + (x + c)^2 + y^2.$$

By multiplying out and combining, we obtain

$$\sqrt{(x + c)^2 + y^2} = a + \frac{c}{a} x.$$

We square again to find

$$(x + c)^2 + y^2 = a^2 + 2cx + \frac{c^2}{a^2} x^2,$$

and this becomes

$$\frac{x^2}{a^2} + \frac{y^2}{a^2 - c^2} = 1.$$

Since $a > c$, we can introduce a new quantity,

$$b = \sqrt{a^2 - c^2},$$

and we can write

$$\frac{x^2}{a^2} + \frac{y^2}{b^2} = 1.$$

This is **the equation of an ellipse**.

We have shown that every point on the locus satisfies this equation. Conversely (by reversing the steps*), it can be shown that each point which satisfies this equation is on the ellipse. A sketch of such an equation is shown in Fig. 6–15. From the equation, we see immediately that the curve is symmetric with respect to both the x and y axes.

DEFINITIONS. The line passing through the two foci F_1 and F_2 is called the **major axis**. The perpendicular bisector of the line segment F_1F_2 is called the **minor axis**. The intersection of the major and minor axes is called the **center**.

The equation of the ellipse as we have obtained it indicates that the ellipse intersects the x axis at $(a, 0)$, $(-a, 0)$. These points are called the **vertices** of the ellipse. The distance between them, $2a$, is called the *length of the major axis*. (Some-

---

* In the process of reversing the steps we must verify (because of the squaring operations) that

$$a + \frac{c}{a} x \geq 0 \quad \text{and} \quad 2a - \sqrt{(x + c)^2 + y^2} \geq 0.$$

However,

$$2a - \sqrt{(x + c)^2 + y^2} = 2a - \left(a + \frac{c}{a} x \right) = a - \frac{c}{a} x.$$

Since $-a \leq x \leq a$ for every point satisfying the equation, and since $c < a$, the verification follows.

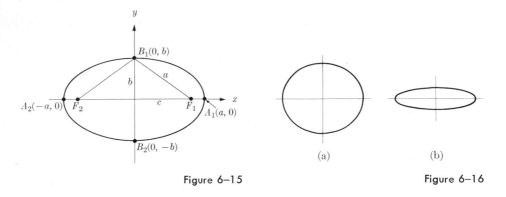

Figure 6–15   Figure 6–16

times this segment is called simply the major axis.) The ellipse intersects the
y axis at $(0, b)$ and $(0, -b)$. The distance between them, $2b$, is called the *length
of the minor axis*. (Sometimes this segment is called the minor axis.)

The **eccentricity** e of an ellipse is defined as

$$e = \frac{c}{a}.$$

Note that since $c < a$, the eccentricity is always between 0 and 1. Further, it is
dimensionless, being the ratio of two lengths.   The eccentricity measures the
flatness of an ellipse. If a is kept fixed and c is very "small," then e is close to
zero. But this means that the foci are close together, and the ellipse is almost a
circle, as in Fig. 6–16(a). On the other hand, if c is close to a, then e is near 1
and the ellipse is quite flat, as shown in Fig. 6–16(b). The limiting position of an
ellipse as $e \rightarrow 0$ is a circle of radius a. The limiting position as $e \rightarrow 1$ is a line
segment of length $2a$.

In developing the equation of an ellipse we placed the foci at $(c, 0)$ and $(-c, 0)$.
If instead we put them on the y axis at $(0, c)$ and $(0, -c)$ and carry through the
same argument, then the equation of the ellipse we obtain is

$$\frac{x^2}{b^2} + \frac{y^2}{a^2} = 1.$$

The y axis is now the major axis, the x axis is the minor axis, and the vertices are
at $(0, a)$ and $(0, -a)$. The quantity $b = \sqrt{a^2 - c^2}$ and the eccentricity $e = c/a$
are defined as before. Figure 6–17 shows an ellipse with foci on the y axis.

EXAMPLE 1. Given the ellipse with equation $9x^2 + 25y^2 = 225$, find the major and
minor axes, the eccentricity, the coordinates of the foci and the vertices.   Sketch the
ellipse.

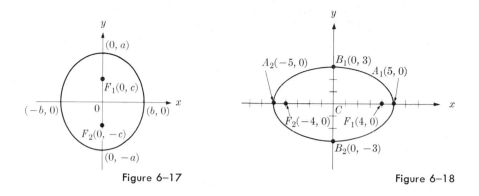

Figure 6–17   Figure 6–18

Solution. We put the equation in "standard form" by dividing by 225. We obtain

$$\frac{x^2}{25} + \frac{y^2}{9} = 1,$$

which tells us that $a = 5, b = 3$. Since

$$a^2 = b^2 + c^2,$$

we find that $c = 4$. The eccentricity $e = c/a = \frac{4}{5}$. The major axis is along the x axis, the minor axis is along the y axis, the vertices are at $(\pm 5, 0)$, and the foci are at $(\pm 4, 0)$. The ellipse is sketched in Fig. 6–18.

EXAMPLE 2. Given the ellipse with equation $16x^2 + 9y^2 = 144$, find the major and minor axes, the eccentricity, the coordinates of the foci, and the vertices. Sketch the ellipse.

Solution. We divide by 144 to get

$$\frac{x^2}{9} + \frac{y^2}{16} = 1.$$

We note that *the number under the y term is larger.* Therefore $a = 4, b = 3, c^2 = a^2 - b^2$, and $c = \sqrt{7}$. The eccentricity $e = \sqrt{7}/4$, the major axis is along the y axis, the minor axis is along the x axis, the foci are at $(0, \pm\sqrt{7})$, and the vertices at $(0, \pm 4)$. The ellipse is sketched in Fig. 6–19.

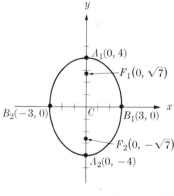

Figure 6–19

The foregoing examples illustrate the fact that when we have an equation of the form

$$\frac{x^2}{(\ )^2} + \frac{y^2}{(\ )^2} = 1,$$

the equation represents an ellipse (or a circle if the denominators are equal in size), and the *larger denominator* determines whether the foci, vertices, and major axes are along the x axis or the y axis.

EXAMPLE 3.   Find the equation of the ellipse with axes along the x axis and y axis which passes through the points $P(6, 2)$ and $Q(-4, 3)$.

Solution. Let the equation be

$$\frac{x^2}{A^2} + \frac{y^2}{B^2} = 1;$$

we don't know whether A is larger than B or B is larger than A. Since the points P and Q are on the ellipse, we have

$$36\left(\frac{1}{A^2}\right) + 4\left(\frac{1}{B^2}\right) = 1,$$

$$16\left(\frac{1}{A^2}\right) + 9\left(\frac{1}{B^2}\right) = 1,$$

which gives us two equations in the unknowns $1/A^2$, $1/B^2$. Solving simultaneously, we obtain

$$\frac{1}{A^2} = \frac{1}{52}, \qquad \frac{1}{B^2} = \frac{1}{13},$$

and the resulting ellipse is

$$\frac{x^2}{52} + \frac{y^2}{13} = 1.$$

The major axis is along the x axis.

PROBLEMS

In problems 1 through 10, find the lengths of the major and minor axes, the coordinates of the foci and vertices, and the eccentricity. Sketch the curve.

1. $4x^2 + 9y^2 = 36$

2. $9x^2 + 25y^2 = 225$

3. $9x^2 + 4y^2 = 144$

4. $16x^2 + 25y^2 = 100$

5. $4x^2 + 3y^2 = 12$

6. $2x^2 + 3y^2 = 6$

7. $5x^2 + 3y^2 = 15$

8. $3x^2 + 7y^2 = 21$

9. $3x^2 + 2y^2 = 13$

10. $4x^2 + 11y^2 = 17$

In problems 11 through 21, find the equation of the ellipse satisfying the given conditions.

11. Vertices at $(\pm 5, 0)$, foci at $(\pm 3, 0)$

12. Vertices at $(0, \pm 10)$, foci at $(0, \pm 8)$

13. Vertices at $(0, \pm 10)$, eccentricity $\frac{3}{5}$

14. Vertices at $(\pm 9, 0)$, eccentricity $\frac{2}{3}$

15. Foci at $(0, \pm 3)$, eccentricity $\frac{3}{5}$

16. Vertices at $(0, \pm 4)$, passing through $(-1, 3)$

17. Axes along the coordinate axes, passing through $(-3, 4)$, $(-4, -1)$

18. Axes along the coordinate axes, passing through $(-2, -5)$ and $(-7, 3)$

19. Foci at $(\pm 4, 0)$, passing through $(5, 3)$

20. Eccentricity $\frac{3}{4}$, foci along the x axis, center at origin, and passing through $(2, 4)$

21. Eccentricity $\frac{3}{4}$, foci on the y axis, center at origin, and passing through $(2, 4)$

22. Find the locus of a point which moves so that the sum of its distances from $(8, 0)$ and $(-8, 0)$ is always 20.

23. Find the locus of a point which moves so that the sum of its distances from $(3, 0)$ and $(9, 0)$ is always 12.

24. A square is inscribed in the ellipse

$$\frac{x^2}{16} + \frac{y^2}{9} = 1.$$

Find the coordinates of the vertices; find the perimeter and area of the square.

25. The orbit in which the earth travels about the sun is approximately an ellipse with the sun at one focus. The major axis of the ellipse is 186 million miles and the eccentricity is 0.017, approximately. Find the maximum and minimum distance from the earth to the sun.

▶4. GEOMETRICAL PROPERTIES OF AN ELLIPSE. DIRECTRICES. TANGENTS

A mechanical construction of an ellipse can be made directly from the facts given in the definition, as shown in Fig. 6–20. Select a piece of string of length $2a$. Fasten its ends with thumbtacks at the foci, insert a pencil so as to draw the string taut, and trace out an arc. The curve will be an ellipse.

The ellipse has many interesting geometrical properties. The paths of the planets about the sun and those of the man-made satellites about the earth are approximately elliptical; whispering galleries are ellipsoidal (an ellipse revolved about the major axis) in shape. If a signal (in the form of a light, sound, or other type of wave) has its source at one focus of an ellipse, *all* reflected waves will pass through the other focus.

We have seen that a parabola has a focus and a directrix. The ellipse has two foci and also two directrices. If the ellipse has foci at $(\pm c, 0)$ and vertices at $(\pm a, 0)$, *the* **directrices** *are the lines*

$$x = \frac{a^2}{c} = \frac{a}{e}$$

and

$$x = -\frac{a^2}{c} = -\frac{a}{e}.$$

To draw an analogy with the parabola, we can show that such an ellipse is the locus of all

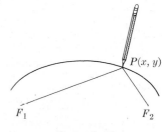

Figure 6–20

points whose distances from a fixed point (F_1) are equal to e $(0 < e < 1)$ times their distances from a fixed line $(x = a/e)$. As Fig. 6–21 shows, the conditions of the locus are

$$d_1 = ed_2 \Leftrightarrow \sqrt{(x - c)^2 + y^2} = e\left|x - \frac{a}{e}\right|.$$

If we square both sides and simplify, we obtain $(x^2/a^2) + (y^2/b^2) = 1$, which we know to be the equation of an ellipse. The same situation holds true if the focus $F_2(-c, 0)$ and the directrix $x = -a/e$ are used.

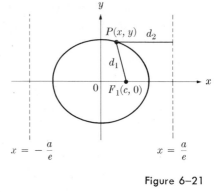

Figure 6–21

EXAMPLE 1.   Find the equation of the ellipse with foci at $(\pm 5, 0)$ and with the line $x = 36/5$ as one directrix.

Solution.   We have $c = 5$, $a/e = 36/5$. Since $e = c/a$, we obtain $a^2/c = 36/5$, or $a = 6$. Then $b = \sqrt{36 - 25} = \sqrt{11}$, and the equation is

$$\frac{x^2}{36} + \frac{y^2}{11} = 1.$$

A line in the plane may intersect an ellipse in two points, in one point, or not at all. A line is said to be **tangent to an ellipse** if the line and ellipse have exactly one point in common.

If we are given an ellipse and a point, the method for finding the line tangent to the ellipse passing through the given point is similar to the one used for finding tangents to a parabola. We write the equation of the family of lines through this point; then we select the particular member of the family (assuming there is one) which has but one point of contact with the ellipse. This member is the desired tangent line. An example illustrates the method.

EXAMPLE 2.   Find the equation of the line tangent to the ellipse

$$S = \{(x, y): 9x^2 + 5y^2 - 81 = 0\}$$

which passes through the point $P_0(2, 3)$.

Solution.   The point P_0 is on S. We write the family of lines

$$l(m) = \{(x, y): y - 3 = m(x - 2)\},$$

and we seek the particular value of m such that $l(m) \cap S = \{P_0\}$. By substitution of the equation of the line into the equation of the ellipse, we find

$$9x^2 + 5(3 + mx - 2m)^2 - 81 = 0.$$

After squaring out the term in the parentheses and collecting terms, we get

$$(9 + 5m^2)x^2 - 10m(2m - 3)x + 20m^2 - 60m - 36 = 0.$$

This quadratic equation in x has one solution if and only if its discriminant vanishes. That is, we must have

$$100m^2(2m - 3)^2 - 4(9 + 5m^2)(20m^2 - 60m - 36) = 0.$$
$$\Leftrightarrow 25m^2 + 60m + 36 = 0 \quad \Leftrightarrow \quad (5m + 6)^2 = 0.$$

The only solution is $m = -\frac{6}{5}$, and the desired tangent line has equation

$$y - 3 = -\tfrac{6}{5}(x - 2) \quad \Leftrightarrow \quad 6x + 5y - 27 = 0.$$

Remarks. If the point P_0 is *outside* the ellipse rather than on the ellipse, then the above analysis yields two values of m corresponding to the two lines which may be drawn tangent to an ellipse from a point outside the ellipse. If the point P_0 is inside the ellipse, the analysis above leads to complex values of m, since no tangent line can be drawn through such a point.

The procedure above is useful for finding the equations of the tangents drawn to an ellipse from an outside point. But it turns out that there is a simple formula for the slope and a simple equation for the tangent to an ellipse at a point on the ellipse. We now derive these formulas.

Let (x_1, y_1) be a point on the ellipse whose equation is

$$\frac{x^2}{a^2} + \frac{y^2}{b^2} = 1. \tag{1}$$

The equation of the family of lines through (x_1, y_1) is

$$y - y_1 = m(x - x_1)$$
$$\Leftrightarrow y = mx + (y_1 - mx_1).$$

Substituting $mx + (y_1 - mx_1)$ for y in Eq. (1) and multiplying by a^2b^2 leads to the equation

$$b^2x^2 + a^2[m^2x^2 + 2m(y_1 - mx_1)x + (y_1 - mx_1)^2] - a^2b^2 = 0$$
$$\Leftrightarrow (a^2m^2 + b^2)x^2 + 2a^2m(y_1 - mx_1)x + a^2[(y_1 - mx_1)^2 - b^2] = 0.$$

The roots of this last equation will coincide if and only if its discriminant vanishes. This condition is

$$4a^4m^2(y_1 - mx_1)^2$$
$$\qquad - 4a^2[a^2m^2(y_1 - mx_1)^2 + b^2(y_1 - mx_1)^2 - a^2b^2m^2 - b^4] = 0$$
$$\Leftrightarrow 4a^2b^2[a^2m^2 + b^2 - y_1^2 + 2mx_1y_1 - m^2x_1^2] = 0$$
$$\Leftrightarrow (a^2 - x_1^2)m^2 + 2x_1y_1m + (b^2 - y_1^2) = 0. \tag{2}$$

Now, since $(x_1^2/a^2) + (y_1^2/b^2) = 1$, we find that

$$a^2 - x_1^2 = \frac{a^2 y_1^2}{b^2}$$

and

$$b^2 - y_1^2 = \frac{b^2 x_1^2}{a^2}.$$

(3)

Substituting (3) into (2), we conclude that (2) holds

$$\Leftrightarrow \left(\frac{a y_1 m}{b} + \frac{b x_1}{a} \right)^2 = 0,$$

from which we conclude that *the slope of the tangent at* (x_1, y_1) *is*

$$m = - \frac{b^2 x_1}{a^2 y_1}.$$

(4)

Substituting this value for m in the equation $y - y_1 = m(x - x_1)$ and using the fact that (x_1, y_1) satisfies (1), we find that *the equation of the tangent at* (x_1, y_1) *to the ellipse whose equation is* (1) *is*

$$\frac{x_1 x}{a^2} + \frac{y_1 y}{b^2} = 1.$$

(5)

The details of the derivation of this equation are left to the student (see problem 15 below).

The reflected-wave property of ellipses is exhibited by the following procedure. Let $P(x_1, y_1)$ be a point on the ellipse (not a vertex), as shown in Fig. 6–22. Construct a tangent line M at this point. Then the angle made by M and the line PF_1 (α in the figure) is equal to the angle made by M and the line PF_2 (β in the figure). This result may be established by setting up the formulas for $\tan \alpha$ and $\tan \beta$ as the angles between two lines and showing them to be equal.

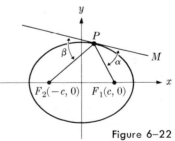

Figure 6–22

PROBLEMS

In problems 1 through 6, find the equation of the ellipse satisfying the given conditions.

1. Foci at $(0, \pm 3)$, directrices $y = \pm 25/3$ 2. Eccentricity $\frac{2}{3}$, directrices $x = \pm 6$
3. Foci at $(\pm 4, 0)$, directrices $x = \pm 16$ 4. Foci at $(0, \pm 4)$, directrices $y = \pm 16$
5. Vertices at $(0, \pm 6)$, directrices $y = \pm 8$ 6. Eccentricity $\frac{2}{3}$, directrices $y = \pm 6$

7. Find the locus of all points whose distances from (4, 0) are always $\frac{1}{2}$ of their distances from the line $x = 16$.

8. Find the locus of all points whose distances from (0, 3) are always $\frac{3}{4}$ of their distances from the line $y = 6$.

9. Find the locus of all points whose distances from (4, 0) are always $\frac{2}{5}$ of their distances from the line $x = 6$. Locate the major and minor axes. (Center not at (0, 0).)

10. Find the locus of all points whose distances from (0, 5) are always $\frac{2}{3}$ of their distances from the line $y = 10$. Locate the major and minor axes. (Center not at (0, 0).)

11. Find the equation of the line tangent to the ellipse

$$\frac{x^2}{10} + \frac{y^2}{6} = 1$$

at the point $P(-2, 6/\sqrt{10})$.

12. Find the equation of the line tangent to the ellipse $2x^2 + 8y^2 = 1$ at the point $P(\frac{1}{2}, \frac{1}{4})$.

13. Find the equation of the line normal to the ellipse $9x^2 + 4y^2 = 36$ at the point $P(-\frac{2}{3}\sqrt{5}, 2)$.

14. Find the equation of the line normal to the ellipse $2x^2 + 5y^2 = 13$ at the point $P(2, 1)$.

15. Using the formula (4) for the slope, derive Eq. (5) for the tangent.

16. Prove the reflected-wave property of an ellipse, as described in this section.

17. Prove that the segment of a tangent to an ellipse between the point of contact and a directrix subtends a right angle at the corresponding focus.

18. A line through a focus parallel to a directrix is called a **latus rectum**. The length of the segment of this line cut off by the ellipse is called the *length of the latus rectum*. Show that the length of the latus rectum of

$$\frac{x^2}{a^2} + \frac{y^2}{b^2} = 1$$

is $2b^2/a$.

19. Prove that the line joining a point P_0 of an ellipse to the center and the line through a focus and perpendicular to the tangent at P_0 intersect on the corresponding directrix.

20. The tangent to an ellipse at a point P_0 meets the tangent at a vertex in a point Q. Prove that the line joining the other vertex to P_0 is parallel to the line joining the center to Q.

21. Given the ellipse $(x^2/4) + (y^2/2) = 1$, find the equations of the tangent lines passing through $P(1, 4)$.

22. Given the ellipse $x^2 + (y^2/9) = 1$, find the equations of the tangent lines passing through $P(2, 3)$.

23. With a fixed, the equation $(x^2/a^2) + (y^2/b^2) = 1$ represents a family of ellipses as b varies. Describe the way the location of each focus changes when $b \to 0$; also when $b \to \infty$.

24. The vertices of a triangle are at $P(-2, 0)$ and $Q(3, 1)$. If the perimeter of the triangle is a fixed number p, show that the third vertex must lie on an ellipse. Find its equation.

25. A highway overpass is in the shape of half an ellipse with the base 200 feet wide. The highway is 120 feet wide and the minimum clearance is 40 ft. Find the height ofthe overpass at its center.

▶**5.   THE HYPERBOLA**

A study of the hyperbola will complete our discussion of second degree curves in the plane.

DEFINITIONS. A **hyperbola** is the locus of all points the difference of whose distances from two fixed points is a positive constant. The two fixed points are called the **foci**.

A typical situation is shown in Fig. 6–23, where the two fixed points (foci) are labeled F_1 and F_2. For a point P to be on the locus, the distance $|PF_1|$ minus the distance $|PF_2|$ must be equal to a positive constant. Also, a point, say Q, will be on the locus if the distance $|QF_2|$ minus $|QF_1|$ is this same constant. We remark that the definition of a hyperbola, like that of a parabola and an ellipse, is purely geometric and makes no mention of coordinate systems. It is true that hyperbolas are second degree curves.

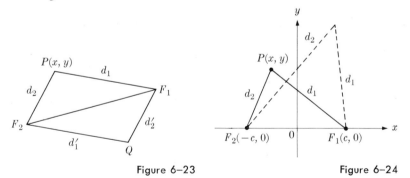

Figure 6–23   Figure 6–24

Suppose that the distance between the foci is $2c$ ($c > 0$). We place the foci at convenient points, F_1 at $(c, 0)$ and F_2 at $(-c, 0)$. Let $P(x, y)$ be a point on the hyperbola; then the conditions of the locus assert that $d_1 - d_2$ or $d_2 - d_1$ is always a positive constant (Fig. 6–24), which we label $2a$ ($a > 0$). We write the conditions of the locus as

$$d_1 - d_2 = 2a \qquad \text{or} \qquad d_2 - d_1 = 2a,$$

which may be combined as

$$d_1 - d_2 = \pm 2a.$$

Using the distance formula, we obtain

$$\sqrt{(x - c)^2 + y^2} - \sqrt{(x + c)^2 + y^2} = \pm 2a.$$

We transfer one radical to the right side and square both sides to get

$$(x - c)^2 + y^2 = (x + c)^2 + y^2 \pm 4a\sqrt{(x + c)^2 + y^2} + 4a^2.$$

Some terms may be canceled to yield

$$\pm 4a\sqrt{(x + c)^2 + y^2} = 4a^2 + 4cx.$$

We divide by $4a$ and once again square both sides, to find

$$x^2 + 2cx + c^2 + y^2 = a^2 + 2cx + \frac{c^2}{a^2} x^2$$

$$\Leftrightarrow \left(\frac{c^2}{a^2} - 1\right) x^2 - y^2 = c^2 - a^2.$$

We now divide through by $c^2 - a^2$, with the result that we have

$$\frac{x^2}{a^2} - \frac{y^2}{c^2 - a^2} = 1.$$

We shall show that the quantity c must be larger than a. We recall that the sum of the lengths of any two sides of a triangle must be greater than the length of the third side. When we refer to Fig. 6–24, we observe that $2c + d_2 > d_1$ and $2c + d_1 > d_2$. Therefore $2c > d_1 - d_2$ and $2c > d_2 - d_1$; combined, these inequalities give $2c > |d_1 - d_2|$. But $|d_1 - d_2| = 2a$, and so $c > a$.

We define the positive number b by the relation

$$b = \sqrt{c^2 - a^2},$$

and we write *the equation of the hyperbola:*

$$\frac{x^2}{a^2} - \frac{y^2}{b^2} = 1.$$

Conversely, by showing that all the steps are reversible, we could have demonstrated that every point which satisfies the above equation is on the locus. (We shall not enter into the discussion required for the proof of this fact.)

The equation of the hyperbola in the above form shows at once that it is symmetric with respect to both the x axis and the y axis.

DEFINITIONS. The line passing through the foci F_1 and F_2 is called the **transverse axis.** The perpendicular bisector of the segment F_1F_2 is called the **conjugate axis.** The intersection of these axes is called the **center.**

[See Fig. 6–25, where the foci are at $(c, 0)$ and $(-c, 0)$, the transverse axis is the x axis, the conjugate axis is the y axis, and the center is at the origin.]

The points of intersection of a hyperbola with the transverse axis are called its **vertices.** In Fig. 6–25, corresponding to the equation

$$\frac{x^2}{a^2} - \frac{y^2}{b^2} = 1,$$

these vertices occur at $(a, 0)$ and $(-a, 0)$. The length $2a$ is called the *length of the transverse axis.* Even though the points $(0, b)$ and $(0, -b)$ are not on the locus, the length $2b$ is a useful quantity. It is called the *length of the conjugate axis.*

The **eccentricity** of a hyperbola is defined as

$$e = \frac{c}{a}.$$

Note that since $c > a$, *the eccentricity of a hyperbola is always larger than* 1.

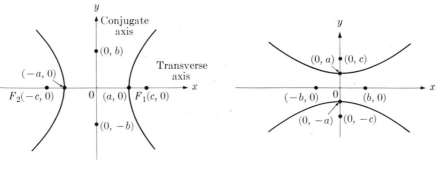

Figure 6–25   Figure 6–26

If the foci of the hyperbola are placed along the y axis at the points $(0, c)$ and $(0, -c)$, the equation takes the form

$$\frac{y^2}{a^2} - \frac{x^2}{b^2} = 1,$$

where, as before,

$$b = \sqrt{c^2 - a^2}.$$

The curve has the general appearance of the curve shown in Fig. 6–26.

In contrast with the equation of an ellipse, the equation for a hyperbola indicates that the relative sizes of a and b play no role in determining where the foci and axes are. An equation of the form

$$\frac{x^2}{(\ )^2} - \frac{y^2}{(\ )^2} = 1$$

always has its transverse axis (and foci) on the x axis, while an equation of the form

$$\frac{y^2}{(\ )^2} - \frac{x^2}{(\ )^2} = 1$$

always has its transverse axis (and foci) on the y axis. In the first case, the quantity under the x^2 term is a^2 and that under the y^2 term is b^2. In the second case the quantity under the y^2 term is a^2 and that under the x^2 term is b^2. In *both cases* we have

$$c^2 = a^2 + b^2.$$

EXAMPLE 1. Given the hyperbola with equation $9x^2 - 16y^2 = 144$. Find the axes, the coordinates of the vertices and the foci, and the eccentricity. Sketch the curve.

Solution. Dividing by 144, we find that

$$\frac{x^2}{16} - \frac{y^2}{9} = 1,$$

and, therefore,

$$a = 4, b = 3, c^2 = a^2 + b^2 = 25, c = 5.$$

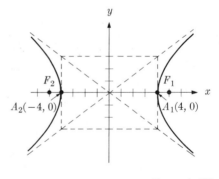

The transverse axis is along the x axis; the conjugate axis is along the y axis. The vertices are at $(\pm4, 0)$. The foci are at $(\pm5, 0)$. The eccentricity $e = c/a = 5/4$. The curve is sketched in Fig. 6–27, where we have constructed a rectangle by drawing parallels to the axes through

Figure 6–27

the points $(a, 0)$, $(-a, 0)$, $(0, b)$, and $(0, -b)$. This is called the **central rectangle**. Note that the diagonal from the origin to one corner of this rectangle has length c.

EXAMPLE 2. Find the equation of the hyperbola with vertices at $(0, \pm6)$ and eccentricity $\frac{5}{3}$. Locate the foci.

Solution. Since the vertices are along the y axis, the equation is of the form

$$\frac{y^2}{a^2} - \frac{x^2}{b^2} = 1,$$

with $a = 6$, $e = c/a = \frac{5}{3}$, and $c = (5 \cdot 6)/3 = 10$. Therefore $b = \sqrt{100 - 36} = 8$, and the equation is

$$\frac{y^2}{36} - \frac{x^2}{64} = 1.$$

The foci are at the points $(0, \pm 10)$.

PROBLEMS

In problems 1 through 10, find the lengths of the transverse and conjugate axes, the coordinates of the foci and the vertices, and the eccentricity. Sketch the curve.

1. $9x^2 - 4y^2 = 36$

2. $9x^2 - 4y^2 + 36 = 0$

3. $9x^2 - 16y^2 = 144$

4. $9x^2 - 16y^2 + 576 = 0$

5. $3x^2 - 4y^2 - 12 = 0$

6. $3x^2 - 4y^2 + 12 = 0$

7. $5x^2 - 3y^2 = 15$

8. $5x^2 - 6y^2 + 30 = 0$

9. $2x^2 - y^2 = 4$

10. $4x^2 - 3y^2 - 8 = 0$

In problems 11 through 20, find for each case the equation of the hyperbola (or hyperbolas) satisfying the given conditions.

11. Vertices at $(\pm 3, 0)$, foci at $(\pm 5, 0)$

12. Vertices at $(\pm 6, 0)$, foci at $(\pm 10, 0)$

13. Vertices at $(0, \pm 6)$, eccentricity $\frac{3}{2}$

14. Foci at $(\pm 3, 0)$, eccentricity $\frac{4}{3}$

15. Eccentricity $\sqrt{3}$, foci on the x axis, center at origin, passing through the point $(2, 1)$

16. Vertices at $(0, \pm 2)$, passing through $(-2, 3)$

17. Foci at $(0, \pm \sqrt{5})$, passing through $(\sqrt{2}, 2)$

18. Axes along the coordinate axes, passing through $(-2, 4)$ and $(-7, -6)$

19. Axes along the coordinate axes, passing through $(-4, -3)$, and $(-6, 5)$

20. Foci at $(\pm 4, 0)$, length of conjugate axis 6

21. Find the locus of a point such that the difference of its distances from $(3, 0)$ and $(-3, 0)$ is always equal to 2.

22. Find the locus of a point such that the difference of its distances from $(0, 5)$ and $(0, -5)$ is always equal to 4.

23. Find the locus of a point such that the difference of its distances from $(7, 0)$ and $(1, 0)$ is always 2.

24. Find the locus of a point such that the difference of its distances from $(2, 0)$ and $(2, 12)$ is always equal to 3.

25. A focal radius is the line segment from a point on a hyperbola to one of the foci. The foci of a hyperbola are at $(4, 0)$ and $(-4, 0)$. The difference in the lengths of the focal radii from any point is always ± 6. Find the equation of the hyperbola.

26. A hyperbola has its foci at $(c, 0)$ and $(-c, 0)$. The point $P(2, 4)$ is on the hyperbola and the focal radii (see Problem 25) from P are perpendicular. Find the equation of the hyperbola.

▶ 6. THE HYPERBOLA: DIRECTRICES, TANGENTS, ASYMPTOTES

The hyperbola, like the ellipse, has two directrices. For the hyperbola

$$\frac{x^2}{a^2} - \frac{y^2}{b^2} = 1,$$

the directrices are the lines

$$x = \pm \frac{a}{e},$$

and for the hyperbola

$$\frac{y^2}{a^2} - \frac{x^2}{b^2} = 1,$$

the directrices are the lines

$$y = \pm \frac{a}{e}.$$

A hyperbola has the property that it is the locus of all points $P(x, y)$ whose distances from a fixed point (F_1) are equal to e ($e > 1$) times their distances from a fixed line (say $x = a/e$). As Fig. 6–28 shows, the conditions of the locus are

$$d_1 = ed_2 \iff \sqrt{(x - c)^2 + y^2} = e \left| x - \frac{a}{e} \right|.$$

If we square both sides and simplify, we obtain $(x^2/a^2) - (y^2/b^2) = 1$. The same is true if we start with $F_2(-c, 0)$ and the line $x = -a/e$.

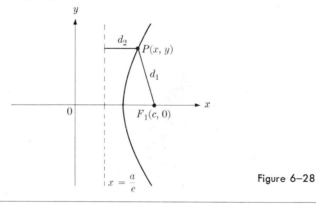

Figure 6–28

EXAMPLE 1. Find the equation of the hyperbola if the foci are at the points $(\pm 6, 0)$ and one directrix is the line $x = 3$.

Solution. We have $c = 6$ and $a/e = 3$. Since $e = c/a$, we may also write $a^2/c = 3$ and $a^2 = 18$, $a = \sqrt{18}$. In addition, $c^2 = a^2 + b^2$, from which we obtain $b^2 = 18$. The equation is

$$\frac{x^2}{18} - \frac{y^2}{18} = 1.$$

In studying the hyperbola

$$\frac{x^2}{a^2} - \frac{y^2}{b^2} = 1 \qquad (1)$$

we shall show that the lines

$$y = \frac{b}{a}x \qquad \text{and} \qquad y = -\frac{b}{a}x$$

are of special interest. For this purpose we consider a point $\bar{P}(\bar{x}, \bar{y})$ on the hyperbola (1) and compute the distance from $\bar{P}$ to the line $y = bx/a$. According to the formula for the distance from a point to a line (see Section 7 of Chapter 3), we have for the distance d:

$$d = \frac{|b\bar{x} - a\bar{y}|}{\sqrt{a^2 + b^2}} .$$

Suppose we select a point $\bar{P}$ which has both of its coordinates positive as shown in Fig. 6–29. Since $\bar{P}$ is a point on the hyperbola, its coordinates satisfy the equation $b^2\bar{x}^2 - a^2\bar{y}^2 = a^2b^2$, which we may write

$$(b\bar{x} - a\bar{y})(b\bar{x} + a\bar{y}) = a^2b^2 \iff b\bar{x} - a\bar{y} = a^2b^2/(b\bar{x} + a\bar{y}).$$

Substituting this last relation in the above expression for d, we find

$$d = \frac{a^2b^2}{\sqrt{a^2 + b^2}\,|b\bar{x} + a\bar{y}|} .$$

Now a and b are fixed positive numbers and we selected $\bar{x}, \bar{y}$ positive. As $\bar{P}$ moves out along the hyperbola farther and farther away from the vertex, $\bar{x}$ and $\bar{y}$ increase without bound. Therefore, the number d tends to zero as $\bar{x}$ and $\bar{y}$ get larger and larger. That is, the distance between the hyperbola (1) and the line $y = bx/a$ shrinks to zero as $\bar{P}$ tends to infinity. We define the line $y = bx/a$ as an **asymptote of the hyperbola** $(x^2/a^2) - (y^2/b^2) = 1$. Analogously, the line $y = -bx/a$ (see Fig. 6–29) is called an asymptote of the same hyperbola. We have only to recall the symmetry of the hyperbola to verify this statement.

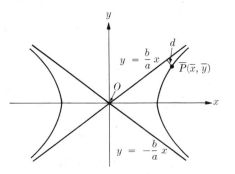

Figure 6–29

Now we can see one of the uses of the central rectangle. *The asymptotes are the lines which contain the diagonals of the central rectangle.* In an analogous way the hyperbola $(y^2/a^2) - (x^2/b^2) = 1$ has the asymptotes

$$y = (a/b)x \quad \text{and} \quad y = -(a/b)x.$$

These lines contain the diagonals of the appropriately constructed central rectangle.

EXAMPLE 2. Given the hyperbola $25x^2 - 9y^2 = 225$, find the foci, vertices, eccentricity, directrices, and asymptotes. Sketch the curve.

Solution. When we divide by 225, we obtain $(x^2/9) - (y^2/25) = 1$, and therefore $a = 3$, $b = 5$, $c^2 = 9 + 25$, and $c = \sqrt{34}$. The foci are at $(\pm\sqrt{34}, 0)$, vertices at $(\pm 3, 0)$; the eccentricity is $e = c/a = \sqrt{34}/3$, and the directrices are the lines $x = \pm a/e = \pm 9/\sqrt{34}$. To find the asymptotes we merely write $y = \pm(b/a)x = \pm\frac{5}{3}x$. Since the asymptotes are a great help in sketching the curve, we draw them first (Fig. 6–30); knowing the vertices and one or two additional points helps us obtain a fairly accurate graph.

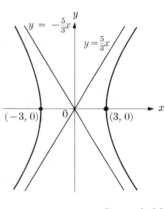

Figure 6–30

An easy way to remember the equations of the asymptotes is to recognize that if the hyperbola is

$$\frac{x^2}{a^2} - \frac{y^2}{b^2} = 1, \quad \text{the asymptotes are} \quad \frac{x^2}{a^2} - \frac{y^2}{b^2} = 0,$$

and if the hyperbola is

$$\frac{y^2}{a^2} - \frac{x^2}{b^2} = 1, \quad \text{the asymptotes are} \quad \frac{y^2}{a^2} - \frac{x^2}{b^2} = 0.$$

A line parallel to an asymptote is easily seen to intersect a hyperbola in exactly one point. Any *other* line which intersects a hyperbola in one point is called a **tangent line to the hyperbola.** To determine these tangent lines, we use the same method we employed for getting tangent lines to parabolas and ellipses. Suppose $P_1(x_1, y_1)$ is a given point and we wish to obtain the equations of all the tangent lines to a given hyperbola through this point. First we write the equation of the line $l(m)$ through P_1 with slope m:

$$l(m) = \{(x, y): y - y_1 = m(x - x_1)\}.$$

Then we choose the value or values of $m(\neq \pm b/a)$ for which the line $l(m)$ intersects the hyperbola in exactly one point. Any such line $l(m)$ is a tangent line. An example illustrates the method.

EXAMPLE 3. Given the hyperbola $(x^2/4) - (y^2/3) = 1$ and the point $P_1(1, 3)$, find the lines (if any) tangent to the hyperbola which pass through P_1.

Solution. We form the family of lines through P_1:

$$l(m) = \{(x, y): y - 3 = m(x - 1)\}.$$

Solving the equation of this line simultaneously with that of the hyperbola, we get

$$3x^2 - 4(mx - m + 3)^2 - 12 = 0.$$

Squaring out the term in parentheses and collecting terms, we find

$$(3 - 4m^2)x^2 - 8m(3 - m)x - 4(m^2 - 6m + 12) = 0.$$

In order that the above quadratic equation in x have one solution, its discriminant must vanish. Therefore we have

$$4m^2(3 - m)^2 + (3 - 4m^2)(m^2 - 6m + 12) = 0$$
$$\Leftrightarrow m^2 + 2m - 4 = 0$$
$$\Leftrightarrow m = -1 \pm \sqrt{5}.$$

The equations of the tangent lines are

$$y - 3 = (-1 + \sqrt{5})(x - 1) \quad \text{and} \quad y - 3 = (-1 - \sqrt{5})(x - 1).$$

In case $P_1(x_1, y_1)$ is *on* the hyperbola

$$\frac{x^2}{a^2} - \frac{y^2}{b^2} = 1,$$

the method just illustrated can be used to find the slope of the tangent to the hyperbola at that point as was done in Section 4. The result is

$$m = \frac{b^2 x_1}{a^2 y_1}.$$

Using this result, the equation of the tangent can be found to be

$$\frac{x_1 x}{a^2} - \frac{y_1 y}{b^2} = 1.$$

The analysis leading to these results is omitted.

PROBLEMS

In problems 1 through 8, find the vertices, foci, eccentricity, equations of the asymptotes, and directrices. Construct the central rectangle and sketch the hyperbola.

1. $4x^2 - y^2 = 16$ 2. $4x^2 - y^2 = 4$ 3. $9x^2 - y^2 = 9$
4. $9x^2 - 16y^2 = 144$ 5. $16x^2 - 9y^2 = 576$ 6. $x^2 - y^2 = 9$
7. $x^2 - y^2 + 9 = 0$ 8. $x^2 - 9y^2 = 36$

In problems 9 through 17, find in each case the equation of the hyperbola (or hyperbolas) satisfying the given conditions.

9. Asymptotes $y = \pm\frac{3}{2}x$, vertices $(\pm4, 0)$

10. Asymptotes $y = \pm\frac{3}{2}x$, vertices $(0, \pm6)$

11. Eccentricity $\frac{3}{2}$, directrices $x = \pm6$

12. Foci at $(\pm5, 0)$, directrices $x = \pm3$

13. Asymptotes $y = \pm2x$, foci at $(\pm\sqrt{5}, 0)$

14. Directrices $y = \pm6$, asymptotes $y = \pm2x$

15. Asymptotes $y = \pm\frac{4}{3}x$, passing through $(-5, 4)$

16. Directrices $x = \pm2$, passing through $(-3, 1)$, center at origin

17. Directrices $y = \pm\frac{3}{2}$, passing through $(3, 2\sqrt{3})$, center at origin

18. Find the equations of the tangent and normal lines to the hyperbola

$$\frac{x^2}{9} - \frac{y^2}{4} = 1$$

at the point $(-3\sqrt{2}, 2)$.

19. Find the point of intersection of the lines tangent to the hyperbola

$$\frac{x^2}{36} - \frac{y^2}{9} = 1$$

at the points $(2\sqrt{10}, 1)$, $(-10, -4)$.

20. Prove that the lines $y = \pm(a/b)x$ are asymptotes of the hyperbola

$$(y^2/a^2) - (x^2/b^2) = 1.$$

21. A line through a focus of a hyperbola and parallel to the directrix is called a **latus rectum.** The length of the segment cut off by the hyperbola is called the *length of the latus rectum.* Show that for the hyperbola $(x^2/a^2) - (y^2/b^2) = 1$ the length of the latus rectum is $2b^2/a$.

22. Find the locus of a point such that its distance from $(5, 0)$ is always $\frac{5}{3}$ times its distance from the line $x = \frac{9}{5}$.

23. Find the locus of a point such that its distance from the point $(0, \sqrt{41})$ is always $\sqrt{41}/5$ times its distance from the line $y = 25/\sqrt{41}$.

24. Find the locus of a point such that its distance from the point $(8, 0)$ is always 3 times its distance from the line $x = 1$.

25. Show that an asymptote, a directrix, and the line through the corresponding focus perpendicular to the asymptote pass through a point.

26. Given the hyperbola $(x^2/4) - y^2 = 1$. Find the equations of the lines tangent to this hyperbola passing through the point $(1, 2)$.

27. Given the hyperbola $(y^2/4) - (x^2/6) = 1$. Decide whether or not tangents to this hyperbola can be drawn which pass through any of the points $P(0, 8)$, $Q(1, 5)$, $R(4, 0)$, and $S(-8, -9)$. Find the equations of the lines, wherever possible.

28. An ellipse and a hyperbola have their foci at $(3, 0)$ and $(-3, 0)$. If their eccentricities are $\frac{2}{3}$ and $\frac{3}{2}$, find their points of intersection.

29. The hyperbolas

$$\frac{x^2}{a^2} - \frac{y^2}{b^2} = 1 \quad \text{and} \quad \frac{x^2}{a^2} - \frac{y^2}{b^2} = -1$$

are called **conjugate**. Show that the lines $y = \pm bx/a$ are the only ones which do not intersect either hyperbola.

*30. Two hyperbolas are called **confocal** if they have the same foci. Show that confocal hyperbolas cannot intersect each other.

31. A hyperbola which has the lines $y = \pm x$ as asymptotes is called **equilateral**. Given the curve $xy = k$, where k is any constant ($\neq 0$). Show that the curve is an equilateral hyperbola. [*Hint*: Let $x = x' - y'$ and $y = x' + y'$ and plot the graph in the x', y' coordinate system.]

▶7. TRANSLATION OF AXES

Coordinate systems were not employed in our definitions and descriptions of the principal properties of second degree curves. However, when we want to find equations for such curves and when we wish to use the methods of analytic geometry, coordinate systems become essential.

When we use the method of coordinate geometry we place the axes at a position "convenient" with respect to the curve under consideration. In the examples of ellipses and hyperbolas which we studied, the foci were located on one of the axes and were situated symmetrically with respect to the origin. But now suppose that we have a problem in which the curve (hyperbola, parabola, ellipse, etc.) is *not* situated so conveniently with respect to the axes. We would then like to change the coordinate system in order to have the curve at a convenient and familiar location. The process of making this change is called a **transformation of coordinates.**

The first type of transformation we shall consider is one of the simplest, and is called the **translation of axes.** Figure 6–31 shows the usual Cartesian xy coordinate system. We now introduce an additional coordinate system, with axis x' parallel to x and k units away, and axis y' parallel to y and h units away. This means that the origin O' of the new coordinate system has coordinates (h, k) in the original system. The positive x' and y' directions are taken to be the same as the positive x and y directions. When P is any point in the plane, what are its coordinates in each of the systems? Suppose P has coordinates (x, y) in the original system and (x', y') in the new system and *suppose the new origin has coordinates (h, k) in the original system.* An inspection of Fig. 6–32 suggests that if $x > h > 0$

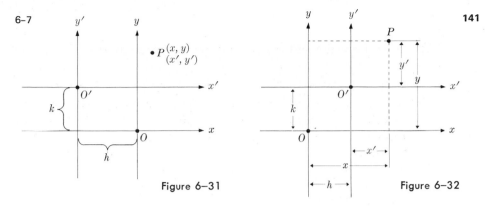

Figure 6–31

Figure 6–32

and $y > k > 0$, then

$$x = x' + h \quad \text{and} \quad y = y' + k \tag{1}$$

or equivalently

$$x' = x - h \quad \text{and} \quad y' = y - k.$$

To prove that these formulas hold for any positions of O, O', and P, we proceed as follows: Let us use the notation $\mathbf{v}[\overrightarrow{AB}]$ to denote that vector having $\overrightarrow{AB}$ as a representative. Let I and J have the coordinates $(1, 0)$ and $(0, 1)$, respectively, in the original system and let I' and J' have these respective coordinates in the new system (see Fig. 6–33). Let

$$\mathbf{i} = \mathbf{v}[\overrightarrow{OI}], \qquad \mathbf{j} = \mathbf{v}[\overrightarrow{OJ}],$$
$$\mathbf{i}' = \mathbf{v}[\overrightarrow{O'I'}], \qquad \mathbf{j}' = \mathbf{v}[\overrightarrow{O'J'}].$$

Then since $\overrightarrow{OI} \approx \overrightarrow{O'I'}$ and $\overrightarrow{OJ} \approx \overrightarrow{O'J'}$, we conclude that

$$\mathbf{i}' = \mathbf{i} \quad \text{and} \quad \mathbf{j}' = \mathbf{j}.$$

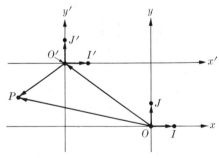

Figure 6–33

From the representation theorem for vectors (Theorem 2, Chapter 4), we obtain

$$\mathbf{v}[\overrightarrow{OP}] = x\mathbf{i} + y\mathbf{j}, \qquad \mathbf{v}[\overrightarrow{OO'}] = h\mathbf{i} + k\mathbf{j}, \qquad \mathbf{v}[\overrightarrow{O'P}] = x'\mathbf{i}' + y'\mathbf{j}' = x'\mathbf{i} + y'\mathbf{j}.$$

Since $\mathbf{v}[\overrightarrow{OP}] = \mathbf{v}[\overrightarrow{OO'}] + \mathbf{v}[\overrightarrow{O'P}]$, we conclude that

$$x\mathbf{i} + y\mathbf{j} = (h + x')\mathbf{i} + (k + y')\mathbf{j}.$$

The formulas (1) follow by equating coefficients of $\mathbf{i}$ and $\mathbf{j}$.

The most general equation of the second degree has the form

$$Ax^2 + Bxy + Cy^2 + Dx + Ey + F = 0,$$

where A, B, C, D, E, and F represent numbers. We assume that A, B, and C are not all zero, since then the equation would be in the first degree. We recognize that the equations we have been discussing are all special cases of the above equation. For example, the circle $x^2 + y^2 - 16 = 0$ has $A = 1$, $B = 0$, $C = 1$, $D = E = 0$, $F = -16$. The ellipse $(x^2/9) + (y^2/4) = 1$ has $A = \frac{1}{9}$, $B = 0$, $C = \frac{1}{4}$, $D = E = 0$, $F = -1$. Similarly, we see that parabolas and hyperbolas are of the above form.

An interesting case is $A = 1$, $B = 0$, $C = -1$, $D = E = F = 0$. We get $x^2 - y^2 = 0$ or $y = \pm x$, and the locus is two intersecting lines. Second degree curves whose loci reduce to lines or points are frequently called **degenerate**.

We now illustrate how to use translation of axes to reduce an equation of the form

$$Ax^2 + Cy^2 + Dx + Ey + F = 0$$

to an equation of the same form but with new letters (x', y'), and with D and E both equal to zero (with certain exceptions). Although presented in a new way, the equation will be easily recognized as one of the second degree curves we have studied. *The principal tool in this process is "completing the square."*

EXAMPLE 1. Given the equation

$$9x^2 + 25y^2 + 18x - 100y - 116 = 0,$$

by using a translation of axes determine whether the locus of the equation is a parabola, ellipse, or hyperbola. Determine foci (or focus), vertices (or vertex), directrices (or directrix), eccentricity, and asymptotes, if any. Sketch the curve.

Solution. To complete the square in x and y, we write the equation in the form

$$9(x^2 + 2x \quad\quad) + 25(y^2 - 4y \quad\quad) = 116.$$

We add 1 in the parentheses for x, which means adding 9 to the left side, and we add 4 in the parentheses for y, which means adding 100 to the left side. We obtain

$$9(x^2 + 2x + 1) + 25(y^2 - 4y + 4) = 116 + 9 + 100$$
$$\Leftrightarrow 9(x + 1)^2 + 25(y - 2)^2 = 225.$$

Now we have the clue for translation of axes. We define

$$x' = x + 1 \quad\quad \text{and} \quad\quad y' = y - 2.$$

That is, the translation is made with $h = -1$, $k = 2$. The equation becomes

$$9x'^2 + 25y'^2 = 225.$$

Dividing by 225, we find

$$\frac{x'^2}{25} + \frac{y'^2}{9} = 1,$$

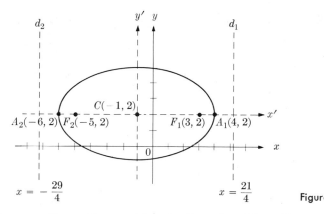

$$x = -\frac{29}{4} \qquad\qquad x = \frac{21}{4} \qquad\qquad \text{Figure 6-34}$$

which we recognize as an ellipse with $a = 5, b = 3, c^2 = a^2 - b^2 = 16, c = 4, e = \frac{4}{5}$. In the $x'y'$ system, we have: center $(0, 0)$; vertices $(\pm5, 0)$; foci $(\pm4, 0)$; directrices, $x' = \pm\frac{25}{4}$.

In the xy system, we use the relations $x = x' - 1, y = y' + 2$ to obtain: center $(-1, 2)$; vertices $(4, 2), (-6, 2)$; foci $(3, 2), (-5, 2)$; directrices, $x = \frac{21}{4}, x = -\frac{29}{4}$. The curve is sketched in Fig. 6–34.

EXAMPLE 2. Discuss the properties of the locus of the equation

$$x^2 + 4x + 4y - 4 = 0.$$

Solution. Here we have $A = 1, C = 0, D = 4, E = 4, F = -4$. There is only one second degree term, and we complete the square to obtain

$$(x^2 + 4x + 4) = -4y + 4 + 4$$
$$\Leftrightarrow (x + 2)^2 = -4(y - 2).$$

We read off the appropriate translation of axes. It is

$$x' = x + 2, \qquad y' = y - 2,$$

and we have

$$x'^2 = -4y',$$

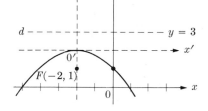

Figure 6–35

which we recognize as a parabola. In the $x'y'$ system, the vertex is at $(0, 0)$, focus at $(0, -1)$ (since $p = 2$); directrix, $y' = 1$. In the xy system (since $h = -2, k = 2$), we have the vertex at $(-2, 2)$, focus at $(-2, 1)$, and directrix $y = 3$. The curve is sketched in Fig. 6–35.

EXAMPLE 3. Find the equation of the ellipse with eccentricity $\frac{1}{2}$ and foci at $(4, 2)$ and $(2, 2)$.

Solution. The center of the ellipse is halfway between the foci and therefore is at the point $(3, 2)$. The major axis of the ellipse is along the line $y = 2$. To move the origin to the center of the ellipse by a translation of coordinates, we let

$$x' = x - 3, \qquad y' = y - 2.$$

In this system the foci are at $(\pm 1, 0)$; $e = c/a = \frac{1}{2}$ and, since $c = 1$, we have $a = 2$. Also, $c^2 = a^2 - b^2$ gives us $b^2 = 3$. In the $x'y'$ system, the equation is

$$\frac{x'^2}{4} + \frac{y'^2}{3} = 1.$$

In the xy system, the equation is

$$\frac{(x - 3)^2}{4} + \frac{(y - 2)^2}{3} = 1,$$

or

$$3(x - 3)^2 + 4(y - 2)^2 = 12$$
$$\Leftrightarrow 3x^2 + 4y^2 - 18x - 16y + 31 = 0.$$

We are now in a position to discuss all the curves that an equation

$$Ax^2 + Cy^2 + Dx + Ey + F = 0$$

can possibly represent, presented here in the form of a theorem:

Theorem 1. (a) *If A and C are both positive or both negative, then the locus is an ellipse, a circle (if $A = C$), a point, or nothing.* (b) *If A and C are of opposite signs, the locus is a hyperbola or two intersecting lines.* (c) *If either A or C is zero, the locus is a parabola, two parallel lines, one line, or nothing.*

We shall not prove this theorem, but some of its possible uses will be illustrated by examples.

The equation $3x^2 + y^2 + 5 = 0$ has no locus, since the sum of positive quantities can never add up to zero. This exhibits the last case under (a) of Theorem 1. Under (c), the equation $x^2 - 2x = 0$ ($A = 1$, $C = 0$, $D = -2$, $E = 0$, $F = 0$) is an example of two parallel lines, while $x^2 - 2x + 1 = 0$ ($A = 1$, $C = 0$, $D = -2$, $E = 0$, $F = 1$) is an example of one line.

A proof of this theorem is within the scope of any curious student.

PROBLEMS

1. A Cartesian system of coordinates is translated to a new $x'y'$ system whose origin is at $(-4, -3)$. The points P, Q, R, and S have coordinates $(-3, 4)$, $(-7, -2)$, $(-\frac{3}{2}, -6)$, and $(2, -5)$, respectively, in the original system. Find the coordinates of these points in the new system.

2. A trasnlation of coordinates moves the origin to the point of intersection of the lines $-3x + 2y = 0$ and $-4x + y - 1 = 0$. Find the translation of coordinates and the equations of these lines in the new system.

3. A translation of coordinates moves the origin to the point of intersection of the lines $7x + 3y + 1 = 0$ and $5x + 2y - 6 = 0$. Find the equations of these lines in the new coordinate system.

In problems 4 through 17, translate the coordinates so as to eliminate the first degree terms (or one first degree term in the case of parabolas), describe the principal properties (as in Example 1) and sketch the curves.

4. $x^2 + 6x + 8y + 1 = 0$

5. $16x^2 + 25y^2 + 64x + 50y - 311 = 0$

6. $16x^2 + 9y^2 + 32x - 36y - 92 = 0$

7. $4x^2 - 9y^2 - 16x + 18y - 29 = 0$

8. $y^2 - 6x - 4y + 16 = 0$

9. $9x^2 + 4y^2 + 36x + 8y + 4 = 0$

10. $4x^2 + 9y^2 - 8x - 18y + 4 = 0$

11. $x^2 - 10x - 4y + 5 = 0$

12. $9x^2 - 4y^2 - 18x + 8y + 4 = 0$

13. $4x^2 + 3y^2 - 8x - 12y + 4 = 0$

14. $2x^2 - 3y^2 + 8x + 6y + 17 = 0$

15. $y^2 + 8x + 6y + 1 = 0$

16. $3x^2 + 2y^2 + 6x - 8y + 11 = 0$

17. $3x^2 - 4y^2 - 12x + 8y + 8 = 0$

In problems 18 through 23, find the equation of the locus indicated. Sketch the curve.

18. Parabola: vertex at $(-1, 2)$; directrix $x = -3$

19. Ellipse: vertices at $(3, 2)$ and $(-5, 2)$; foci at $(2, 2)$ and $(-4, 2)$

20. Ellipse: foci at $(-3, -2)$ and $(-3, 4)$; directrices $y = -5$ and $y = 7$

21. Hyperbola: vertices at $(-1, -3)$ and $(-1, 5)$; passing through $(-3, -5)$

22. Hyperbola: asymptotes $3x + 2y + 7 = 0$ and $-3x + 2y + 1 = 0$; passing through $(-5, 1)$

23. Parabola: axis $x = 1$; directrix $y = 2$; focus $(1, 5)$

In problems 24 through 28, find the equation of the given locus. Identify the curve.

24. The locus of all points whose distances from $(-3, 4)$ equal their distances from the line $y = 6$

25. The locus of all points whose distances from $(-1, 2)$ are equal to one-half their distances from the line $x = 2$

26. The locus of all points whose distances from $(-2, -1)$ are equal to twice their distances from the line $y = -4$

27. The locus of all points whose distances from $(3, 2)$ are equal to twice their distances from $(-3, -4)$

28. The locus of all points whose distances from $(-2, 1)$ are equal to their distances from the line $4x + 3y - 5 = 0$.

▶8. ROTATION OF AXES. THE GENERAL EQUATION OF THE SECOND DEGREE

Suppose we make a transformation of coordinates from an xy system to an $x'y'$ system in the following way. The origin is kept fixed and the x' and y' axes are obtained by rotating the x and y axes counterclockwise by an amount θ, as shown in Fig. 6–36. Every point P will have coordinates (x, y) with respect to the original system and coordinates (x', y') with respect to the new system. We now find the relationship between (x, y) and (x', y'). Let I and J have the respective coordinates $(1, 0)$ and $(0, 1)$ in the original system, let I' and J' have the same respective coordinates in the new system, and let $\mathbf{i} = \mathbf{v}[\overrightarrow{OI}], \mathbf{j} = \mathbf{v}[\overrightarrow{OJ}], \mathbf{i}' = \mathbf{v}[\overrightarrow{OI'}],$ $\mathbf{j}' = \mathbf{v}[\overrightarrow{OJ'}]$. Then I' and J' have the respective coordinates $(\cos \theta, \sin \theta)$ and

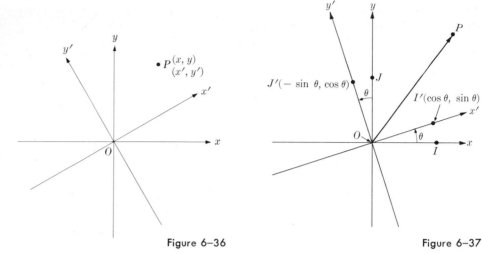

Figure 6–36   Figure 6–37

$(\cos [\theta + 90°], \sin [\theta + 90°]) = (-\sin \theta, \cos \theta)$ in the original system (see Fig. 6–37), so that

$$\mathbf{i}' = (\cos \theta)\mathbf{i} + (\sin \theta)\mathbf{j},$$
$$\mathbf{j}' = (-\sin \theta)\mathbf{i} + (\cos \theta)\mathbf{j}.$$

Then by the representation theorem for vectors (Theorem 2, Chapter 4)

$$\mathbf{v}[\overrightarrow{OP}] = x\mathbf{i} + y\mathbf{j}$$
$$= x'\mathbf{i}' + y'\mathbf{j}' = x'[(\cos \theta)\mathbf{i} + (\sin \theta)\mathbf{j}] + y'[(-\sin \theta)\mathbf{i} + (\cos \theta)\mathbf{j}]$$
$$= (x' \cos \theta - y' \sin \theta)\mathbf{i} + (x' \sin \theta + y' \cos \theta)\mathbf{j}.$$

Equating coefficients of $\mathbf{i}$ and $\mathbf{j}$, we get

$$x = x' \cos \theta - y' \sin \theta, \qquad y = x' \sin \theta + y' \cos \theta.$$

Solving for x' and y' in terms of x and y, we get

$$x' = x \cos \theta + y \sin \theta,$$
$$y' = -x \sin \theta + y \cos \theta.$$

EXAMPLE 1. A Cartesian coordinate system is rotated 60°. Find the coordinates of the point $P(3, -1)$ in the new system. What is the equation of the line $2x - 3y + 1 = 0$ in the rotated system?

Solution. We have $\sin 60° = \frac{1}{2}\sqrt{3}$, $\cos 60° = \frac{1}{2}$. The equations relating the xy system and the $x'y'$ system are

$$x' = \tfrac{1}{2}x + \tfrac{1}{2}\sqrt{3}\, y, \qquad y' = -\tfrac{1}{2}\sqrt{3}\, x + \tfrac{1}{2}y.$$

The coordinates of P are $x' = \frac{3}{2} - \frac{1}{2}\sqrt{3}$, $y' = -\frac{3}{2}\sqrt{3} - \frac{1}{2}$. The relationships giving the xy system in terms of the $x'y'$ system are

$$x = \tfrac{1}{2}x' - \tfrac{1}{2}\sqrt{3}\,y', \qquad y = \tfrac{1}{2}\sqrt{3}\,x' + \tfrac{1}{2}y';$$

and so the equation of the line in the new system is

$$2(\tfrac{1}{2}x' - \tfrac{1}{2}\sqrt{3}\,y') - 3(\tfrac{1}{2}\sqrt{3}\,x' + \tfrac{1}{2}y') + 1 = 0$$

$$\Leftrightarrow (1 - \tfrac{3}{2}\sqrt{3})x' - (\tfrac{3}{2} + \sqrt{3})y' + 1 = 0.$$

We recall that the most general equation of the second degree has the form

$$Ax^2 + Bxy + Cy^2 + Dx + Ey + F = 0 \quad (A, B, C \text{ not all zero}).$$

In Section 7 we showed (when $B = 0$) how the axes could be translated so that in the new system $D = E = 0$ (except in the case of parabolas, when we could make only one of them zero).

Now we shall show that *it is always possible to rotate the coordinates in such a way that in the new system there is no $x'y'$ term.* To do this we take the equations

$$x = x' \cos\theta - y' \sin\theta,$$
$$y = x' \sin\theta + y' \cos\theta,$$

and we substitute in the general equation of the second degree. Then we have

$$Ax^2 = A(x' \cos\theta - y' \sin\theta)^2,$$
$$Bxy = B(x' \cos\theta - y' \sin\theta)(x' \sin\theta + y' \cos\theta),$$
$$Cy^2 = C(x' \sin\theta + y' \cos\theta)^2,$$
$$Dx = D(x' \cos\theta - y' \sin\theta),$$
$$Ey = E(x' \sin\theta + y' \cos\theta),$$
$$F = F.$$

We add these equations and obtain (after multiplying out the right side)

$$Ax^2 + Bxy + Cy^2 + Dx + Ey + F$$
$$= A'x'^2 + B'x'y' + C'y'^2 + D'x' + E'y' + F' = 0,$$

where A', B', C', D', E', and F' are the abbreviations of

$$A' = A \cos^2\theta + B \sin\theta \cos\theta + C \sin^2\theta,$$
$$B' = 2(C - A) \sin\theta \cos\theta + B(\cos^2\theta - \sin^2\theta),$$
$$C' = A \sin^2\theta - B \sin\theta \cos\theta + C \cos^2\theta,$$
$$D' = D \cos\theta + E \sin\theta,$$
$$E' = -D \sin\theta + E \cos\theta,$$
$$F' = F.$$

Our purpose is to select θ so that the $x'y'$ term is missing, in order that B' will be zero. Let us set it equal to zero and see what happens:

$$2(C - A) \sin \theta \cos \theta + B(\cos^2 \theta - \sin^2 \theta) = 0.$$

We recall from trigonometry the double-angle formulas

$$\sin 2\theta = 2 \sin \theta \cos \theta, \qquad \cos 2\theta = \cos^2 \theta - \sin^2 \theta,$$

and we write

$$(C - A) \sin 2\theta + B \cos 2\theta = 0$$

$$\Leftrightarrow \cot 2\theta = \frac{A - C}{B}.$$

In other words, if we select θ so that $\cot 2\theta = (A - C)/B$ we will obtain $B' = 0$. There is always an angle 2θ between 0 and π which solves this equation.

The next example will illustrate the process.

EXAMPLE 2. Given the equation $8x^2 - 4xy + 5y^2 = 36$. Choose new axes by rotation so as to eliminate the $x'y'$ term; sketch the curve and locate the principal quantities.

Solution. We have $A = 8$, $B = -4$, $C = 5$, $D = E = 0$, $F = -36$. We select $\cot 2\theta = (A - C)/B = (8 - 5)/(-4) = -\frac{3}{4}$. This means that 2θ is in the second quadrant and $\cos 2\theta = -\frac{3}{5}$.

We remember from trigonometry the half-angle formulas

$$\sin \theta = \sqrt{(1 - \cos 2\theta)/2}, \qquad \cos \theta = \sqrt{(1 + \cos 2\theta)/2},$$

and we use these to get

$$\sin \theta = \sqrt{\tfrac{4}{5}}, \qquad \cos \theta = \sqrt{\tfrac{1}{5}},$$

which gives the rotation

$$x = \frac{1}{\sqrt{5}} x' - \frac{2}{\sqrt{5}} y' = \frac{x' - 2y'}{\sqrt{5}},$$

$$y = \frac{2}{\sqrt{5}} x' + \frac{1}{\sqrt{5}} y' = \frac{2x' + y'}{\sqrt{5}}.$$

Substituting in the given equation, we obtain

$$\tfrac{8}{5}(x' - 2y')^2 - \tfrac{4}{5}(x' - 2y')(2x' + y') + \tfrac{5}{5}(2x' + y')^2 = 36;$$

when we multiply out, we find that

$$4x'^2 + 9y'^2 = 36 \qquad \text{or} \qquad \frac{x'^2}{9} + \frac{y'^2}{4} = 1.$$

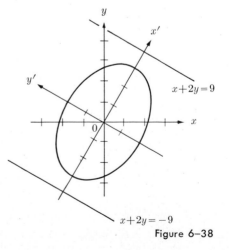

$x + 2y = 9$

$x + 2y = -9$

Figure 6–38

The locus is an ellipse, which is sketched in Fig. 6–38. The $x'y'$ coordinates of the vertices are at $(\pm 3, 0)$, and the foci are at $(\pm\sqrt{5}, 0)$. The eccentricity is $\sqrt{5}/3$; the directrices are the lines $x' = \pm 9/\sqrt{5}$. In the original system, the vertices are at $(3/\sqrt{5}, 6/\sqrt{5})$, $(-3/\sqrt{5}, -6/\sqrt{5})$. The foci are at $(1, 2)$, $(-1, -2)$. The directrices are the lines $x + 2y = \pm 9$.

In Example 2 the coefficients D and E are zero. When a problem arises with B, D, and E all different from zero, we may eliminate them by performing in succession a rotation (eliminating B) and a translation (eliminating the D and E terms). Here we have an example of the way processes develop in mathematics: a complicated structure is erected by combining simple building blocks. The technique of performing several transformations in succession occurs frequently in mathematical problems.

EXAMPLE 3. Given the equation

$$2x^2 - 4xy - y^2 + 20x - 2y + 17 = 0.$$

Reduce the equation to standard form by eliminating B, D, and E. Identify the curve and locate the vertices and foci.

Solution. We choose θ so that

$$\cot 2\theta = \frac{A - C}{B} = \frac{2 + 1}{-4} = -\frac{3}{4}.$$

This puts 2θ in the second quadrant and $\cos 2\theta = -\frac{3}{5}$ and, as in Example 2, $\cos \theta = \sqrt{\frac{1}{5}}$, $\sin \theta = \sqrt{\frac{4}{5}}$. The rotation is

$$x = \frac{1}{\sqrt{5}}(x' - 2y'), \qquad y = \frac{1}{\sqrt{5}}(2x' + y').$$

Substituting into the given equation, we obtain

$$\tfrac{2}{5}(x' - 2y')^2 - \tfrac{4}{5}(x' - 2y')(2x' + y') - \tfrac{1}{5}(2x' + y')^2$$
$$+ \frac{20}{\sqrt{5}}(x' - 2y') - \frac{2}{\sqrt{5}}(2x' + y') + 17 = 0.$$

Simplification yields

$$2x'^2 - 3y'^2 - \frac{16}{\sqrt{5}}x' + \frac{42}{\sqrt{5}}y' - 17 = 0,$$

which we recognize as a hyperbola. To eliminate the coefficients of x' and y', we proceed as in the preceding section and translate the coordinates. Completing the square is the standard method, and we write

$$2\left(x'^2 - \frac{8}{\sqrt{5}}x'\qquad\right) - 3\left(y'^2 - \frac{14}{\sqrt{5}}y'\qquad\right) = 17.$$

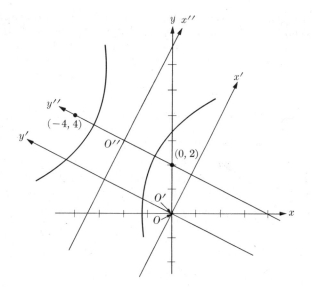

Figure 6–39

Then we have

$$2\left(x'^2 - \frac{8}{\sqrt{5}}x' + \frac{16}{5}\right) - 3\left(y'^2 - \frac{14}{\sqrt{5}}y' + \frac{49}{5}\right) = 17 + \frac{32}{5} - \frac{147}{5}$$

$$\Leftrightarrow 2\left(x' - \frac{4}{\sqrt{5}}\right)^2 - 3\left(y' - \frac{7}{\sqrt{5}}\right)^2 = -6.$$

The translation of axes is read off from this equation. It is

$$x'' = x' - \frac{4}{\sqrt{5}}, \qquad y'' = y' - \frac{7}{\sqrt{5}};$$

the equation becomes

$$\frac{y''^2}{2} - \frac{x''^2}{3} = 1.$$

Figure 6–39 shows the transformations. In the $x''y''$ system, the vertices are at $(0, \pm\sqrt{2})$ and the foci are at $(0, \pm\sqrt{5})$. This puts the vertices at $(4/\sqrt{5}, \pm\sqrt{2} + 7/\sqrt{5})$ and the foci at $(4/\sqrt{5}, \pm\sqrt{5} + 7/\sqrt{5})$ in the $x'y'$ system. Substituting in the equations

$$x = \frac{1}{\sqrt{5}}(x' - 2y') \qquad \text{and} \qquad y = \frac{1}{\sqrt{5}}(2x' + y'),$$

we get the vertices at

$$\left(-\frac{10 + 2\sqrt{10}}{5}, \frac{15 + \sqrt{10}}{5}\right), \qquad \left(\frac{-10 + 2\sqrt{10}}{5}, \frac{15 - \sqrt{10}}{5}\right),$$

and the foci at $(-4, 4)$, $(0, 2)$, in the xy system.

The next example illustrates a somewhat different situation.

EXAMPLE 4. Given the equation

$$4x^2 - 12xy + 9y^2 - 52x + 26y + 81 = 0,$$

reduce it to standard form by eliminating B, D, and E. Identify the curve and locate the principal quantities.

Solution. We first rotate and choose θ so that

$$\cot 2\theta = \frac{A - C}{B} = \frac{4 - 9}{-12} = \frac{5}{12}.$$

Then 2θ is in the first quadrant and $\cos 2\theta = \frac{5}{13}$. From the half-angle formulas (as in Example 2),

$$\cos \theta = \sqrt{\tfrac{9}{13}}, \qquad \sin \theta = \sqrt{\tfrac{4}{13}}.$$

The desired rotation of coordinates is

$$x = \frac{1}{\sqrt{13}}(3x' - 2y'), \qquad y = \frac{1}{\sqrt{13}}(2x' + 3y').$$

Substituting in the given equation, we obtain

$$\tfrac{4}{13}(9x'^2 - 12x'y' + 4y'^2) - \tfrac{12}{13}(6x'^2 + 5x'y' - 6y'^2)$$
$$+ \tfrac{9}{13}(4x'^2 + 12x'y' + 9y'^2) - 4\sqrt{13}(3x' - 2y')$$
$$+ 2\sqrt{13}(2x' + 3y') + 81 = 0,$$

and, after simplification,

$$13y'^2 - 8\sqrt{13}\, x' + 14\sqrt{13}\, y' + 81 = 0.$$

Note that in the process of eliminating the $x'y'$ term we also eliminated the x'^2 term. We now realize that the curve must be a parabola. To translate the coordinates properly, we complete the square in y. This yields

$$13(y'^2 + \tfrac{14}{13}\sqrt{13}\, y' \quad) = 8\sqrt{13}\, x' - 81$$
$$\Leftrightarrow 13\left(y' + \frac{7}{\sqrt{13}}\right)^2 = 8\sqrt{13}\left(x' - \frac{4}{\sqrt{13}}\right).$$

The translation of axes,

$$x'' = x' - \frac{4}{\sqrt{13}}, \qquad y'' = y' + \frac{7}{\sqrt{13}},$$

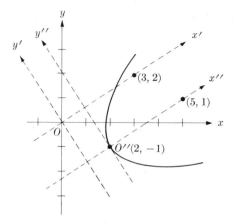

Figure 6-40

leads to the equation

$$y''^2 = \frac{8}{\sqrt{13}} x''.$$

In the $x''y''$ system, $p = 4/\sqrt{13}$, the vertex is at the origin, the focus is at $(2/\sqrt{13}, 0)$, the directrix is the line $x'' = -2/\sqrt{13}$, and the x'' axis is the axis of the parabola. In the $x'y'$ system, the focus is at $(6/\sqrt{13}, -7/\sqrt{13})$, and the directrix is the line $x' = 2/\sqrt{13}$. As for the original xy system, we find that the focus is at $(\frac{32}{13}, -\frac{9}{13})$. The directrix is the line $3x + 2y = 2$. The parabola and all sets of axes are sketched in Fig. 6-40.

In the rotation of the coordinates, the general equation of the second degree,

$$Ax^2 + Bxy + Cy^2 + Dx + Ey + F = 0,$$

goes into the equation

$$A'x^2 + B'xy + C'y^2 + D'x + E'y + F = 0,$$

with

$$A' = A \cos^2 \theta + B \sin \theta \cos \theta + C \sin^2 \theta,$$
$$B' = 2(C - A) \sin \theta \cos \theta + B(\cos^2 \theta - \sin^2 \theta),$$
$$C' = A \sin^2 \theta - B \sin \theta \cos \theta + C \cos^2 \theta.$$

The quantity $A' + C'$, when calculated, is

$$A' + C' = A(\cos^2 \theta + \sin^2 \theta) + C(\sin^2 \theta + \cos^2 \theta) = A + C.$$

In other words, even though A changes to A' and C changes to C' when a rotation through *any* angle is made, the quantity $A + C$ does not change at all. We say that $A + C$ is **invariant** under a rotation of coordinates.

If we compute the expression $B'^2 - 4A'C'$ (a tedious computation) and use some trigonometry, we find that

$$B'^2 - 4A'C' = B^2 - 4AC.$$

In other words, $B^2 - 4AC$ is also an *invariant* under rotation of axes.

It can readily be checked that $A + C$ and $B^2 - 4AC$ are invariant under translation of axes, and therefore we can formulate the following theorem for *general equations of the second degree.*

Theorem 2. (a) *If $B^2 - 4AC < 0$, the curve is an ellipse, a circle, a point, or there is no curve.* (b) *If $B^2 - 4AC > 0$, the curve is a hyperbola or two intersecting straight lines.* (c) *If $B^2 - 4AC = 0$, the curve is a parabola, two parallel lines, one line, or there is no curve.*

The circle, ellipse, parabola, and hyperbola are often called **conic sections,** for all of them can be obtained as sections cut from a right circular cone by planes. The cone is thought of as extending indefinitely on both sides of its vertex; the part of the cone on one side of the vertex is called a **nappe.**

If the plane intersects only one nappe, as in Fig. 6–41(a), the curve of the intersection is an ellipse. (A circle is a special case of the ellipse, and occurs when the plane is perpendicular to the axis of the cone.) If the plane is parallel to one of the generators of the cone, the intersection is a parabola, as shown in Fig. 6–41(b). If the plane intersects both nappes, the curve is a hyperbola, one branch coming from each nappe, as in Fig. 6–41(c).

Certain degenerate cases also occur; the locus is two intersecting lines when the plane intersects both nappes and also passes through the vertex. If the plane

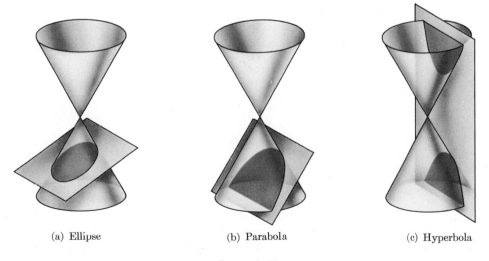

(a) Ellipse   (b) Parabola   (c) Hyperbola

Figure 6–41

contains one of the generators, the locus of the intersection is a single straight line. Finally, the locus is a single point if the plane contains the vertex and does not intersect either nappe of the cone. The degenerate locus of two parallel lines cannot be obtained as a plane section of a cone.

PROBLEMS

In problems 1 through 12, in each case change from an xy system to an $x'y'$ system such that the $x'y'$ term is missing. For the ellipses and hyperbolas, find the coordinates of the vertices. For each parabola find the xy coordinates of the focus and the xy equation of the directrix. If the locus consists of lines, find their xy equations.

1. $4x^2 - 4xy + y^2 = 9$
2. $4x^2 + 4xy + 7y^2 = 240$
3. $2x^2 - 3xy - 2y^2 + 25 = 0$
4. $x^2 - 6xy - 7y^2 = 0$
5. $x^2 - 2xy + y^2 - 8x - 8y = 0$
6. $x^2 - 4xy + y^2 + 32 = 0$
7. $13x^2 - 12xy + 8y^2 = 884$
8. $4x^2 + 4xy + y^2 + 20x - 40y = 0$
9. $2xy + y^2 = 10$
10. $xy = -4$
11. $2xy + 3x^2 + 5 = 0$
12. $xy = 3$

In problems 13 through 22, in each case rotate the axes to an $x'y'$ system such that the $x'y'$ term is missing. Then translate the axes so that the first degree terms are absent. Sketch the loci and identify the principal quantities.

13. $3x^2 - 10xy + 3y^2 + 14x - 2y - 5 = 0$
14. $2x^2 - 8xy - 4y^2 - 4x - 20y - 15 = 0$
15. $x^2 - 4xy + 4y^2 - 38x - 24y - 139 = 0$
16. $9x^2 + 24xy + 16y^2 + 42x + 56y + 49 = 0$
17. $19x^2 - 24xy + 12y^2 + 40x - 12y + 31 = 0$
18. $5x^2 + 4\sqrt{5}\,xy + 4y^2 + 4\sqrt{5}\,x + 8y - 21 = 0$
19. $21x^2 - 10\sqrt{3}\,xy + 31y^2 - (168 - 20\sqrt{3})x - (124 - 40\sqrt{3})y + 316 - 80\sqrt{3} = 0$
20. $x^2 - 2xy + y^2 + 4x - 4y + 4 = 0$
21. $\sqrt{3}\,xy + y^2 + 3x + 2\sqrt{3}\,y - 3 = 0$
22. $4x^2 + 3xy - 2x - y - 1 = 0$

23. Prove that a second degree equation with an xy term in it can never represent a circle.

24. Prove that a second degree equation with both D and E absent (i.e., no x and y terms) cannot be a parabola.

25. Prove that $B^2 - 4AC$ is invariant under rotation and translation of axes.

26. Given the transformation of coordinates

$$x' = ax + by, \qquad y' = cx + dy,$$

with a, b, c, and d numbers such that $ad - bc$ is positive. If the general equation of the second degree undergoes such a transformation, what can be said about $B^2 - 4AC$?

Graphs of Algebraic Relations 7

▶ 1. LOCI OF ALGEBRAIC RELATIONS

In Chapters 5 and 6 we discussed the loci in the plane of equations of the second degree. After completing the study of conics, i.e., equations of the second degree, we naturally wish to determine the loci of equations of the third degree, the fourth degree, and so on. Unfortunately, there is no systematic method for plotting the graphs of equations of degree higher than two. However, in certain special cases, it is possible to make a preliminary determination of the domain, the range, the intercepts, and the symmetry properties of the solution set of the equation and, on the basis of these, to plot the graph.

We recall that the **domain** of a relation is the set of all numbers x_0 such that the vertical line $x = x_0$ intersects the locus. The **range** of a relation is the set of all y_0 such that the horizontal line $y = y_0$ intersects the locus. To determine the domain of a relation involving x and y we perform the following two steps: (i) *We solve for y in terms of x.* (We caution that this step is often not possible for equations of higher degree.) (ii) *In the resulting expression or expressions in x, we determine those values of x for which at least one of the expressions has meaning.* The domain is the totality of such values of x. To find the range, we perform steps (i) and (ii) with the roles of x and y interchanged. Two examples illustrate the method.

EXAMPLE 1. Find the domain and range of the relation defined by

$$y^4 = x^2 - 4x + 3.$$

Solution. To find the domain we first solve for y in terms of x. Setting $f(x) = x^2 - 4x + 3$, we see that

$$y^2 = +\sqrt{f(x)}, \qquad y^2 = -\sqrt{f(x)},$$

and

$$y = +\sqrt{+\sqrt{f(x)}}, \quad y = -\sqrt{+\sqrt{f(x)}}, \quad y = +\sqrt{-\sqrt{f(x)}}, \quad y = -\sqrt{-\sqrt{f(x)}}.$$

Excluding the places where $f(x) = 0$, we conclude that the last two expressions for y in terms of x yield no locus since the square root of a negative quantity is always present. The first two expressions show that the domain consists of all x for which $f(x) \geq 0$.

To determine when $f(x)$ is nonnegative, we apply the method of Chapter 1, Section 5, for the solution of inequalities by factoring. We have

$$f(x) \equiv x^2 - 4x + 3 \equiv (x - 1)(x - 3)$$

and we make the table:

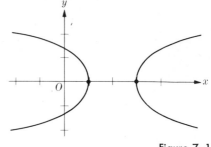

x	0	1	2	3	4
$f(x)$	3	0	-1	0	3

Figure 7–1

Therefore $f(x)$ is positive for x in the interval $(-\infty, 1)$ and in the interval $(3, \infty)$. The domain is $(-\infty, 1] \cup [3, \infty)$. To find the range we solve for x in terms of y. From the relation $x^2 - 4x + (3 - y^4) = 0$, we obtain (by the quadratic formula)

$$x = 2 \pm \sqrt{1 + y^4}.$$

Since $1 + y^4$ is positive for all y, the range is $(-\infty, \infty)$. A graph of the relation is sketched in Fig. 7–1.

EXAMPLE 2. Given the curve with equation

$$y(x^2 - 1) = 2,$$

find the intercepts, test for symmetry, and sketch the graph.

Solution. Setting $x = 0$, we get $y = -2$, which is the y intercept; $y = 0$ yields the impossible statement $0 = 2$, and the curve has no x intercept.

To test for symmetry with respect to the x axis, we replace y by $-y$, getting

$$-y(x^2 - 1) = 2,$$

which means that the curve is not symmetric with respect to the x axis. Replacing x by $-x$, we find

$$y[(-x)^2 - 1] = 2 \qquad \text{or} \qquad y(x^2 - 1) = 2,$$

which is the same as the original equation, and the curve is symmetric with respect to the y axis. The curve is not symmetric with respect to the origin, since

$$(-y)[(-x)^2 - 1] = 2$$

is not the same as the original equation.

We make up a table of values for positive x:

x	0	1	2	3	4	5
y	-2	$-$	$\frac{2}{3}$	$\frac{1}{4}$	$\frac{2}{15}$	$\frac{1}{12}$

We see that as x increases, the values of y get closer and closer to zero. On the other hand, there is no value of y corresponding to $x = 1$. In order to get a closer look at

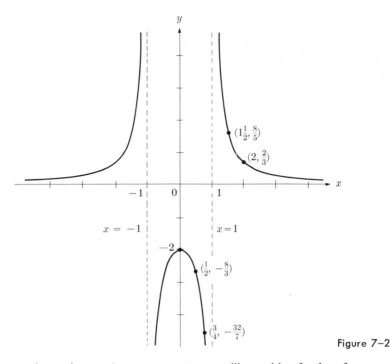

Figure 7–2

what happens when x is near 1, we construct an auxiliary table of values for x near 1. We obtain the following set of values:

x	$\frac{1}{2}$	$\frac{3}{4}$	$\frac{7}{8}$	0.9	1.1	1.2	1.3	$1\frac{1}{2}$
y	$-\frac{8}{3}$	$-\frac{32}{7}$	$-\frac{128}{15}$	$-\frac{200}{19}$	$\frac{200}{21}$	$\frac{50}{11}$	$\frac{200}{69}$	$\frac{8}{5}$

As x moves closer to 1 from the left, the corresponding values of y become large negative numbers. As x approaches 1 from the right, the values of y become large positive numbers. The curve is sketched in Fig. 7–2. The portion to the left of the y axis is sketched from our knowledge of the symmetry property; therefore a table of values for negative x is not needed.

In Example 2, the vertical line through the point $(1, 0)$ plays a special role. The curve to the right of the line gets closer and closer to this line as the curve becomes higher and higher. In fact, the distance between the curve and the line tends to zero as the curve continues upward beyond all bound. Such a line is called a *vertical asymptote* to the curve. Similarly, the x axis is called a *horizontal asymptote*, since the distance between the curve and the x axis tends to zero as x increases beyond all bound. A knowledge of the location of the vertical and horizontal asymptotes is of great help in sketching the curve. We now give a rule (which works in many cases) for finding the asymptotes.

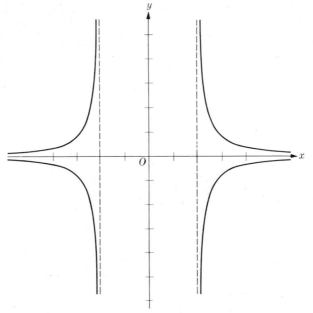

Figure 7–3

RULE. *To locate the vertical asymptotes, solve the equation for y in terms of x. If the result is a quotient of two expressions involving x, find all those values of x for which the denominator vanishes and the numerator does not. If a is such a value, the vertical line through the point (a, 0) will be a vertical asymptote. To locate the horizontal asymptotes, solve for x in terms of y, and find those values of y for which the denominator vanishes (and the numerator does not). If b is such a value, the horizontal line through the point (0, b) is a horizontal asymptote.*

Two examples illustrate the technique.

EXAMPLE 3.   Find the intercepts, symmetries, domain, range, and asymptotes, and sketch a graph of the equation

$$(x^2 - 4)y^2 = 1.$$

Solution. (a) *Intercepts:* The value $y = 0$ yields no x intercept; when $x = 0$, we have $y^2 = -\frac{1}{4}$, and there are no y intercepts.

(b) *Symmetry:* Locus is symmetric with respect to both axes (and therefore with respect to the origin).

(c) *Domain:* Solving for y in terms of x:

$$y = \pm \frac{1}{\sqrt{x^2 - 4}} \cdot$$

The domain is all x with $|x| > 2$; that is, the set $(-\infty, -2) \cup (2, +\infty)$.

(d) *Range:* Solving for x in terms of y: $x = \pm \sqrt{1 + 4y^2}/y$. The range is all y except $y = 0$.

(e) *Asymptotes:* To find the vertical asymptotes, we use the expression in part (c) for the domain. Setting the denominator equal to zero, we have the vertical lines $x = 2$, $x = -2$ as asymptotes. To find the horizontal asymptotes, we use the expression in part (d) for the range. Setting the denominator equal to zero, we have the horizontal line $y = 0$ as an asymptote. The graph is sketched in Fig. 7–3.

EXAMPLE 4.   Find the intercepts, symmetry, domain, range, and asymptotes, and sketch the locus of the equation

$$x^2 y = x - 3.$$

Solution. (a) *Intercepts:* If $y = 0$, then $x = 3$ and the x intercept is 3. Setting $x = 0$ yields no y intercept.

(b) *Symmetry:* All symmetry tests fail.

(c) *Domain:* Solving for y in terms of x, we have

$$y = \frac{x - 3}{x^2}.$$

The domain is all $x \neq 0$.

(d) *Range:* To solve for x, we write the equation $yx^2 - x + 3 = 0$. The solution of this quadratic in x is

$$x = \frac{1 \pm \sqrt{1 - 12y}}{2y} \qquad \text{if} \quad y \neq 0.$$

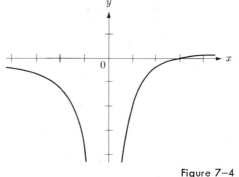

Figure 7–4

The original equation shows that $x = 3$ when $y = 0$. The range is all $y \leq \frac{1}{12}$; that is, $(-\infty, \frac{1}{12}]$.

(e) *Asymptotes:* The line $x = 0$ is a vertical asymptote. The line $y = 0$ is a horizontal asymptote.

To draw the locus we construct the table:

x	-4	-3	-2	-1	1	2	3	4	5
y	$-\frac{7}{16}$	$-\frac{2}{3}$	$-\frac{5}{4}$	-4	-2	$-\frac{1}{4}$	0	$\frac{1}{16}$	$\frac{2}{25}$

The graph is sketched in Fig. 7–4.

PROBLEMS

In problems 1 through 25, find in each case the intercepts, symmetries, domain, range, and asymptotes. Also sketch the graph.

1. $y = 9 - x^2$

2. $x^2 y = 9$

3. $4x^2 + 3y^2 = 12$

4. $y^2 = x^2 + 9$

5. $4x^2 - 3y^2 = 12$

6. $4x^2 - 9y^2 + 36 = 0$

7. $4x^2 + 2xy + y^2 = 12$ 8. $x^2 - 2xy + 4 = 0$

9. $y^2(x - 2) = 3$ 10. $y(x^2 - 4) = 3$

11. $y^2(x^2 - 9) = 28$ 12. $x^2(y + 1) = 8$

13. $y^2(x^2 + 4) = 8$ 14. $x^2(y - 3) + 8 = 0$

15. $xy(x - 2) = 16$ 16. $xy^2(x - 2) = 64$

17. $x(9 - y^2) = 6y$ 18. $x^2(9 - y^2) = 6y$

19. $x(y^2 - 9) = 6y$ 20. $y^4 = 4y^2 - x^2$

21. $y^4 = x^2 - y^2$ 22. $x^2y^2 = x - 3$

23. $x^2(y^2 - 1) = x + 2$ 24. $y^2 = x^4 - 4x^2$

25. $x^2 + 5x + 6 = 0$

26. Show that if a relation is symmetric with respect to the x axis and with respect to the origin, it must be symmetric with respect to the y axis.

▶2. NONLINEAR INEQUALITIES IN THE PLANE

We discussed *linear* inequalities in the plane in Section 11 of Chapter 3. There we saw that the solution set of a single linear inequality is a half-plane. The solution set of several simultaneous linear inequalities was found to be the *intersection* of the solution sets of the individual inequalities.

We shall now develop methods for the solution of nonlinear inequalities, a subject requiring more complicated techniques than does linear inequalities. For example, suppose we wish to determine the solution set in the plane of the inequality

$$x^2 > a, \qquad a \text{ is a real number.}$$

If a is negative, then the solution set is all of R_2. If $a = 0$, the solution set consists of all points in R_2 for which $x \neq 0$, that is, all points not on the y axis. If $a > 0$, then

$$x^2 > a$$

$$\Leftrightarrow x > \sqrt{a} \quad \text{or} \quad x < -\sqrt{a}.$$

Denoting by S_1, S_2, and S, the sets

$$S_1 = \{(x, y): x > \sqrt{a}\},$$

$$S_2 = \{(x, y): x < -\sqrt{a}\},$$

$$S = \{(x, y): x^2 > a\},$$

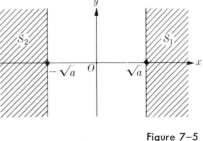

Figure 7–5

we see that S is the *union* of the half-planes S_1 and S_2; that is, $S = S_1 \cup S_2$. The solution set is the shaded region shown in Fig. 7–5.

Having obtained the solution set of the inequality $x^2 > a$, we set out to solve the inequality $x^2 < a$. If $a \leq 0$, the solution set is empty. If $a > 0$, then

$$x^2 < a \Leftrightarrow x < \sqrt{a} \quad \text{and} \quad x > -\sqrt{a}.$$

Let S_1, S_2, and S be the sets

$$S_1 = \{(x, y): x < \sqrt{a}\},$$

$$S_2 = \{(x, y): x > -\sqrt{a}\},$$

$$S = \{(x, y): x^2 < a\}.$$

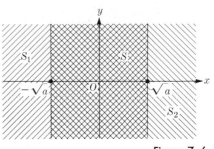

In contrast with the preceding case, we see
that S is the *intersection* of the half-planes
S_1 and S_2; that is, $S = S_1 \cap S_2$. The
solution set is the doubly-shaded region

Figure 7–6

shown in Fig. 7–6. The set S may also be described by the double inequality
$-\sqrt{a} < x < \sqrt{a}$. Here we see the importance of distinguishing problems in the
plane from those on the line. The inequality $-\sqrt{a} < x < \sqrt{a}$ considered as a
problem in R_1 represents the points in an open interval, while the same inequality,
considered in R_2, consists of the points in an infinite strip.

We work an example and then state a principle for the solution of certain classes
of inequalities.

EXAMPLE 1. Determine the solution set of
the inequality

$$x^2 - y^2 > 4.$$

Solution. We see that

$$x^2 - y^2 > 4 \Leftrightarrow x^2 > 4 + y^2.$$

This holds if and only if

$$x > \sqrt{4 + y^2} \quad \text{or} \quad x < -\sqrt{4 + y^2}. \quad (1)$$

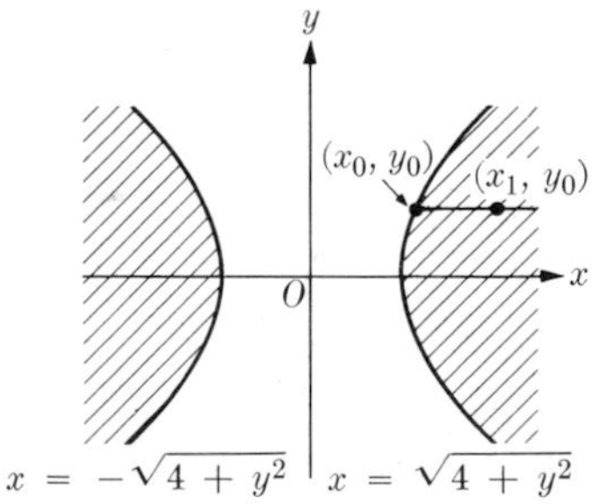

Figure 7–7

Suppose that (x_0, y_0) is a point such that $x_0 = \sqrt{4 + y_0^2}$. Then any point (x_1, y_0) on
the line $y = y_0$ to the right of (x_0, y_0) will satisfy the inequality $x_1 > \sqrt{4 + y_0^2}$, and
so will satisfy the first of the two inequalities in (1). The locus of the inequality $x >
\sqrt{4 + y^2}$ is the shaded region to the right of the curve $x = \sqrt{4 + y^2}$ as shown in Fig.
7–7. Similarly, the locus of $x < -\sqrt{4 + y^2}$ is the shaded region to the left of the curve
$x = -\sqrt{4 + y^2}$. The entire locus of $x^2 - y^2 > 4$ is the *union* of the two shaded
regions.

In general, the solution set of an inequality such as

$$x^2 > f(y)$$

may be found as the union of the solution sets of the inequalities

$$x > \sqrt{f(y)}, \quad x < -\sqrt{f(y)}.$$

The solution set of the inequality

$$x^2 < f(y)$$

is the *intersection* of the solution sets of

$$x < \sqrt{f(y)}, \qquad x > -\sqrt{f(y)}.$$

Analogous statements hold for inequalities of the type $y^2 > g(x)$ and $y^2 < g(x)$.

As in the case of linear inequalities, the solution set of two or more simultaneous inequalities (some of which may be nonlinear) is the intersection of the solution sets of the individual inequalities.

EXAMPLE 2. Describe and draw a graph of the solution set of the simultaneous inequalities

$$-1 \leq x \leq 2 \quad \text{and} \quad x^2 \leq y \leq x + 6.$$

Solution. The solution set of the *two* inequalities $-1 \leq x \leq 2$ is the infinite strip between the vertical lines $x = -1$ and $x = 2$. We note that the bounding lines are also in this set, which we denote by S_1; that is,

$$S_1 = \{(x, y): -1 \leq x \leq 2\}.$$

To solve the inequality $y \geq x^2$ we first construct the parabola $y = x^2$ and consider a point (x_0, y_0) on this curve (see Fig. 7–8). Then any point (x_0, y_1) on the vertical line $x = x_0$ which is *above* the parabola satisfies $y_1 > x_0^2$. The set $S_2 = \{(x, y): y \geq x^2\}$ is the shaded region shown in Fig. 7–8. Finally, the linear inequality $y \leq x + 6$ has as its solution set, denoted by S_3, the points on the line $y = x + 6$ and in the half-plane below this line. Solving simultaneously the equations $y = x^2$ and $y = x + 6$, we find that the parabola and the line intersect at $(-2, 4)$ and $(3, 9)$. Consequently, the solution set we seek is the shaded region shown in Fig. 7–9. Denoting this region by S, we obtain it as the intersection $S = S_1 \cap S_2 \cap S_3$. We observe that the boundary of the shaded region is included in S.

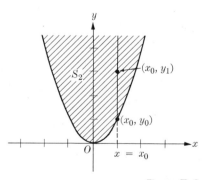

Figure 7–8

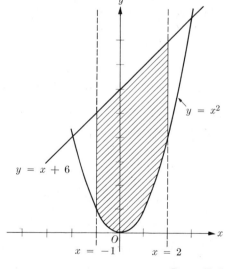

Figure 7–9

EXAMPLE 3. Describe in set notation and draw a graph of the bounded region whose boundary consists of (parts of) the loci of the equations

$$y = 1,$$

$$y = x,$$

and

$$xy^2 = 8.$$

Solution. The loci of the equations are plotted in Fig. 7–10. The loci of $x = y$ and $xy^2 = 8$ intersect at the point (2, 2). The region we seek is the shaded region shown in Fig. 7–10. It is bounded on the right by the curve $xy^2 = 8$ or $x = 8/y^2$. Therefore the points in the shaded region must satisfy $x < 8/y^2$. These points also satisfy $x > y$, since the region is bounded on the left by the line $x = y$. The region is bounded above and below by the lines $y = 2$ and $y = 1$, respectively. Thus the solution set is described by

$$\{(x, y): 1 \leq y \leq 2 \text{ and } y \leq x \leq 8/y^2\}.$$

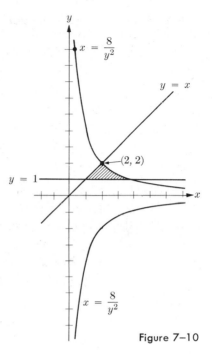

Figure 7–10

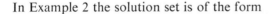

In Example 2 the solution set is of the form

$$\{(x, y): a \leq x \leq b \text{ and } f_1(x) \leq y \leq f_2(x)\},\qquad(2)$$

while that in Example 3 has the form

$$\{(x, y): c \leq y \leq d \text{ and } g_1(y) \leq x \leq g_2(y)\}.\qquad(3)$$

Regions of both these types occur frequently in calculus. For this reason, students who study analytic geometry as a preparation for calculus should develop great facility in determining the exact nature of regions described by a number of inequalities. The type of a given region depends critically on whether the bounding curves are of the form $y = f(x)$ or $x = g(y)$.

PROBLEMS

In problems 1 through 6, in each case plot the graph of the solution set of the given inequality. Also describe the set in words.

1. $2x < y^2$

2. $y < 1 - x^2$

3. $x > 9 - y^2$

4. $x^2 + 2y - 4 < 0$

5. $4x^2 - y^2 \leq 4$

6. $x^2 + y^2 \leq 9$

In problems 7 through 14, in each case draw a graph of the given set. Also describe the set in words.

7. $\{(x, y): -1 \le x \le 2, 0 \le y \le x + 2\}$

8. $\{(x, y): 1 \le x \le 3, 1 \le y \le x^2\}$

9. $\{(x, y): -1 \le y \le 2, 1 \le x \le y + 2\}$

10. $\{(x, y): 0 \le x \le 2, x^2 + 1 \le y \le 5x\}$

11. $\{(x, y): -1 \le x \le 1, x^2 \le y \le x + 1\}$

12. $\{(x, y): -2 \le y \le 1, y^2 + 1 \le x \le 3 - y\}$

13. $\{(x, y): -1 \le x \le 1, -\sqrt{9 - x^2} \le y \le \sqrt{9 - x^2}\}$

14. $\{(x, y): y^2 \le 2x \text{ and } x \le 2\}$

In problems 15 through 26, sketch the region bounded by the given curves and describe it using set notation of one of the forms (2) or (3) on page 163.

15. $x = -2, x = 2, y = x^2 - 8, y = -x$

16. $x = -1, x = 2, y = x^2 - 4, y = 4 - x^2$

17. $y = -1, y = 1, x = y^2, x = 2 - y$

18. $x = 1, x = 2, y = \frac{1}{2}x, y = \sqrt{x + 2}$

19. $x = 0, x = 3, y = x^2, y = 12x$

20. $y = 1, y = 2, x = 1, x = y^2$

21. $y = 0, y = \frac{3}{4}, x = y^2, x = y$

22. $x = -2, x = 0, y = x^3, y = -x$

23. $x = 1, x = 2, x^2y = 2, x + y = 4$

24. $x = 1, x = 2, y^2 = x$

25. $y = 1, y = 2, y = x^2, y = 3x$

26. $y = x^2, y = \sqrt[3]{x}$

The Transcendental Functions 8

▶ 1. GRAPHS OF TRIGONOMETRIC FUNCTIONS

In the study of trigonometry, although we occasionally have use for a short table of the values of trigonometric functions in terms of **radian** measure, the most common measure used for angles is the degree, with most tables for the sine, cosine, etc., expressed in degrees, minutes, and sometimes seconds.

In calculus, however, the situation is reversed; the *natural unit* for the study of trigonometric and related functions is the *radian*, because measurement of angles in degrees is awkward for problems in calculus. Those students who are studying analytic geometry preparatory to a course in calculus will find it advantageous to get practice in using radians to graph trigonometric functions. In the process they will become familiar with many of the properties of these functions. As an additional aid to the student there is a brief review of trigonometry in the Appendix 3.

The graph of the function $y = \sin x$ in a Cartesian coordinate system is made by drawing up a table of values and sketching the curve, as shown in Fig. 8–1.

x	0	$\pi/6$	$\pi/4$	$\pi/3$	$\pi/2$	$2\pi/3$	$3\pi/4$	$5\pi/6$	π
y	0	$1/2$	$\sqrt{2}/2$	$\sqrt{3}/2$	1	$\sqrt{3}/2$	$\sqrt{2}/2$	$1/2$	0

x	$7\pi/6$	$5\pi/4$	$4\pi/3$	$3\pi/2$	$5\pi/3$	$7\pi/4$	$11\pi/6$	2π
y	$-1/2$	$-\sqrt{2}/2$	$-\sqrt{3}/2$	-1	$-\sqrt{3}/2$	$-\sqrt{2}/2$	$-1/2$	0

It is necessary to make a table of values only between 0 and 2π. From the relation $\sin(2\pi + x) = \sin x$, we know that the curve then repeats indefinitely to the left and right.

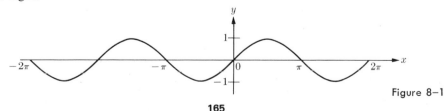

Figure 8–1

165

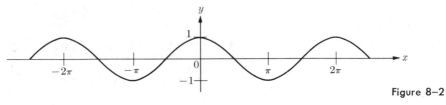

Figure 8-2

The graph of $y = \cos x$ is found in the same way. A table of values between 0 and 2π and a knowledge of the periodic nature of the function yield the graph shown in Fig. 8-2. Note that the curves are identical except that the cosine function is shifted to the left by an amount $\pi/2$. This fact is not surprising if we recall the relation $\sin [x + (\pi/2)] = \cos x$.

A function f is called **periodic** if there is a number l such that

$$f(x + l) = f(x)$$

for all x, in which case the number l is called a **period** of the function f. For the sine function, 2π is a period, but so are 4π, 6π, 8π, etc., since $\sin (x + 4\pi) = \sin x$, $\sin (x + 6\pi) = \sin x$, and so on. For periodic functions there is always a smallest number l for which $f(x + l) = f(x)$. This number is called the **least period** of f. The least period of $\sin x$ is 2π; the same is true for $\cos x$. The function $\tan x$ has period 2π, but this is not the least period. A graph of $y = \tan x$ is given in Fig. 8-3 which shows clearly that π is a period of $\tan x$. This is the least period for the tangent function. We note that the lines $x = (\pi/2) + n\pi$, $n = 0, \pm 1, \pm 2, \ldots$, are all vertical asymptotes of the locus of $y = \tan x$.

The function $y = \sin kx$ repeats when kx changes by an amount 2π. If k is a constant, then $\sin kx$ repeats when x changes by an amount $2\pi/k$. The least period of the function $\sin kx$ is $2\pi/k$. The same is true for $y = \cos kx$. For example, $y = \cos 5x$ has as its least period $2\pi/5$.

A periodic function f, which never becomes larger than some number b in value and never becomes smaller than some number c, is said to have **amplitude** $\frac{1}{2}(b - c)$,

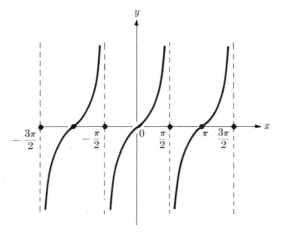

Figure 8-3

provided that there are points where f actually takes on the values b and c. For example, the function $y = 3 \sin 2x$ is never larger than 3 or smaller than -3. The amplitude is 3. Note that $y = 3$ when $x = \pi/4$. The tangent function has no amplitude.

EXAMPLE 1. Find the amplitude and period of $y = 4 \cos (3x/2)$. Sketch the graph.

Solution. The amplitude is 4; the period is $2\pi/(\frac{3}{2}) = 4\pi/3$. The graph is sketched in Fig. 8–4.

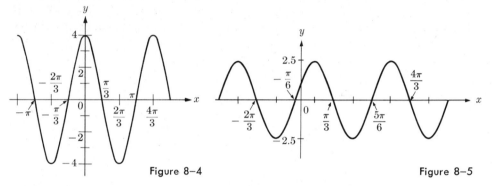

Figure 8–4   Figure 8–5

EXAMPLE 2. Find the amplitude and period of $y = 2.5 \sin [2x + (\pi/3)]$ and sketch its graph.

Solution. This function is similar to $2.5 \sin 2x$, except that it is shifted horizontally. We write $y = 2.5 \sin 2[x + (\pi/6)]$. When $x = -\pi/6$, $y = 0$, and we see that the "zero point" is moved to the left an amount $\pi/6$. The period is $2\pi/2 = \pi$, and the amplitude is 2.5. With this information it is now easy to sketch the curve, as shown in Fig. 8–5.

Combinations of trigonometric functions may be sketched quickly by a method known as **addition of ordinates.** If the sum of two functions is to be sketched, we

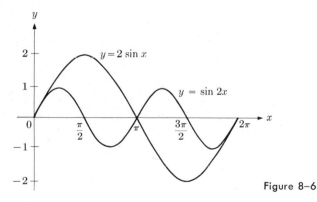

Figure 8–6

start by sketching each function separately. Then, at a particular value of x, the y values of each of the functions are added to give the result. Incidentally, this method is quite general and may be used for the sum or difference of any functions. An example illustrates the technique.

EXAMPLE 3. Sketch the graph of

$$y = 2 \sin x + \sin 2x.$$

Solution. The function $2 \sin x$ has period 2π and amplitude 2, while $\sin 2x$ has period π and amplitude 1. We sketch the graph of each of these functions on the same coordinate system from 0 to 2π, as shown in Fig. 8–6. At convenient values of x, the y values of the two graphs are added and the results plotted either on the same graph or on a similar one, as shown in Fig. 8–7.

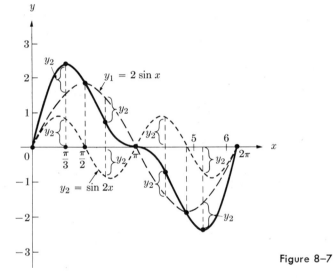

Figure 8–7

PROBLEMS

In problems 1 through 16, find the amplitudes and the periods of the functions defined by the given expressions and sketch their graphs.

1. $3 \sin \frac{1}{3}x$ 2. $4 \cos \frac{4}{3}x$ 3. $2 \cos \frac{1}{3}\pi x$

4. $2 \sin \frac{1}{2}\pi x$ 5. $5 \sin \frac{3}{5}x$ 6. $3 \sin \frac{2}{3}\pi x$

7. $3 \sin (\frac{3}{2}x + \pi/2)$ 8. $2 \cos (\frac{1}{2}\pi x + \frac{3}{4}\pi)$ 9. $-3 \cos 3x$

10. $-2 \sin (\frac{1}{3}x + 4)$ 11. $3 \cos (3x + \frac{1}{4}\pi)$ 12. $4 \sin (3x - \frac{1}{2}\pi)$

13. $-4 \cos (\frac{3}{4}\pi x)$ 14. $\frac{1}{2} \sin (3x + \frac{2}{3}\pi)$ 15. $1 + \cos 2x$

16. $-2 + 3 \cos (x - \frac{1}{2}\pi)$

In problems 17 through 24, find the periods and amplitudes (if any), and sketch the graphs.

17. $\tan \frac{3}{2}x$ 18. $\cot 2x$ 19. $\sec 3x$

20. $\csc 2x$ 21. $2 \cot (\pi x - \frac{1}{2}\pi)$ 22. $-3 \cot (\frac{3}{2}x - \frac{1}{2}\pi)$

23. $\sin 2x + \cos 2x$ [*Hint*: Express this in the form $A \sin (x + \alpha)$.]

24. $2\sqrt{3} \cos 3x + 2 \sin 3x$

In problems 25 through 32, sketch the graphs of the curves, using the method of addition of ordinates.

25. $y = \cos x + \sin 2x$ 26. $y = 2 \sin x + 2 \sin 2x$

27. $y = 2 \cos \frac{1}{2}x - \sin x$ 28. $y = 3 \sin \frac{2}{3}\pi x - \frac{3}{2} \cos \frac{4}{3}\pi x$

29. $y = -2 \sin 3x + 3 \cos 3x$ 30. $y = -3 \cos 2x + 2 \cos 4x$

31. $y = x + \sin x$ 32. $y = 2 - 2x + \sin 2x$

33. Sketch the graph of $y = 2 \cos^2 x$. Show that it has period π. How does it compare with the graph of $\cos 2x$?

34. Show that the graph of $y = \sin nx + \cos nx$ is that of a sine function. [*Hint*: Write $y = \sqrt{2}[(1/\sqrt{2}) \sin nx + (1/\sqrt{2}) \cos nx]$ and use the formula $\sin (A + B) = \sin A \cos B + \cos A \sin B$.]

35. Extend the method of Problem 34 to show that the graph of $y = a \sin nx + b \cos nx$ is that of a sine function.

▶ 2.   INVERSE RELATIONS AND FUNCTIONS

In Section 3 of Chapter 2, we defined a relation as a set of points in the number plane, R_2. Suppose we are given a relation S. Then we define the **inverse relation** of S as the set of all points (x, y) in R_2 such that (y, x) belongs to the relation S. For example, if $(3, 7)$ is in the relation S, then $(7, 3)$ is in the inverse relation of S.

In case a relation is the solution set of an equation in x and y, the inverse relation is the solution set of the equation obtained by interchanging x and y in the given equation. We give four examples:

(i)   If a relation is defined by $x^2 - y^2 = 1$, the inverse relation is the solution set of

$$y^2 - x^2 = 1.$$

(ii) If a relation is defined by $y = x^2$, the inverse relation is the solution set of

$$y^2 = x.$$

(iii) If a relation is defined by $y = (x - 1)^3$, the inverse relation is the solution set of

$$(y - 1)^3 = x \quad \text{or} \quad y = 1 + x^{1/3}.$$

(iv) If a relation is defined by

$$y^2 = x^3,$$

the inverse relation is the solution set of

$$y^3 = x^2 \quad \text{or} \quad y = x^{2/3}.$$

In example (i), neither the original nor the inverse relation is a function. In example (ii), the given relation is a function and the inverse relation is not. In example (iii), both the given relation and its inverse are functions. Finally, in example (iv), the given relation is not a function, but its inverse is. Therefore we see that there is no simple interconnection between inverse relations and functions.

It is a direct consequence of the definition of inverse relation that the *domain* of the inverse relation is the range of the given relation and that the *range* of the inverse is the domain of the original. Also, the inverse of the inverse of a given relation is just the given relation.

Since a function is a special case of a relation, no additional effort is needed to define the **inverse** of a function. As we saw in example (ii) above, the inverse of a function need not be a function. In general, a function f which is the locus of the equation $y = f(x)$ has as its inverse the locus given by

$$x = f(y).$$

The inverse of a function is a relation which may consist of several functions. These functions are called the **branches** of the inverse relation. If we can solve the equation $x = f(y)$ for y in terms of x, then we can obtain explicit expressions for the branches of the inverse of f.

EXAMPLE 1. Given $f(x) = x^2 - 2x + 2$, plot the inverse relation of f. Show that the inverse consists of two branches and find explicit formulas for these branches.

Solution. The inverse relation is the solution set of the equation

$$x = y^2 - 2y + 2.$$

The locus is the parabola sketched in Fig. 8–8. Solving for y, we get the branches

$$y = 1 - \sqrt{x - 1} \equiv g_1(x),$$

$$y = 1 + \sqrt{x - 1} \equiv g_2(x).$$

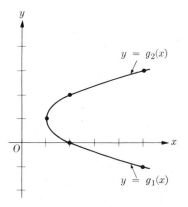

Figure 8–8

Suppose a given function f has an inverse which is also a function. Call this inverse function g. Then we have two equivalent ways of expressing the inverse of f:

$$x = f(y) \quad \text{and} \quad y = g(x). \tag{1}$$

A simple substitution of the second equation in (1) into the first shows that

$$f[g(x)] = x \quad \text{for } x \text{ in the domain of } g. \tag{2}$$

Similarly, substituting the first equation of (1) into the second, we obtain

$$g[f(y)] = y \quad \text{for } y \text{ in the domain of } f.$$

Or, changing notation, we have

$$g[f(x)] = x \quad \text{for } x \text{ in the domain of } f. \tag{3}$$

By using another interpretation of function, we can give formulas (2) and (3) additional significance. We may think of a function as an "operator" which carries one object into another. To see this, we note first that an element in the domain of a given function f is a number in R_1; also, an element in the range of f is a number in R_1. We may say that f "carries" a number x in R_1 (the element of the domain) into a number y in R_1 (the element of the range). The number y is called the **image** of x under the operation* of f. We now interpret (2) and (3) by observing that the inverse function g is that operator which carries the image of an element x

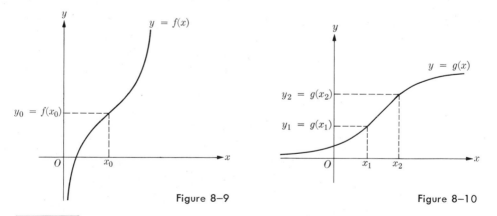

Figure 8–9   Figure 8–10

---

* The term **mapping** is currently in use as a synonym for operation. We say that f maps a number x in the domain into a number y in the range.

in the domain of f back into x itself; g reverses the action of f. For example, if $f(x) = x^3$ and g is the inverse of f, then $g(x) = x^{1/3}$. Equations (2) and (3) in this case are

$$(x^{1/3})^3 = x \quad \text{and} \quad (x^3)^{1/3} = x,$$

respectively.

Suppose that f is a function whose graph increases steadily in height as we proceed along it from left to right. Writing $y = f(x)$, we say that f is an **increasing function** of x. In such a case it is not hard to show that the inverse of f is also a function (which we denote by g) and, in fact, this inverse function is also an increasing function as we go from left to right along its domain. To establish these facts, we let y_0 be in the domain of g. By definition, this means that y_0 is in the range of f. Therefore, there is a number x_0 such that $y_0 = f(x_0)$. (See Fig. 8–9.) The definition that f is increasing asserts that for $x > x_0$, we must have $f(x) > f(x_0)$. Similarly, if $x < x_0$, we must have $f(x) < f(x_0)$. Consequently any **horizontal** line $y = y_0$ can intersect the graph of $y = f(x)$ only once. This statement is equivalent to the assertion that g, the inverse of f, is a function. To see that g is increasing, we write $y = g(x)$ and let x_1 and x_2 be two numbers in the domain of g (see Fig. 8–10). Suppose $x_1 < x_2$ and $y_1 = g(x_1)$, $y_2 = g(x_2)$. We must show $y_1 < y_2$. However, from the definition of inverse function

$$x_1 = f(y_1) \quad \text{and} \quad x_2 = f(y_2).$$

If $y_1 = y_2$, then $x_1 = x_2$. If $y_1 > y_2$, then $x_1 > x_2$ since f is an increasing function. Consequently, the only possibility remaining is $y_1 < y_2$, which is what we set out to prove.

If the graph of a function f decreases steadily as we go from left to right, the corresponding argument shows that the inverse of f is again a function and that this function is also a decreasing function of x.

If f is a function which is neither always increasing nor always decreasing over its whole domain, it may be possible to split the domain into several intervals on each of which f is an increasing or a decreasing function. Figure 8–11 shows a typical situation in which the function f, when restricted to the interval (a_i, a_{i+1}), $i = 1, 2, 3, 4$, is an increasing or a decreasing function. For example, suppose we

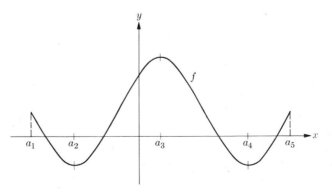

Figure 8–11

define

$$f_1(x) = x^2 \quad \text{on} \quad (-\infty, 0], \qquad f_2(x) = x^2 \quad \text{on} \quad [0, \infty),$$

and $f(x) = x^2$ on $(-\infty, \infty)$ as the union of f_1 and f_2. We observe that f_1 is steadily decreasing on its domain, while f_2 is steadily increasing on its domain. The functions f_1 and f_2 have inverses g_1 and g_2 which are functions. In fact, $g_1(x) = -\sqrt{x}, g_2(x) = \sqrt{x}, x \geq 0$.

PROBLEMS

In problems 1 through 10, in each case draw a graph of the inverse relation of the given function. Indicate its various branches on the graph and find formulas for the branches.

1. $f(x) = 2x + 3$

2. $f(x) = 2 - \frac{1}{3}x$

3. $f(x) = x^2 + 4x + 5$

4. $f(x) = x^2 - 2x - 3$

5. $f(x) = 4 - 3x - x^2$

6. $f(x) = 3 + 2x - x^2$

7. $f(x) = \dfrac{x - 1}{x}$

8. $f(x) = \dfrac{3 - x}{3 + x}$

9. $f(x) = \dfrac{1}{x^2}$

10. $f(x) = \dfrac{2x}{1 + x^2}$

In problems 11 through 15, in each case express the given function as the union of functions each restricted to a domain where it is steadily increasing or steadily decreasing. Find the inverses of these restricted functions.

11. $f(x) = -x^4, (-\infty < x < \infty)$

12. $f(x) = 5 - 2x - x^2, (-\infty < x < \infty)$

13. $f(x) = (x - 2)^3, (-\infty < x < \infty)$

14. $f(x) = x^{1/3}, (-\infty < x < \infty)$

15. $f(x) = x + \sqrt{x}, (0 \leq x < \infty)$

▶3. THE INVERSE TRIGONOMETRIC FUNCTIONS

The inverse of the sine function is the relation defined by the equation

$$x = \sin y.$$

A graph of this relation is sketched in Fig. 8–12 with the curve repeating indefinitely in both upward and downward directions. Since the *range* of the sine function is the interval $[-1, 1]$, this interval is the *domain* of the inverse of the sine function. For each number x on $[-1, 1]$, there are infinitely many values of y such that $\sin y = x$. For example, if $x = \frac{1}{2}$, then

$$y = \frac{\pi}{6} + 2n\pi, \quad y = \frac{5\pi}{6} + 2n\pi, \quad \text{where} \quad n = 0, \pm 1, \pm 2, \ldots$$

Referring to Fig. 8–1 on p. 165, we see that $\sin x$ increases as x goes from $-\pi/2$ to $\pi/2$, decreases from $\pi/2$ to $3\pi/2$, increases from $3\pi/2$ to $5\pi/2$, decreases from $5\pi/2$ to $7\pi/2$, and so on. We define the *increasing* function f_1 by restricting

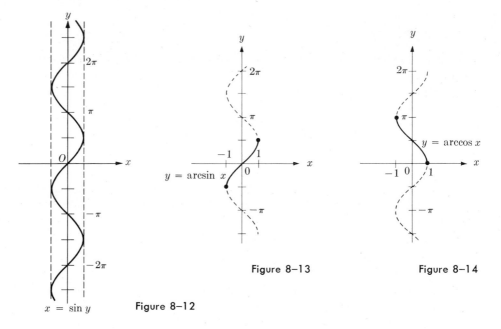

$x = \sin y$ Figure 8–12

Figure 8–13

Figure 8–14

the domain of the sine function to the interval $[-\pi/2, \pi/2]$:

$$f_1(x) = \sin x, \qquad -\frac{\pi}{2} \le x \le \frac{\pi}{2}.$$

We define the **arcsin function** as the inverse of f_1; that is,

$$y = \arcsin x \Leftrightarrow x = \sin y \quad \text{and} \quad -\frac{\pi}{2} \le y \le \frac{\pi}{2}.$$

The graph of this function is indicated by the solid line in Fig. 8–13. Since arcsin is the inverse of f_1, formulas (2) and (3) of Section 2 become

$$\sin (\arcsin x) = x \qquad \text{if} \quad x \text{ is on } [-1, 1],$$

$$\arcsin (\sin x) = x \qquad \text{if} \quad x \text{ is on } \left[-\frac{\pi}{2}, \frac{\pi}{2}\right].$$

The inverse relation of the cosine function is the locus of the equation

$$x = \cos y.$$

The graph is drawn in Fig. 8–14. By restricting $\cos x$ to the interval $[0, \pi]$, we obtain a decreasing function of x. We define

$$f_2(x) = \cos x, \qquad 0 \le x \le \pi,$$

and so obtain the **arccos function** as the inverse of f_2; that is,

$$y = \arccos x \iff x = \cos y \quad \text{and} \quad 0 \le y \le \pi.$$

The graph of this function is indicated by the solid line in Fig. 8–14.

EXAMPLE 1. Sketch the graph of

$$y = 2 \arccos \left(\frac{x}{2}\right).$$

Solution. We have

$$y = 2 \arccos \left(\frac{x}{2}\right)$$

$$\iff \left(\frac{y}{2}\right) = \arccos \left(\frac{x}{2}\right)$$

$$\iff \left(\frac{x}{2}\right) = \cos \left(\frac{y}{2}\right) \quad \text{and} \quad 0 \le \frac{y}{2} \le \pi$$

$$\iff x = 2 \cos \left(\frac{y}{2}\right) \quad \text{and} \quad 0 \le y \le 2\pi.$$

The student may use the methods of Section 1
to obtain the graph in Fig. 8–15.

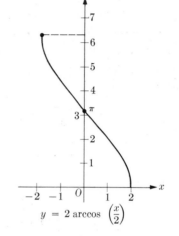

$$y = 2 \arccos \left(\frac{x}{2}\right)$$

Figure 8–15

The inverse relation of the tangent function is the locus of the equation

$$x = \tan y.$$

Figure 8–16 shows a graph of this relation.
We define the **arctan function** by the conditions

$$y = \arctan x$$

$$\iff x = \tan y$$

and

$$-\frac{\pi}{2} < y < \frac{\pi}{2}.$$

The graph of this function is shown by the
solid line in Fig. 8–16.

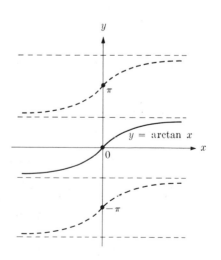

$$y = \arctan x$$

Figure 8–16

EXAMPLE 2. Evaluate (i) arcsin $(-\frac{1}{2})$, (ii) arccos $(-\frac{1}{2})$, (iii) arctan $(-1/\sqrt{3})$.

Solution. We have (i) arcsin $(-\frac{1}{2}) = -\pi/6$, (ii) arccos $(-\frac{1}{2}) = 2\pi/3$, and (iii) arctan $(-1/\sqrt{3}) = -\pi/6$.

Inverses of the cotangent, secant and cosecant functions may be obtained by a procedure similar to the one above for the sine, cosine, and tangent functions. (See problem 11 at the end of this section.) The arcsin, arccos, and arctan functions are often denoted by the symbols $\sin^{-1}$, $\cos^{-1}$, and $\tan^{-1}$, respectively. When this notation is used, *the* -1 *is not an exponent*. We shall usually use the "arc" notation, but the student should be familiar with both since many texts on calculus prefer the -1 symbol.

PROBLEMS

In problems 1 through 4, evaluate each expression.

1. (a) arcsin $(1/2)$, (b) arccos $(\sqrt{3}/2)$
2. (a) arctan $(-\sqrt{3})$, (b) $\cos^{-1}(-1/2)$
3. (a) arctan $(1/\sqrt{3})$, (b) $\sin^{-1}(-\sqrt{3}/2)$
4. (a) $\sin^{-1}(-1/2)$, (b) $\tan^{-1}(-1)$

In problems 5 through 10, plot graphs of the equations.

5. $y = 2$ arcsin $(\frac{1}{2}x)$ 6. $y = 2$ arccos $(2x)$
7. $y = \frac{1}{3}$ arccos $(\frac{1}{3}x)$ 8. $y = 3$ arcsin $(x/3)$
9. $y = 3$ arctan $(2x)$ 10. $y = \frac{1}{2}$ arctan $(x/2)$

11. Sketch the graphs of the inverse relations of the cot, sec, and csc functions. Give appropriate definitions of the arccot, arcsec, and arccsc functions.

►4.   EXPONENTIAL AND LOGARITHMIC CURVES

Let us consider the equation

$$y = 2^x$$

with the intention of sketching its graph. By assigning integer values to x, we obtain the table

x	-3	-2	-1	0	1	2	3
2^x	$\frac{1}{8}$	$\frac{1}{4}$	$\frac{1}{2}$	1	2	4	8

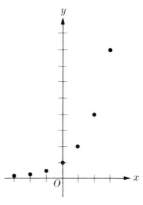

Figure 8–17

Plotting these points as shown in Fig. 8–17, we are tempted to draw a smooth curve through them.

However, to get a more accurate estimate of the graph, we use the facts

$$2^{1/2} = \sqrt{2}, \qquad 2^{3/2} = 2\sqrt{2}, \qquad 2^{5/2} = 4\sqrt{2},$$

$$2^{-1/2} = \tfrac{1}{2}\sqrt{2}, \qquad 2^{-3/2} = \tfrac{1}{4}\sqrt{2}, \qquad 2^{-5/2} = \tfrac{1}{8}\sqrt{2}.$$

In this way, we may adjoin the following table to the one above:

x	$-5/2$	$-3/2$	$-1/2$	$1/2$	$3/2$	$5/2$
2^x	$\tfrac{1}{8}\sqrt{2}$	$\tfrac{1}{4}\sqrt{2}$	$\tfrac{1}{2}\sqrt{2}$	$\sqrt{2}$	$2\sqrt{2}$	$4\sqrt{2}$
Approx.	0.18	0.35	0.71	1.41	2.83	5.66

All the points in both tables appear to lie on the smooth curve drawn in Fig. 8–18. Furthermore, the curve appears to get closer to the x axis as x decreases and it increases without bound as x increases.

Suppose that r is a *rational* number; that is, r is of the form p/q with p and q *integers*. We saw in algebra that expressions of the form a^r are well defined for $a > 0$. It is merely the qth root of a raised to the pth power. We recall the *laws of exponents*, valid for a and b positive and r and s rational numbers:

$$a^{r+s} = a^r \cdot a^s, \qquad a^{r-s} = a^r/a^s, \qquad (a^r)^s = a^{rs},$$
$$(ab)^r = a^r b^r, \qquad (a/b)^r = a^r/b^r. \tag{1}$$

Using these laws we are able to expand indefinitely the tables for $y = 2^x$. For example,

$$2^{1/4} = \sqrt{\sqrt{2}},$$

$$2^{3/4} = \sqrt{2} \cdot \sqrt{\sqrt{2}},$$

$$2^{5/4} = 2 \cdot 2^{1/4},$$

and so on. All these additional points would be seen to lie on the same smooth curve shown in Fig. 8–18. However, there is a difficulty. Suppose we set $x = \sqrt{5}$. The curve in Fig. 8–18 has a point corresponding to this number, but so far no meaning has been assigned to $2^{\sqrt{5}}$. More generally, if x is any *irrational* number and a is positive, the expression a^x has not been defined thus far. We state without

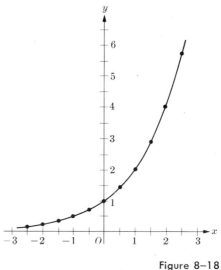

Figure 8–18

proof that it is possible to define a^x (if $a > 0$) for all real numbers x in such a way that

(i)   *a^x increases as x increases if $a > 1$;*
(ii) *a^x decreases as x increases if $0 < a < 1$;*
(iii) *the usual laws of exponents as given by* (1) *hold for all real numbers;*
(iv) *the range of the relation given by $y = a^x$ is the interval $(0, \infty)$.*

In particular, statements (i), (iii), and (iv) assert that Fig. 8–18 portrays an accurate sketch of the graph of $y = 2^x$.

Functions f of the form a^x are called **exponential functions.** Their graphs are **exponential curves.** (We always assume that a is positive and that $a \neq 1$.)

EXAMPLE 1. Given the equation

$$y = 3^{(1-x)},$$

plot its graph.

Solution. We obtain the table

x	-1	0	1	2	3
y	9	3	1	$\frac{1}{3}$	$\frac{1}{9}$

The graph is plotted in Fig. 8–19.

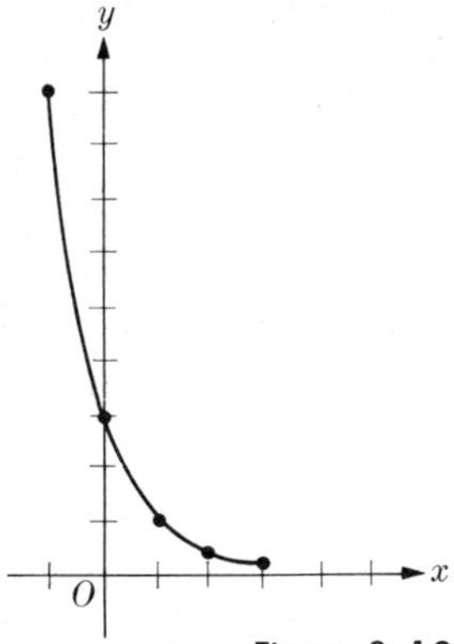

Figure 8–19

Suppose that $f(x) = a^x$, $a > 1$, is an exponential function. Since f is an increasing function (according to (i) above), we know from the results in Section 2 that the inverse of f is a function. The inverse of a^x is called the **logarithm to the base a.** We denote it by $\log_a$. In other words, we define the logarithm (or log) to the base a by the condition

$$y = \log_a x \Leftrightarrow x = a^y, \quad x > 0.$$

Since we showed in Section 2 that the inverse of an increasing function is an increasing function, it follows that $\log_a x$ is an increasing function of x (if $a > 1$). Selecting $f(x) = a^x$ in equations (2) and (3) of Section 2, we find that

$$a^{(\log_a x)} = x \qquad \text{if} \quad x > 0$$

and

$$\log_a (a^x) = x \qquad \text{for all real } x.$$

In words,

> *the logarithm of x to the base a is the power to which a must be raised to obtain x.*

If $a < 1$, then a^x is decreasing as x increases. The results of Section 2 show that the inverse is again a function and, in fact, a decreasing function. We still define the inverse as logarithm to the base a and denote it by $\log_a$.

The student who has already studied logarithms should be familiar with the following laws of logarithms and with their use in computation:

> If $a > 0$ and $a \neq 1$,
>
> (i) $\log_a (x \cdot y) = \log_a x + \log_a y,$
>
> (ii) $\log_a \left(\dfrac{x}{y}\right) = \log_a x - \log_a y,$
>
> (iii) $\log_a (x^y) = y \log_a x,$
>
> (iv) $\log_a 1 = 0,$
>
> (v) $\log_b x = \dfrac{\log_a x}{\log_a b}.$

These rules follow easily from the laws of exponents (1). For example, to prove (i), we let

$$u = \log_a x, \qquad v = \log_a y.$$

Then

$$u = \log_a x \Leftrightarrow x = a^u, \qquad v = \log_a y \Leftrightarrow y = a^v.$$

Therefore

$$x \cdot y = a^u \cdot a^v = a^{u+v} \Leftrightarrow u + v = \log_a (x \cdot y).$$

Hence

$$\log_a (x \cdot y) = u + v = \log_a x + \log_a y.$$

The proofs of laws (ii) through (v) are left to the student. (See problems 21, 22, and 23 at the end of this section.)

EXAMPLE 2. Plot a graph of the equation

$$y = 2 \log_3 x.$$

Solution. We have

$$y = 2 \log_3 x \Leftrightarrow \frac{y}{2} = \log_3 x \Leftrightarrow x = 3^{y/2}.$$

We first construct a table:

x	$\frac{1}{9}$	$\sqrt{3}/9$	$\frac{1}{3}$	$\sqrt{3}/3$	1	$\sqrt{3}$	3	$3\sqrt{3}$	9
x approx.	0.11	0.19	0.33	0.58	1.00	1.73	3.00	5.20	9.00
y	-4	-3	-2	-1	0	1	2	3	4

Then we draw the graph as shown in Fig. 8–20.

EXAMPLE 3. Compute: (i) $\log_3\left(\frac{1}{27}\right)$, (ii) $\log_{(1/2)} 4$.

Solution. (i) Let $y = \log_3\left(\frac{1}{27}\right)$. Then

$$\tfrac{1}{27} = 3^y \quad \text{and} \quad \tfrac{1}{27} = 3^{-3}$$

so

$$y = -3 = \log_3\left(\tfrac{1}{27}\right).$$

(ii) Let $y = \log_{(1/2)} 4$. Then

$$4 = \left(\frac{1}{2}\right)^y = \frac{1}{2^y}$$

$$\Leftrightarrow 2^y = \frac{1}{4} \Leftrightarrow y = -2 = \log_{(1/2)} 4.$$

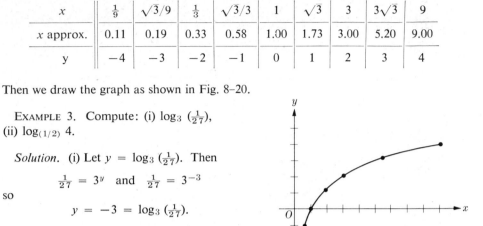

Figure 8–20

Since logarithms reduce problems in multiplication and division to those of addition and subtraction, they are used extensively in numerical computation. Logarithms to the base 10 are known as **common logarithms,** and many accurate tables of these logarithms are available for computational purposes. In calculus, a particular irrational number, for which the symbol e is used, plays an important role. It turns out that logarithms to the base e are quite useful in many calculus problems. Logarithms to this base are known as **natural logarithms.** The quantity e has the value 2.7, approximately; in fact, $e = 2.71828+$.

PROBLEMS

In problems 1 through 16, sketch graphs of the given equations.

1. $y = 5^x$
2. $y = \left(\frac{5}{2}\right)^x$
3. $y = \left(\frac{1}{3}\right)^x$
4. $y = \left(\frac{1}{2}\right)^{2x}$
5. $y = 3^{-x}$
6. $y = 2^{-2x}$
7. $y = 3^{x-1}$
8. $y = 2^{2-x}$
9. $y = 2^{x^2-1}$
10. $y = \log_3 x$
11. $y = \log_5 x$
12. $y = \log_{(1/3)} x$
13. $y = 2 \log_3 x$
14. $y = \frac{1}{2} \log_4 x$
15. $y = \log_2 3x$
16. $y = \log_2 (x - 2)$

In problems 17 through 20, find the values of the given expressions.

17. (a) $\log_2 16$, (b) $\log_9 81$, (c) $\log_2 (\frac{1}{8})$

18. (a) $\log_3 (\frac{1}{27})$, (b) $\log_5 25$, (c) $\log_4 16$

19. (a) $\log_{(1/3)} 1$, (b) $\log_6 (\frac{1}{36})$, (c) $\log_a a$ $(a > 0)$

20. (a) $\log_4 8$, (b) $\log_9 (\frac{1}{27})$, (c) $\log_4 32$

21. Prove that $\log_a (x/y) = \log_a x - \log_a y$ $(a, x, y$ positive, $a \neq 0)$.

22. Prove that $\log_a (x^y) = y \log_a x$ $(a > 0, x > 0, a \neq 1)$.

23. Prove that $\log_b x = (\log_a x)/(\log_a b)$ $(a > 0, b > 0, x > 0, a \neq 1, b \neq 1)$.

Using the result of problem 23, express the following quantities in terms of common logarithms (i.e., logs to the base 10).

24. (a) $\log_4 9$, (b) $\log_5 17$, (c) $\log_3 12$

25. (a) $\log_7 23$, (b) $\log_{(1/3)} 13$, (c) $\log_7 16$

Without using tables, find the values of the following, given that $\log_{10} 2 = 0.3010$ and $\log_{10} 3 = 0.4771$.

26. (a) $\log_{10} 25$, (b) $\log_{10} 40$

27. $\log_3 24$	28. $\log_5 60$	29. $\log_{(1/3)} 10$
30. $\log_5 20$	31. $\log_6 30$	

32. If $1000 is invested at 6% and the interest is compounded continuously, the resulting amount y at the end of x years is given by

$$y = 1000e^{0.06x}.$$

(a) Find, approximately, the amount accumulated after two years and compare this sum with the amount accumulated if the interest is compounded annually. [The amount accumulated by a sum P after n periods at i% interest per period is $P(1 + i/100)^n$.]

(b) Approximately how long will it take for the investment to double?

9 Parametric Equations, Polar Coordinates

▶ 1. PARAMETRIC EQUATIONS

Let us consider the two equations

$$x = t^2 + 2t, \qquad y = t - 2. \tag{1}$$

Each number t determines a value of x and one of y and hence a point (x, y) in R_2. The totality of these points forms a relation. The graph of this relation may be exhibited by the following method. We set up a table of values for t, x, and y by letting t take on various values and then computing x and y from the given equations. The points (x, y) are plotted in the usual way on a Cartesian coordinate system. The table is shown below and the graph is sketched in Fig. 9–1. The t

t	0	1	2	3	4	-1	-2	-3	-4
x	0	3	8	15	24	-1	0	3	8
y	-2	-1	0	1	2	-3	-4	-5	-6

scale is completely separate and does not appear in the graph. The quantity t is called a **parameter,** and the equations (1) are called the parametric equations of the locus shown in Fig. 9–1. We recall that in Section 6 of Chapter 3 we derived parametric equations for a straight line. In general, if f and g are functions and we write

$$x = f(t), \qquad y = g(t) \tag{2}$$

then we say that the two equations form a set of **parametric equations.** (If f and g are linear, they represent a straight line.) We denote the domain of f by S_1 and that of g by S_2. Then both equations in (2) have meaning only when t is in the intersection of

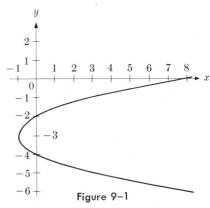

Figure 9–1

182

S_1 and S_2. The domain of the set of parametric equations (2) is the set $S = S_1 \cap S_2$. The locus of (2) is the set of points in the xy plane which result when t takes on all values in S.

Sometimes it is possible to *eliminate the parameter* by solving one of the equations for t and then substituting into the other. In the example we considered, we have

$$t = y + 2$$

and, substituting this value for t in the equation for x in terms of t, we obtain

$$x = (y + 2)^2 + 2(y + 2) = y^2 + 6y + 8.$$

We recognize the equation $x = y^2 + 6y + 8$ as that of a parabola, as indeed Fig. 9–1 shows.

The equations

$$x = t^7 + 3t^2 - 1, \qquad y = 2^t + 2t^2 - 3\sqrt{t},$$

are a pair of parametric equations. Because of the square-root sign, it is clear that the domain for t is restricted to nonnegative real numbers. It is not possible to eliminate the parameter in any simple way, since solving either equation for t gives the appearance of being a herculean task.

EXAMPLE 1. Plot the locus of the equations

$$x = t^2 + 2t - 1, \qquad y = t^2 + t - 2.$$

Eliminate the parameter, if possible.

Solution. We construct the following table:

t	-4	-3	-2	-1	0	1	2	3
x	7	2	-1	-2	-1	2	7	14
y	10	4	0	-2	-2	0	4	10

By plotting these points, we obtain the curve shown in Fig. 9–2. To eliminate the parameter, we first subtract the equations, obtaining

$$x - y = t + 1 \qquad \text{and} \qquad t = x - y - 1.$$

Substituting for t in the second of the given equations, we find that

$$y = (x - y - 1)^2 + (x - y - 1) - 2,$$

and, simplifying,

$$x^2 - 2xy + y^2 - x - 2 = 0.$$

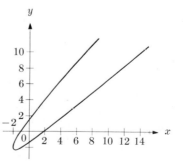

Figure 9–2

EXAMPLE 2. Plot the locus of the equations

$$x = 4 \cos \theta, \qquad y = 3 \sin \theta,$$

and eliminate the parameter, if possible.

Solution. In this problem the parameter is θ. While the domain for θ consists of all real numbers, only those values between 0 and 2π are needed, since both functions have period 2π. Rather than set up a table of values for θ, x, and y, we find it simpler first to write

$$\cos \theta = \frac{x}{4}, \qquad \sin \theta = \frac{y}{3}.$$

Then, by squaring and adding, we get

$$1 = \cos^2 \theta + \sin^2 \theta = \frac{x^2}{16} + \frac{y^2}{9},$$

which we recognize to be an ellipse, as shown in Fig. 9–3.

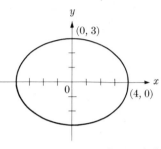

Figure 9–3

EXAMPLE 3. A projectile moves approximately according to the law

$$x = (v_0 \cos \alpha)t, \qquad y = (v_0 \sin \alpha)t - 16t^2,$$

where v_0 and α are constants and t is the time, in seconds, after the projectile is fired. The equations give the Cartesian coordinates x, y (in feet) of the center of the projectile in the vertical plane of motion with the muzzle of the gun at the origin of the coordinate system, the x axis horizontal, and the y axis vertical; v_0 is the muzzle velocity, i.e., the velocity of the projectile at the instant it leaves the gun; α is the angle of inclination of the projectile as it leaves the gun (Fig. 9–4). If $v_0 = 100$ ft/sec, $\cos \alpha = \frac{3}{5}$, and $\sin \alpha = \frac{4}{5}$, what are the coordinates of the center of the projectile at times $t = 1, 2, 3, 4$, and 5? Find the time T when the projectile hits the ground, and find the distance R from the muzzle of the gun to the place where the projectile strikes the ground.

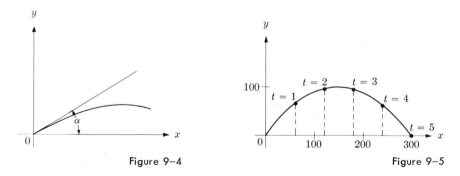

Figure 9–4   Figure 9–5

Solution. We have $v_0 \cos \alpha = 60$, $v_0 \sin \alpha = 80$, and so

$$x = 60t, \qquad y = 80t - 16t^2.$$

The projectile hits the ground when $y = 0$. Therefore $0 = 80t - 16t^2$ and $t = 0, 5$.

Consequently, $T = 5$. To find R, we insert $t = 5$ in $x = 60t$, obtaining $R = 300$ ft. We construct the table:

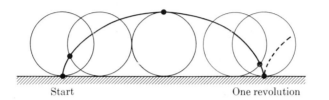

t	0	1	2	3	4	5
x	0	60	120	180	240	300
y	0	64	96	96	64	0

Plotting the points yields the curve shown in Fig. 9–5.

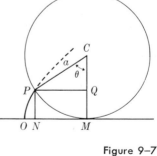

Start   One revolution   Figure 9–6

A circular hoop starts rolling along a stretch of level ground. A point on the rim of the hoop has a mark on it. We wish to find the path traced out by the marked point. Figure 9–6 shows the hoop in a number of different positions as it rolls along. The curve traced out by the marked point, which we denote by P, may be expressed in terms of parametric equations. Let a be the radius of the hoop, and suppose that when the hoop begins rolling the point P is on the ground at the point labeled O in Fig. 9–7. Figure 9–7 shows the position of the point P after the hoop has turned through an angle θ. Since the hoop is assumed to roll without slipping, we have, from the diagram, $\overline{OM} = \text{arc } \widehat{MP} = a\theta$, θ in radians. From $\triangle CPQ$ we read off

$$\overline{PQ} = a \sin \theta, \qquad 0 \leq \theta \leq \frac{\pi}{2},$$

$$\overline{QC} = a \cos \theta, \qquad 0 \leq \theta \leq \frac{\pi}{2}.$$

Denoting the coordinates of P by (x, y) we see that

$$x = \overline{ON} = \overline{OM} - \overline{NM} = \text{arc } \widehat{MP} - \overline{PQ}$$

$$= a\theta - a \sin \theta,$$

$$y = \overline{NP} = \overline{MC} - \overline{QC} = a - a \cos \theta.$$

Figure 9–7

Although these equations were derived for θ between 0 and $\pi/2$, it can be shown that for all values of θ the parametric equations

$$x = a(\theta - \sin \theta), \qquad y = a(1 - \cos \theta)$$

represent the path of the marked point on the rim. This curve is called a **cycloid.**

The cycloid is a particularly good example of a curve which is obtained without too much difficulty by use of parametric equations, while any attempt to find the relation between x and y without resorting to a parameter would lead to an almost unsurmountable problem.

PROBLEMS

In problems 1 through 16, plot the curves as in Example 1; in each case eliminate the parameter and get a relation between x and y.

1. $x = 3t, y = -4t$

2. $x = t + 1, y = t^2 - 1$

3. $x = t + 1, y = \dfrac{2}{t + 1}$

4. $x = 1 - s^2, y = -2 + \dfrac{1}{4} s^3$

5. $x = 1 + 3 \cos \theta, y = -2 + 3 \sin \theta$

6. $x = 2 \cos \theta, y = 3 \sin \theta$

7. $x = 4 \sin^3 \theta, y = 4 \cos^3 \theta$

8. $x = 3 \tan 2t, y = 4 \sec 2t$

9. $x = -2 + 5 \sec t, y = 3 + 3 \tan t$

10. $x = 3 + 4 \sin (2t), y = 1 + 3 \cos (2t)$

11. $x = t^2 - t - 2, y = t^2 + t + 2$

12. $x = t^2 + 2t - 3, y = 2t^2 + t - 1$

13. $x = \dfrac{20t}{4 + t^2}$,

$y = \dfrac{5(4 - t^2)}{4 + t^2}$

14. $x = \dfrac{(3 - t)^2(3 + t)}{3(t^2 + 3)}$,

$y = \dfrac{(3 - t)(3 + t)^2}{3(t^2 + 3)}$

15. $x = 3^t, y = 3^{1-t}$

16. $x = 2^{2t}, y = 2^{-3t}$

17. Assuming that the equations of Example 3 hold, and given that $v_0 = 500$ ft/sec, $\cos \alpha = \frac{4}{5}$, $\sin \alpha = \frac{3}{5}$, find x and y for $t = 2, 4, 6, 8,$ and 10. Also find T and R and plot the trajectory.

18. Assuming that the equations of Example 3 hold, find T and R in terms of v_0 and α. Find y in terms of x (i.e., eliminate the parameter).

19. A wheel of radius a rolls without slipping along a level stretch of ground. A point on one of the spokes of the wheel is marked. Find the path traced by this point if it is at a distance b from the center. This curve is called a **prolate cycloid**.

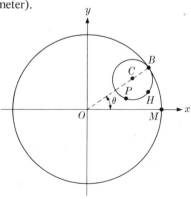

20. Find the locus of a point P on a circle of radius a which rolls without slipping on the inside of a circle of radius $4a$. Choose the center of the large circle as origin, choose the positive x axis through a point where P touches the large circle, and choose the parameter θ as the angle xOC, where C is the center of the small circle (Fig. 9–8).

Figure 9–8

(*Hint:* Note that arc $\overset{\frown}{BHP}$ = arc $\overset{\frown}{MB}$.)

Answer: $x = 3a \cos \theta + a \cos 3\theta$, $y = 3a \sin \theta - a \sin 3\theta$. This curve is called a **hypocycloid.**

21. A string is wound about a circle of radius a. The path traced by the end of the string as it is unwound is called the **involute of the circle.** Refer to Fig. 9–9 and show that the equations of the involute are

$$x = a \cos \theta + a \theta \sin \theta$$

$$y = a \sin \theta - a \theta \cos \theta$$

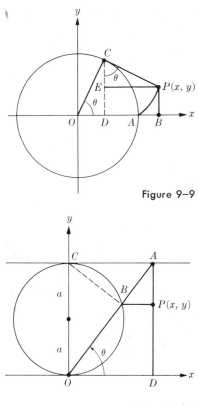

Figure 9–9

22. Plot the graph of the equations $x = a \sin \theta$, $y = a \tan \theta(1 + \sin \theta)$. Eliminate the parameter and show that the line $x = a$ is an asymptote. This curve is called a **strophoid.**

23. A circle of radius a is drawn tangent to the x axis as shown in Fig. 9–10. The line OA intersects the circle at point B. Then the projection of AB on the vertical line through A is the segment AP. The locus of the point P as B moves around the circle is called the **Witch of Agnesi.** Show that the curve is given by $x = 2a \cot \theta$, $y = 2a \sin^2 \theta$. Eliminate the parameter and show that the x axis is an asymptote.

Figure 9–10

▶2. POLAR COORDINATES

A coordinate system in the plane allows us to associate an ordered pair of numbers with each point in the plane. Until now we have considered Cartesian coordinate systems exclusively. However, there are many problems for which other systems of coordinates are more advantageous than are Cartesian coordinates. With this in mind we now describe the system known as **polar coordinates.**

We begin by selecting a point in the plane which we call the **pole** or **origin** and label it O. From this point we draw a half-line starting at the pole and extending indefinitely in one direction. This line is usually drawn horizontally and to the right of the pole, as shown in Fig. 9–11. It is called the **initial line** or **polar axis.**

Let P be any point in the plane. Its position will be determined by its distance from the pole and by the angle that the line OP makes with the initial line. We measure angles θ from the initial line as in trigonometry—positive in a counterclockwise direction and negative in a clockwise direction. The distance r from the origin to the point P will be taken as positive. The coordinates of P (Fig. 9–12) in the polar coordinate system are (r, θ). There is a sharp distinction

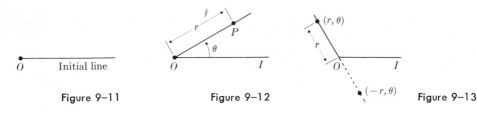

Figure 9–11   Figure 9–12   Figure 9–13

between Cartesian and polar coordinates, in that a point P may be represented in just one way by a pair of Cartesian coordinates, but it may be represented in many ways by polar coordinates. For example, the point Q with polar coordinates $(2, \pi/6)$ also has polar coordinates

$$\left(2, 2\pi + \frac{\pi}{6}\right), \quad \left(2, 4\pi + \frac{\pi}{6}\right), \quad \left(2, 6\pi + \frac{\pi}{6}\right),$$

$$\left(2, -2\pi + \frac{\pi}{6}\right), \quad \left(2, -4\pi + \frac{\pi}{6}\right), \quad \text{etc.}$$

In other words, there are infinitely many representations of the same point. Furthermore, it is convenient to allow r, the distance from the origin, to take on negative values. We establish the convention that a pair of coordinates such as $(-3, \theta)$ is simply another representation of the point with coordinates $(3, \theta + \pi)$. Figure 9–13 shows the relationship of the points (r, θ) and $(-r, \theta)$.

EXAMPLE 1. Plot the points whose polar coordinates are

$$P\left(3, \frac{\pi}{3}\right), \quad Q\left(-2, \frac{2\pi}{3}\right), \quad R\left(-2, \frac{\pi}{4}\right), \quad S\left(2, \frac{3\pi}{4}\right), \quad T\left(3, -\frac{\pi}{6}\right).$$

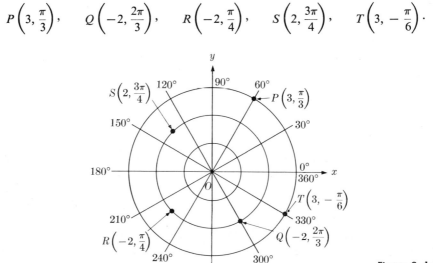

Figure 9–14

Solution. The points are plotted in Fig. 9–14.

It is important to know the connection between Cartesian and polar coordinate systems. To find this relationship, let us consider a plane with one system superimposed on the other in such a way that the origin of the Cartesian system is at the pole and the positive x axis coincides with the initial line (Fig. 9–15). The relationship between the Cartesian coordinates (x, y) and the polar coordinates (r, θ) of a point P is given by the equations

$$x = r \cos \theta, \qquad y = r \sin \theta.$$

When we are given r and θ, these equations tell us how to find x and y. We also have the formulas

$$r = \pm\sqrt{x^2 + y^2},$$

$$\tan \theta = \frac{y}{x},$$

which give us r and θ when the Cartesian coordinates are known.

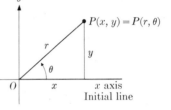

Figure 9–15

EXAMPLE 2. The Cartesian coordinates of a point are $(\sqrt{3}, -1)$. Find a set of polar coordinates for this point.

Solution.   We have

$$r = \sqrt{3 + 1} = 2 \qquad \text{and} \qquad \tan \theta = -1/\sqrt{3}.$$

Since the point is in the fourth quadrant, we select for θ the value $-\pi/6$ (or $11\pi/6$). The answer is $(2, -\pi/6)$.

PROBLEMS

In problems 1 through 4, the polar coordinates of points are given. Find the Cartesian coordinates of the same points and plot them on graphs.

1. $(3, \pi/4)$, $(4, 2\pi/3)$, $(4, \pi)$, $(2, 0)$, $(-3, \pi)$

2. $(2, \pi/3)$, $(3, \pi/4)$. $(4, \pi/2)$, $(2, 3\pi/2)$, $(-2, 4\pi/3)$

3. $(-2, 0)$, $(3, -\pi/4)$, $(2, -\pi/6)$, $(-4, 5\pi/6)$, $(0, \pi)$

4. $(3, -\pi/3)$, $(-2, -\pi/2)$, $(4, -3\pi/2)$, $(-3, -\pi/3)$, $(0, 3\pi/4)$

In problems 5 through 8, the Cartesian coordinates of points are given. Find a pair of polar coordinates for each of the points and plot on graphs.

5. $(2, 2\sqrt{3})$, $(0, 2)$, $(-\sqrt{3}, 1)$, $(0, -2)$, $(3, 0)$

6. $(-\sqrt{3}, -1)$, $(-3, 0)$, $(1, \sqrt{3})$, $(2, -2)$, $(0, 4)$

7. $(-2, 2)$, $(3, -\sqrt{3})$, $(\sqrt{3}, 1)$, $(2, 2\sqrt{3})$, $(-2, 2\sqrt{3})$

8. $(3, 0)$, $(0, 0)$, $(4, 4)$, $(-\sqrt{2}, \sqrt{6})$, $(\frac{5}{2}, -\frac{5}{2})$

9. Describe the locus of all points which, in polar coordinates, satisfy the condition $r = 5$; do the same for the condition $\theta = \pi/3$ and for the condition $\theta = -5\pi/6$. What can be said about the angle of intersection of the curve $r = $ const, with $\theta = $ const?

10. Find the distance between the points with polar coordinates $(3, \pi/4)$, $(2, \pi/3)$.

11. Find the distance between the points with polar coordinates $(1, \pi/2)$, $(4, 5\pi/6)$.

12. Find a formula for the distance between the points (r_1, θ_1) and (r_2, θ_2).

▶3.   GRAPHS IN POLAR COORDINATES

Suppose that r and θ are connected by some equation such as $r = 3\cos 2\theta$ or $r^2 = 4\sin 3\theta$. *We define the* **locus of an equation in polar coordinates** (r, θ) *as the set of all points P each of which has at least one pair of polar coordinates* (r, θ) *which satisfies the given equation.* To plot the locus of such an equation we must find all ordered pairs (r, θ) which satisfy the given equation and then plot the points obtained. We can obtain a good approximation of the locus in polar coordinates, as we can in Cartesian coordinates, by making a sufficiently complete table of values, plotting the points, and connecting them by a smooth curve.

EXAMPLE 1. Draw the graph of the equation $r = 3\cos\theta$.

Solution. We construct the table:

θ	0	$\pi/6$	$\pi/3$	$\pi/2$	$2\pi/3$	$5\pi/6$
$\theta°$	0	30	60	90	120	150
r	3	$\frac{3}{2}\sqrt{3}$	$\frac{3}{2}$	0	$-\frac{3}{2}$	$-\frac{3}{2}\sqrt{3}$
r approx.	3.00	2.60	1.50	0	-1.50	-2.60

θ	π	$7\pi/6$	$4\pi/3$	$3\pi/2$	$5\pi/3$	$11\pi/6$
$\theta°$	180	210	240	270	300	330
r	-3	$-\frac{3}{2}\sqrt{3}$	$-\frac{3}{2}$	0	$\frac{3}{2}$	$\frac{3}{2}\sqrt{3}$
r approx.	-3.00	-2.60	-1.50	0	1.50	2.60

The graph is shown in Fig. 9-16. It can be shown to be a circle with center at $(\frac{3}{2}, 0)$ and radius $\frac{3}{2}$. It is worth noting that even though there are 12 entries in the table, only 6 points are plotted. The curve is symmetric with respect to the initial line, and we could have saved effort if this fact had been taken into account. Since cos $(-\theta) = \cos \theta$ for all values of θ, we could have obtained the points on the graph for values of θ between 0 and $-\pi$ without extra computation.

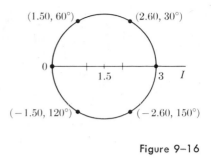

Figure 9–16

We shall set forth the rules of symmetry, as we did in Chapter 2, Section 3. These rules, which are useful as aids in graphing, are described in terms of symmetries with respect to x and y axes in Cartesian coordinates. That is, we suppose that the positive x axis coincides with the initial line of the polar coordinate system.

RULE I. *If the substitution of $(r, -\theta)$ for (r, θ) yields the same equation, the locus is symmetric with respect to the x axis.*

RULE II. *If the substitution of $(r, \pi - \theta)$ for (r, θ) yields the same equation, the locus is symmetric with respect to the y axis.*

RULE III. *If the substitution of $(-r, \theta)$ or of $(r, \theta + \pi)$ for (r, θ) yields the same equation, the locus is symmetric with respect to the pole.*

It is true, as we have already seen, that if any two of the three **symmetries** hold, the remaining one holds automatically. However, it is possible for a locus to have certain symmetry properties which the rules above will fail to exhibit.

EXAMPLE 2. Discuss for symmetry and plot the locus of $r = 3 + 2 \cos \theta$.

Solution. The locus is symmetric with respect to the x axis, since cos $(-\theta) = \cos \theta$ and Rule I applies. We construct the table:

θ	0	$\pm\pi/6$	$\pm\pi/3$	$\pm\pi/2$	$\pm2\pi/3$	$\pm5\pi/6$	$\pm\pi$
$\theta°$	0	$\pm30°$	$\pm60°$	$\pm90°$	$\pm120°$	$\pm150°$	$\pm180°$
r	5	$3 + \sqrt{3}$	4	3	2	$3 - \sqrt{3}$	1
r approx.	5.00	4.73	4.00	3.00	2.00	1.27	1.00

We use values of θ from $-\pi$ to $+\pi$ only, since cos $(\theta + 2\pi) = \cos \theta$ and no new points could possibly be obtained with higher values of θ. The graph, called a **limaçon**, is shown in Fig. 9–17.

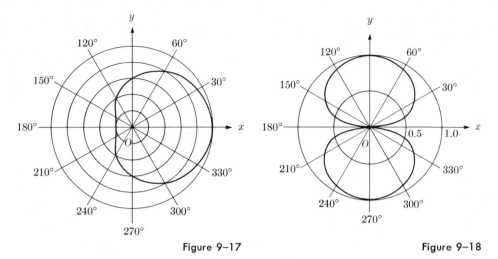

Figure 9–17   Figure 9–18

EXAMPLE 3. Given that $r^2 = \sin \theta$, discuss for symmetry and plot the locus of the equation.

Solution. When we replace r by $-r$ we get the same equation and, therefore, by Rule III, the locus is symmetric with respect to the pole. If we replace θ by $\pi - \theta$ we see that $\sin(\pi - \theta) = \sin \theta$ and, by Rule II, the locus is symmetric with respect to the y axis. Since two of the three *symmetries* hold, the curve is also symmetric with respect to the x axis. But notice that *Rule I does not hold* since replacing (r, θ) by $(r, -\theta)$ leads to the equation

$$r^2 = - \sin \theta,$$

which is different from the original equation (but turns out to have the same locus). If $\sin \theta$ is negative we have no locus, and so we must restrict θ to the interval $0 \le \theta \le \pi$. We make the table:

θ	0	$\pi/6$	$\pi/3$	$\pi/2$	$2\pi/3$	$5\pi/6$	π
$\theta°$	0°	30°	60°	90°	120°	150°	180°
r	0	$\pm\sqrt{2}/2$	$\pm\sqrt{\sqrt{3}/2}$	± 1	$\pm\sqrt{\sqrt{3}/2}$	$\pm\sqrt{2}/2$	0
r approx.	0	± 0.71	± 0.93	± 1	± 0.93	± 0.71	0

The graph is shown in Fig. 9–18.

EXAMPLE 4. Test for symmetry and plot the graph of the equation $r = 2 \cos 2\theta$.

Solution. If we replace θ by $-\theta$ the equation is unchanged and, therefore, by Rule I, we have symmetry with respect to the x axis. When we replace θ by $\pi - \theta$ we get

$$\cos 2(\pi - \theta) = \cos 2\pi \cos 2\theta + \sin 2\pi \sin 2\theta = \cos 2\theta$$

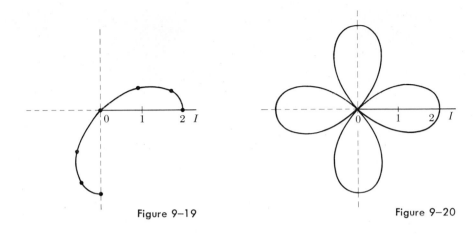

Figure 9–19   Figure 9–20

and, by Rule II, the locus is symmetric with respect to the y axis. Consequently, the locus is symmetric with respect to the pole. We construct the table:

θ	0	$\pi/12$	$\pi/6$	$\pi/4$	$\pi/3$	$5\pi/12$	$\pi/2$
2θ	0	$\pi/6$	$\pi/3$	$\pi/2$	$2\pi/3$	$5\pi/6$	π
$2\theta°$	0°	30°	60°	90°	120°	150°	180°
r	2	$\sqrt{3}$	1	0	-1	$-\sqrt{3}$	-2
r approx.	2.00	1.73	1.00	0	-1.00	-1.73	-2.00

We plot the points as shown in Fig. 9–19. Now, making use of the symmetries, we can easily complete the graph (Fig. 9–20). This curve is called a four-leaved rose. Equations of the form

$$r = a \sin n\theta, \qquad r = a \cos n\theta,$$

where n is a positive integer, have loci which are called **rose** or **petal** curves. The number of petals is equal to n if n is an odd integer and is equal to $2n$ if n is an even integer. If $n = 1$, there is one petal and it is circular.

The loci of equations of the form

$$r = a \pm b \cos \theta \qquad \text{or} \qquad r = a \pm b \sin \theta$$

are called **limaçons**; the locus in Example 2 is a limaçon. In the special cases where $a = b$, the loci are called **cardioids** (see problems 6, 7, and 8 below). The loci of equations of the form

$$r^2 = a^2 \cos 2\theta \qquad \text{or} \qquad r^2 = a^2 \sin 2\theta$$

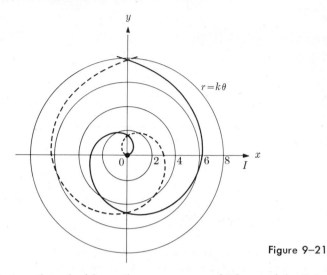

Figure 9–21

are called **lemniscates**; these loci have the appearance of "figure eights" (see problems 15 and 16 below). Polar coordinates are particularly adapted to the study of certain curves called **spirals**. The so-called **spiral of Archimedes** has an equation of the form

$$r = k\theta$$

and its graph is drawn in Fig. 9–21 (the dashed portion arises from negative values of θ). The **logarithmic spiral** has an equation of the form

$$\log_b r = \log_b a + k\theta \quad \text{or} \quad r = a \cdot b^{k\theta}$$

and its graph is drawn in Fig. 9–22.

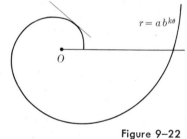

Figure 9–22

PROBLEMS

In problems 1 through 32, discuss for symmetry and plot the loci of the equations.

1. $r = 4 \cos \theta$
2. $r = -2 \sin \theta$
3. $r = 3 \cos (\theta + \pi/3)$
4. $r = 3 \sin (\theta - \pi/3)$
5. $r = 3(1 + \cos \theta)$
6. $r = 4(1 - \sin \theta)$
7. $r = 4(1 - \cos \theta)$
8. $r = 3 - \cos \theta$
9. $r = 4 + 2 \sin \theta$
10. $r = 2 + 3 \cos \theta$
11. $r = 3 - 5 \cos \theta$
12. $r \cos \theta = 4$
13. $r \sin \theta = 2$
14. $r^2 = 4 \cos \theta$
15. $r^2 = 4 \cos 2\theta$
16. $r = 4 \cos 2\theta$
17. $r = 3 \sin 2\theta$
18. $r^2 = 6 \sin 2\theta$
19. $r = 4 \cos 3\theta$
20. $r = 4 \cos^2 \frac{1}{2}\theta$
21. $r = 3 \sin \frac{1}{2}\theta$
22. $r = 2 \tan \theta$
23. $r = 6 \sin 4\theta$
24. $r^2 = \cot \theta$

25. $r = 4 \csc \theta$ 26. $r = -3 \sec \theta$ 27. $r^2 = 4 \sec \theta$

28. $r(3 - 2 \cos \theta) = 6$ 29. $r(1 + \cos \theta) = 3$ 30. $r(2 + 3 \cos \theta) = 4$

31. $r(1 + 2 \sin \theta) = -3$ 32. $r^2 = 4 \cos \theta$

In each of the problems 33 through 40, sketch the curves and find the coordinates of the points of intersection, if any.

33. $r = 3 \cos \theta, r = 2 \sin \theta$

34. $r = \sin \theta, r = \sin 2\theta$

35. $r = 2 \cos \theta, r \cos \theta = 1$

36. $r = \cos \theta, r^2 = \cos 2\theta$

37. $r = \tan \theta, r = 2 \sin \theta$

38. $r = 2 + \cos \theta, r = 6 \cos \theta$

39. $r = \sin 2\theta, r = \sin 4\theta$

40. $r = 1 + \cos \theta, r = 1 + \sin \theta$

▶4. EQUATIONS IN CARTESIAN AND POLAR COORDINATES

The curves discussed in Section 3 find their natural setting in polar coordinates. However, the equations of some curves may be simpler in appearance or their properties more transparent in one coordinate system than in another. For this reason it is useful to be able to transform an equation given in one coordinate system into the corresponding equation in another system.

Suppose that we are given the equation of a curve in the form $y = f(x)$ in a Cartesian system. If we simply make the substitution

$$x = r \cos \theta, \qquad y = r \sin \theta,$$

we have the equation of the same curve in a polar coordinate system. We can also go the other way, so that if we are given $r = g(\theta)$ in polar coordinates, the substitution

$$r^2 = x^2 + y^2, \qquad \theta = \arctan \frac{y}{x}, \qquad \left(\text{i.e., } \tan \theta = \frac{y}{x}\right)$$

transforms the relation into one in Cartesian coordinates. In this connection, it is frequently useful to make the substitutions

$$\sin \theta = \frac{y}{\sqrt{x^2 + y^2}}, \qquad \cos \theta = \frac{x}{\sqrt{x^2 + y^2}}, \qquad \tan \theta = \frac{y}{x},$$

rather than use the formula for θ itself.

EXAMPLE 1. Find the equation in polar coordinates which corresponds to the locus of the equation

$$x^2 + y^2 - 3x = 0$$

in Cartesian coordinates.

Solution. Substituting $x = r \cos \theta$, $y = r \sin \theta$, we obtain

$$r^2 - 3r \cos \theta = 0 \Leftrightarrow r(r - 3 \cos \theta) = 0.$$

Therefore the locus is $r = 0$ or $r - 3 \cos \theta = 0$. Because the pole is part of the locus of $r - 3 \cos \theta = 0$ (since $r = 0$ when $\theta = \pi/2$), the result is

$$r = 3 \cos \theta.$$

We recognize this locus as the circle of Example 1, Section 2. (In Cartesian coordinates, the equation is $(x - \frac{3}{2})^2 + y^2 = \frac{9}{4}$.)

EXAMPLE 2. Given the polar coordinate equation

$$r = \frac{1}{1 - \cos \theta},$$

find the corresponding equation in Cartesian coordinates.

Solution. We write

$$r - r \cos \theta = 1 \Leftrightarrow r = 1 + r \cos \theta.$$

Substituting for r and $\cos \theta$, we get

$$\pm\sqrt{x^2 + y^2} = 1 + x.$$

Squaring both sides (a dangerous operation, since this may introduce extraneous solutions), we obtain

$$x^2 + y^2 = 1 + 2x + x^2 \quad \text{or} \quad y^2 = 2x + 1.$$

When we squared both sides we introduced the extraneous locus $r = -(1 + r \cos \theta)$. This locus is the same as the locus of the equation

$$r = \frac{-1}{1 + \cos \theta}$$

as is seen by solving for r. But this last equation can be put in the form

$$(-r) = \frac{1}{1 - \cos (\theta + \pi)},$$

which is obtained from the original by replacing (r, θ) by $(-r, \theta + \pi)$. Since we have seen that (r, θ) and $(-r, \theta + \pi)$ are polar coordinates of the same point, the extraneous locus coincides with the original.

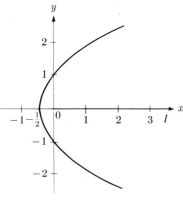

Figure 9–23

Therefore the correct result is $y^2 = 2x + 1$. The graph is drawn in Fig. 9–23.

PROBLEMS

In problems 1 through 12, in each case find a polar coordinate equation of the locus whose corresponding equation is given in Cartesian coordinates. Sketch.

1. $x = 4$ 2. $y = 3$ 3. $x - y = 0$

4. $3x + 2y = 0$ 5. $2x + 2\sqrt{3}\,y = 4$ 6. $x + y = 6$

7. $xy = 5$ 8. $y^2 = 4x$ 8. $x^2 = 2y$

10. $x^2 + y^2 - 2x = 0$ 11. $x^2 + y^2 - 4x + 2y = 0$

12. $x^2 + y^2 - 2x + 2y - 6 = 0$

In problems 13 through 32, in each case find a polynomial equation in Cartesian coordinates whose locus corresponds to the equation given in polar coordinates. Discuss possible extraneous solutions. Sketch.

13. $r = 4$ 14. $\theta = 2\pi/3$ 15. $r = 5\cos\theta$

16. $r = 2\sin\theta$ 17. $r\cos\theta + 2 = 0$ 18. $r\sin\theta = 2$

19. $r\cos(\theta + \pi/3) = 4$ 20. $r\cos(\theta - \pi/4) = 4$ 21. $r^2\cos 2\theta = 6$

22. $r^2\sin 2\theta = -4$ 23. $r^2 = 2\cos 2\theta$ 24. $r^2 = 2\sin 2\theta$

25. $r = 2\sec\theta\tan\theta$ 26. $r = 3\csc\theta\cot\theta$ 27. $r(3 - 2\cos\theta) = 4$

28. $r = 4\cos 3\theta$ 29. $r = 3\sin 3\theta$ 30. $r = 2 - 2\cos\theta$

31. $r = 1 + 2\cos\theta$ 32. $r = 4\sin 2\theta$

▶5.   STRAIGHT LINES, CIRCLES, AND CONICS

The straight line was discussed in detail in Chapter 3. We found that every straight line has an equation (in Cartesian coordinates) of the form

$$Ax + By + C = 0.$$

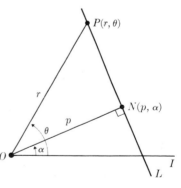

To obtain the equation in polar coordinates we can simply substitute $x = r\cos\theta$, $y = r\sin\theta$. However, it is also possible to find the desired equation directly. In Fig. 9–24, we consider a line L (not passing through the pole), and we draw a perpendicular from the pole to L. Call the point of intersection N and suppose that the coordinates of N are (p, α). The quantities p and

Figure 9–24

α are sufficient to determine the line L completely. To see this, we let $P(r, \theta)$ be any point on L. and we note, according to Fig. 9–24, that $\angle PON = \theta - \alpha$. Therefore we have

$$r\cos(\theta - \alpha) = p.$$

When α and p are given, this is the equation of a straight line. The student may verify that the same equation results regardless of the positions of L and p. The above equation can also be written as

$$r \cos \theta \cos \alpha + r \sin \theta \sin \alpha = p,$$

and changing to rectangular coordinates yields

$$x \cos \alpha + y \sin \alpha = p,$$

which is known as the **normal form of the equation of a straight line.** Whenever the coefficients of x and y are the cosine and sine of an angle, respectively, the constant term gives the distance from the origin. Any linear equation may be put in normal form. For example, $3x + 4y = 7$ may be written (upon division by $\sqrt{3^2 + 4^2} = 5$) as

$$\tfrac{3}{5}x + \tfrac{4}{5}y = \tfrac{7}{5}.$$

This equation is now in normal form, and the distance from the origin to the line is $\tfrac{7}{5}$ units. (Note that this procedure is the same as the one using the formula for the distance from a point to a line, as given in Chapter 3, Section 7).

A line through the pole has $p = 0$, and its polar coordinate equation is

$$\cos(\theta - \alpha) = 0.$$

This is equivalent to the statement

$$\theta = \text{const},$$

a condition which we already know represents straight lines passing through the pole.

We shall employ a direct method for obtaining the equation of a circle in polar coordinates. Let the circle have radius a and center at the point C with coordinates (c, α), as shown in Fig. 9–25. If $P(r, \theta)$ is any point on the circle, then in $\triangle POC$, side $|OP| = r$, side $|OC| = c$, side $|PC| = a$, and $\angle POC = \theta - \alpha$. We use the law of cosines in triangle POC, obtaining

$$r^2 - 2cr \cos(\theta - \alpha) + c^2 = a^2.$$

If c, α, and a are given, the equation of the circle is completely determined. It is instructive to draw a figure similar to 9–25 but with $a > c$ and to repeat the proof with $P(r, \theta)$ in various positions.

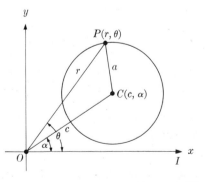

Figure 9–25

EXAMPLE 1. A circle has center at $(5, \pi/3)$ and radius 2. Find its equation in polar coordinates.

Solution. We have $c = 5, \alpha = \pi/3, a = 2$. Therefore

$$r^2 - 10r \cos\left(\theta - \frac{\pi}{3}\right) + 25 = 4$$

$$\Leftrightarrow r^2 - 10r \cos\left(\theta - \frac{\pi}{3}\right) + 21 = 0.$$

If a circle passes through the origin, so that $c = \pm a$, the equation reduces to

$$r = 2c \cos(\theta - \alpha).$$

(With $c = \frac{3}{2}$ and $\alpha = 0$, we get the equation of Example 1, Section 3.)

We now recall the focus-directrix definition of parabolas, ellipses, and hyperbolas, as given in Chapter 6. It states that *the locus of a point which moves so that the ratio of its distance from a fixed point (the focus) to its distance from a fixed line (the directrix) remains constant is*

  a parabola   if the ratio is 1,

  an ellipse   if the ratio is between 0 and 1,

  a hyperbola if the ratio is larger than 1.

Polar coordinates are particularly suited to conics, since all three types have equations of the same form in this system. To derive the equation, we place the focus at the pole and let the directrix L be a line perpendicular to the initial line and p units to the left of the pole (Fig. 9–26). If $P(r, \theta)$ is a point on the locus, then the conditions say that $|OP|/|PD| = $ a constant which we call the eccentricity, e, and we write (since $|OP| = r$),

$$r = e|PD|.$$

We see from Fig. 9–26 that

$$|PD| = |PE| + |ED| = r \cos\theta + p.$$

Therefore we have

$$r = er \cos\theta + ep,$$

or

$$r = \frac{ep}{1 - e \cos\theta}. \qquad (1)$$

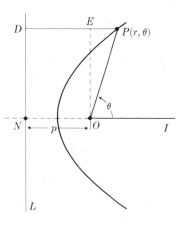

Equation (1) represents an ellipse if $0 < e < 1$, a parabola if $e = 1$, and a hyperbola if $e > 1$.

Figure 9–26

EXAMPLE 2. Find the equation of the conic with focus at the origin, eccentricity $\frac{1}{2}$ and directrix the line $r \cos \theta = -3$.

Solution. We have $p = 3$, and so

$$r = \frac{\frac{1}{2}(3)}{1 - \frac{1}{2} \cos \theta} = \frac{3}{2 - \cos \theta}.$$

The curve is an ellipse.

EXAMPLE 3. Given the conic $r = 7/(4 - 5 \cos \theta)$, find the eccentricity and locate the directrix. Identify the conic.

Solution. We write the equation in the form

$$r = \frac{\frac{7}{4}}{1 - \frac{5}{4} \cos \theta}.$$

Then we read off the eccentricity $e = \frac{5}{4}$ and $ep = \frac{7}{4}$. This gives $p = \frac{7}{5}$, and the directrix is the line $r \cos \theta = -\frac{7}{5}$. The curve is a hyperbola.

It can be shown that the equation

$$r = \frac{ep}{1 - e \cos (\theta - \alpha)}$$

is a conic of eccentricity e with focus at the pole and directrix the line

$$r \cos (\theta - \alpha) = -p.$$

PROBLEMS

In problems 1 through 8, in each case the line L passes through the point N and is perpendicular to the line going through N and the pole O. Find the equation of L in polar coordinates when the coordinates of N are as given.

1. $N(2, \pi/3)$ 2. $N(-3, 2\pi/3)$ 3. $N(4, \pi/2)$

4. $N(4, 9\pi)$ 5. $N(-2, 0)$ 6. $N(2, -3\pi/2)$

7. $N(5\pi/4, 3\pi/4)$ 8. $N(-5, \pi)$

In problems 9 through 15, in each case find, in polar coordinates, the equation of the circle satisfying the given conditions.

9. Center $(3, \pi/4)$, radius 4   10. Center $(-3, 3\pi/4)$, radius 2

11. Center $(3, \pi/6)$, radius 3   12. Center $(-4, -3\pi/2)$, radius 4

13. Center $(3, -\pi/3)$, passing through $(5, \pi/6)$

14. Center $(4, \pi/4)$, passing through $(5, 3\pi/4)$

15. Center on the line $\theta = \pi/4$ and passing through $(4, -\pi/2)$ and $(0, 0)$

In problems 16 through 21 find, in polar coordinates, the equation of the conic with focus at the pole and having the given eccentricity and directrix.

16. $e = \frac{3}{2}$, directrix $r \cos \theta = 6$ 17. $e = 1$, directrix $r \cos \theta = -3$

18. $e = \frac{1}{2}$, directrix $r \cos \theta = 3$ 19. $e = \frac{3}{4}$, directrix $r \cos (\theta + \pi/2) = 2$

20. $e = 2$, directrix $r \cos (\theta - \pi/3) = 4$

21. $e = 1$, directrix $r \sin \theta = 4$ [*Hint*: $\sin \theta = \cos (\theta - \pi/2)$.]

22. Given the line with equation $r \cos (\theta - \pi/4) = 2$, find the distance from the point $(3, \pi/3)$ to this line.

23. Given the line with equation $r \cos (\theta - \alpha) = p$, find a formula for the distance d from a point (r_1, θ_1) to this line.

In problems 24 through 31, conics are given. Find the eccentricity and directrix of each and sketch the curve.

24. $r = \dfrac{4}{2 - \cos \theta}$ 25. $r = \dfrac{3}{1 - \cos \theta}$

26. $r = \dfrac{6}{3 - 2 \cos (\theta - \pi/3)}$ 27. $r = \dfrac{3}{2 - \sin \theta}$

28. $r = \dfrac{3}{2 - 3 \sin \theta}$ 29. $r = \dfrac{4}{2 + \cos \theta}$

30. $r = \dfrac{4}{2 + 3 \cos (\theta + \pi/4)}$ 31. $r = \dfrac{3}{2 + 2 \sin (\theta - \pi/6)}$

32. Show that the equation

$$r = \frac{A}{B \cos \theta + C \sin \theta}$$

with $A, B, C \neq 0$ is always the equation of a straight line.

33. (a) Given the hyperbola

$$r = \frac{3}{1 - 3 \cos \theta}.$$

Show that the asymptotes have inclination $\theta = \arccos (\frac{1}{3})$, $\theta = \arccos (-\frac{1}{3})$.

(b) If

$$r = \frac{ep}{1 - e \cos \theta},$$

show that the asymptotes have inclination $\theta = \arccos (1/e)$, $\theta = \arccos (-1/e)$.

34. Show that the loci of $r = a \sec^2 (\theta/2)$ and $r = a \csc^2 (\theta/2)$ are parabolas.

10 Solid Analytic Geometry

▶1. **THE NUMBER SPACE R_3. COORDINATES. THE DISTANCE FORMULA**

In the study of analytic geometry of the plane we were careful to distinguish the geometric plane from the number plane R_2. As we saw in Chapter 2, the number plane consists of the collection of ordered pairs of real numbers. In the study of three-dimensional geometry it is useful to introduce the set of all ordered *triples* of real numbers. Calling this set the **three-dimensional number space,** we denote it by R_3. Each individual number triple is a **point** in R_3. The three elements in each number triple are called its **coordinates.** We now show how three-dimensional number space may be represented on a geometric or Euclidean three-dimensional space.

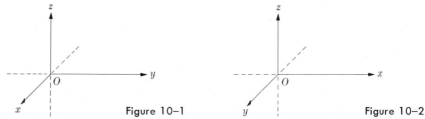

Figure 10–1   Figure 10–2

In three-dimensional space, consider three mutually perpendicular lines which intersect in a point O. We designate these lines the **coordinate axes** and, starting from O, we set up number scales along each of them. If the positive directions of the x, y, and z axes are labeled x, y, and z, as shown in Fig. 10–1, we say the axes form a **right-handed system.** Figure 10–2 illustrates the axes in a **left-handed system.** We shall use a right-handed coordinate system throughout.

Any two intersecting lines in space determine a plane. A plane containing two of the coordinate axes is called a **coordinate plane.** Clearly, there are three such planes.

To each point P in three-dimensional space we can assign a point in R_3 in the following way. Through P construct three planes, each parallel to one of the coordinate planes as shown in Fig. 10–3. We label the intersections of the planes through P with the coordinate axes Q, R, and S, as shown. Then, if Q is x_0 units from the origin O, R is y_0 units from O, and S is z_0 units from O, we assign to P the number triple (x_0, y_0, z_0) and say that the point P has **Cartesian coordinates**

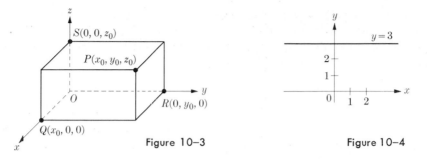

Figure 10–3   Figure 10–4

(x_0, y_0, z_0). To each point in space there corresponds exactly one ordered number triple and, conversely, to each ordered number triple there is associated exactly one point in three-dimensional space. We have just described a **Cartesian** or **rectangular** coordinate system. In Section 9 we shall discuss other coordinate systems.

In studying plane analytic geometry we saw that an equation such as

$$y = 3$$

represents all points lying on a line parallel to the x axis and three units above it (Fig. 10–4). The equation

$$y = 3$$

in the context of three-dimensional geometry represents something entirely different. The locus of points satisfying this equation is a plane parallel to the xz plane (the xz plane is the coordinate plane determined by the x axis and the z axis) and three units from it (Fig. 10–5). The plane represented by $y = 3$ is perpendicular to the y axis and passes through the point $(0, 3, 0)$. Since there is exactly one plane which is perpendicular to a given line and which passes through a given point, we see that the locus of the equation $y = 3$ consists of one and only one such plane. Conversely, from the very definition of a Cartesian coordinate system every point with y coordinate 3 must lie in this plane. Equations such as $x = a$ or $y = b$ or $z = c$ always represent planes parallel to the coordinate planes.

We recall from Euclidean solid geometry that *any two nonparallel planes intersect in a straight line*. Therefore, the locus of all points which simultaneously satisfy the equations

$$x = a \qquad \text{and} \qquad y = b$$

is a line parallel to (or coincident with) the z axis. Conversely, any such line is the locus of a pair of equations of the above form. Since the plane $x = a$ is parallel to the z axis, and the plane $y = b$ is parallel to the z axis, the line of

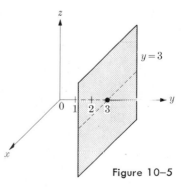

Figure 10–5

intersection must be parallel to the z axis also. (Corresponding statements hold with the axes interchanged.)

A plane separates three-dimensional space into two parts, each of which is called a **half-space.** The inequality

$$x > 5$$

represents all points with x coordinate greater than 5. The set of such points comprises a half-space. Two intersecting planes divide three-space into 4 regions which we call **infinite wedges.** Three intersecting planes divide space into 8 regions (or possibly fewer), four planes into 16 regions (or possibly fewer), and so on. The inequality

$$|y| \le 4$$

represents all points between (and on) the planes $y = -4$ and $y = 4$. Regions in space defined by inequalities are more difficult to visualize than those in the plane. However, **polyhedral domains**—that is, those bounded by a number of planes—are frequently simple enough to be sketched. A polyhedron with six faces in which opposite faces are congruent parallelograms is called a **parallelepiped.** Cubes and rectangular bins are particular cases of parallelepipeds.

Theorem 1. *The distance d between the points $P_1(x_1, y_1, z_1)$ and $P_2(x_2, y_2, z_2)$ is*

$$d = \sqrt{(x_2 - x_1)^2 + (y_2 - y_1)^2 + (z_2 - z_1)^2}.$$

Proof. We make the construction shown in Fig. 10–6. By the Pythagorean theorem we have

$$d^2 = |P_1Q|^2 + |QP_2|^2.$$

Noting that $|P_1Q| = |RS|$, we use the formula for distance in the xy plane to get

$$|P_1Q|^2 = |RS|^2$$
$$= (x_2 - x_1)^2 + (y_2 - y_1)^2.$$

Furthermore, since P_2 and Q are on a line parallel to the z axis, we see that

$$|QP_2|^2 = |T_1T_2|^2 = (z_2 - z_1)^2.$$

Therefore

$$d^2 = (x_2 - x_1)^2 + (y_2 - y_1)^2 + (z_2 - z_1)^2.$$

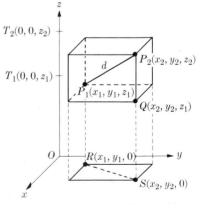

Figure 10–6

The midpoint P of the line segment connecting the point $P_1(x_1, y_1, z_1)$ and $P_2(x_2, y_2, z_2)$ has coordinates $P(\bar{x}, \bar{y}, \bar{z})$ given by the formulas

$$\bar{x} = \frac{x_1 + x_2}{2}, \qquad \bar{y} = \frac{y_1 + y_2}{2}, \qquad \bar{z} = \frac{z_1 + z_2}{2}.$$

If P_1 and P_2 lie in the xy plane—that is, if $z_1 = 0$ and $z_2 = 0$—then so does the midpoint P, and we recognize the formula as the one we learned in plane analytic geometry. The above formula for $\bar{x}$ is proved as in Chapter 3, Section 1, by passing planes through P_1, P, and P_2 perpendicular to the x axis. The formulas for $\bar{y}$ and $\bar{z}$ are established by analogy. (See problem 24 at the end of this section.)

EXAMPLE 1. Find the coordinates of the point Q which divides the line segment from $P_1(1, 4, -2)$ to $P_2(-3, 6, 7)$ in the proportion 3 to 1.

Solution. The midpoint P of the segment P_1P_2 has coordinates $P(-1, 5, \frac{5}{2})$. When we find the midpoint of PP_2 we get $Q(-2, \frac{11}{2}, \frac{19}{4})$.

EXAMPLE 2. One endpoint of a segment P_1P_2 has coordinates $P_1(-1, 2, 5)$. The midpoint P is known to lie in the xz plane, while the other endpoint is known to lie on the intersection of the planes $x = 5$ and $z = 8$. Find the coordinates of P and P_2.

Solution. For $P(\bar{x}, \bar{y}, \bar{z})$ we note that $\bar{y} = 0$, since P is in the xz plane. Similarly, for $P_2(x_2, y_2, z_2)$ we have $x_2 = 5$ and $z_2 = 8$. From the midpoint formula we get

$$\bar{x} = \frac{-1 + 5}{2}, \qquad 0 = \bar{y} = \frac{2 + y_2}{2}, \qquad \bar{z} = \frac{5 + 8}{2}.$$

Therefore the points have coordinates $P(2, 0, \frac{13}{2})$, $P_2(5, -2, 8)$.

PROBLEMS

In problems 1 through 5, find the lengths of the sides of triangle ABC and state whether the triangle is a right triangle, an isosceles triangle, or both.

1. $A(3, 1, 2)$, $B(-2, 3, -1)$, $C(-1, 0, 2)$ 2. $A(-1, 3, -1)$, $B(2, 2, 1)$, $C(2, 6, -1)$

3. $A(1, 4, 3)$, $B(2, 2, 1)$, $C(4, 0, 2)$ 4. $A(0, 0, 0)$, $B(2, 4, 1)$, $C(-1, -5, -5)$

5. $A(-3, 1, 2)$, $B(-1, 4, 3)$, $C(2, 3, 1)$

In problems 6 through 9, find the midpoint of the segment joining the given points A, B.

6. $A(3, 2, 1)$, $B(2, -4, -1)$ 7. $A(1, 4, 6)$, $B(3, 2, -1)$

8. $A(-3, 0, 2)$, $B(6, 1, 4)$ 9. $A(0, -2, 0)$, $B(-1, 0, 4)$

In problems 10 through 13, find the lengths of the medians of the given triangles ABC.

10. $A(3, 2, 1)$, $B(-2, 3, -1)$, $C(-1, 0, 2)$ 11. $A(1, 4, 3)$, $B(2, 2, 1)$, $C(4, 0, 2)$

12. $A(-1, 3, -1)$, $B(1, 1, 2)$, $C(2, 6, -1)$ 13. $A(-3, 1, 2)$, $B(-1, 4, 3)$, $C(2, 3, 1)$

14. One endpoint of a line segment is at $P(-3, 4, 6)$ and the midpoint is at $Q(6, 2, 1)$. Find the other endpoint.

15. One endpoint of a line segment is at $P_1(6, -2, 1)$ and the midpoint Q lies in the plane $y = 3$. The other endpoint, P_2, lies on the intersection of the planes $x = 4$ and $z = -6$. Find the coordinates of P_2 and Q.

In problems 16 through 19, determine whether or not the three given points lie on a line.

16. $A(2, 1, -1)$, $B(3, -1, -4)$, $C(1, 3, 2)$ 17. $A(1, 2, 3)$, $B(5, 4, 6)$, $C(-7, -2, -2)$

18. $A(2, 1, -1)$, $B(4, 3, 3)$, $C(-1, -2, -6)$

19. $A(-3, 1, 2)$, $B(-1, 4, 3)$, $C(2, 3, 1)$

20. Describe the locus of points in R_3 which satisfy the inequalities $-1 \leq y < 2$.

21. Describe the locus of points in R_3 which satisfy the relations $x = 1$, $z = -2$.

22. Describe the locus of points in R_3 which satisfy the inequalities $x \geq 0$, $y \geq 0$, $z \geq 0$.

23. Describe the locus of points in R_3 which satisfy the inequalities $x \leq 0$, $y > 0$, $z < 0$.

24. Derive the formula for determining the midpoint of a line segment.

25. The formula for the coordinates of the point $Q(x_0, y_0, z_0)$ which divides the line segment from $P_1(x_1, y_1, z_1)$ to $P_2(x_2, y_2, z_2)$ in the ratio p to q is

$$x_0 = \frac{px_2 + qx_1}{p + q}, \qquad y_0 = \frac{py_2 + qy_1}{p + q}, \qquad z_0 = \frac{pz_2 + qz_1}{p + q}.$$

Derive this formula.

26. Find the equation of the locus of all points equidistant from the points $(3, 2, -1)$ and $(-1, 3, 1)$. Can you describe the locus?

27. Find the equation of the locus of all points equidistant from the points $(0, 5, 1)$ and $(4, 2, -1)$. Can you describe the locus?

28. The points $A(0, 0, 0)$, $B(1, 0, 0)$, $C(\frac{1}{2}, \frac{1}{2}, 1/\sqrt{2})$, $D(0, 1, 0)$ are the vertices of a four-sided figure. Show that $|AB| = |BC| = |CD| = |DA| = 1$. Prove that the figure is not a rhombus.

29. Prove that the diagonals joining opposite vertices of a rectangular parallelepiped (there are four of them which are interior to the parallelepiped) bisect each other.

▶2. DIRECTION COSINES AND NUMBERS

A **directed line** was defined in Chapter 3, Section 6, as a straight line together with one of the two possible orderings of its points. If no ordering is specified the line is called **undirected.** Although these definitions were given for lines in a plane, it is clear that lines in three-space may be described as directed or undirected in the same way. We indicate the particular ordering of a line by means of an arrow.

Consider a directed line $\vec{L}$ passing through the origin (Fig. 10–7). Denote by α, β, and γ the angles made by the directed line $\vec{L}$ and the positive directions of the x, y, and z axes, respectively. We define these angles to be the **direction angles** of the directed line $\vec{L}$. The *undirected* line L will have two possible sets of direction angles according to the ordering chosen. The two sets are

$$\alpha, \beta, \gamma \quad \text{and} \quad 180° - \alpha, \ 180° - \beta, \ 180° - \gamma.$$

The term "line" without further specification shall mean undirected line.

DEFINITION. If α, β, γ are direction angles of a directed line $\vec{L}$, then $\cos \alpha$, $\cos \beta$, $\cos \gamma$ are called the **direction cosines** of $\vec{L}$.

Since $\cos(180° - \theta) = -\cos \theta$, we see that if λ, μ, ν are direction cosines of a directed line $\vec{L}$, then λ, μ, ν and $-\lambda$, $-\mu$, $-\nu$ are the two sets of direction cosines of the undirected line L.

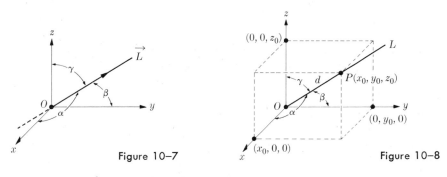

Figure 10–7   Figure 10–8

We shall show that the direction cosines of any line L satisfy the relation

$$\cos^2 \alpha + \cos^2 \beta + \cos^2 \gamma = 1.$$

Let $P(x_0, y_0, z_0)$ be a point on a line L which goes through the origin. Then the distance d of P from the origin is

$$d = \sqrt{x_0^2 + y_0^2 + z_0^2},$$

and (see Fig. 10–8) we have

$$\cos \alpha = \frac{x_0}{d}, \quad \cos \beta = \frac{y_0}{d}, \quad \cos \gamma = \frac{z_0}{d}.$$

Squaring and adding, we get the desired result.

To define the direction cosines of any line L in space, we simply consider the line L' parallel to L which passes through the origin, and assert that *by definition*

L has the same direction cosines as *L'*. Thus *all parallel lines in space have the same direction cosines.*

DEFINITION. Two sets of number triples, *a, b, c* and *a', b', c'*, neither all zero, are said to be **proportional** if there is a number *k* such that

$$a' = ka, \qquad b' = kb, \qquad c' = kc.$$

Remark. The number *k* may be positive or negative but not zero, since by hypothesis neither of the number triples is 0, 0, 0. If none of the numbers *a, b,* and *c* is zero, we may write the proportionality relations as

$$\frac{a'}{a} = k, \qquad \frac{b'}{b} = k, \qquad \frac{c'}{c} = k$$

or, more simply,

$$\frac{a'}{a} = \frac{b'}{b} = \frac{c'}{c}.$$

DEFINITION. Suppose that a line *L* has direction cosines λ, μ, ν. Then a set of numbers *a, b, c* is called a **set of direction numbers** for *L* if *a, b, c* and λ, μ, ν are proportional.

A line *L* has unlimited sets of direction numbers.

Theorem 2. *If $P_1(x_1, y_1, z_1)$ and $P_2(x_2, y_2, z_2)$ are two points on a line L, then*

$$\lambda = \frac{x_2 - x_1}{d}, \qquad \mu = \frac{y_2 - y_1}{d}, \qquad \nu = \frac{z_2 - z_1}{d}$$

is a set of direction cosines of L where d is the distance from P_1 to P_2.

Proof. In Fig. 10–9 we note that the angles α, β, and γ are equal to the direction angles, since the lines P_1A, P_1B, P_1C are parallel to the coordinate axes. We read off from the figure that

$$\cos \alpha = \frac{x_2 - x_1}{d},$$

$$\cos \beta = \frac{y_2 - y_1}{d},$$

$$\cos \gamma = \frac{z_2 - z_1}{d},$$

which is the desired result.

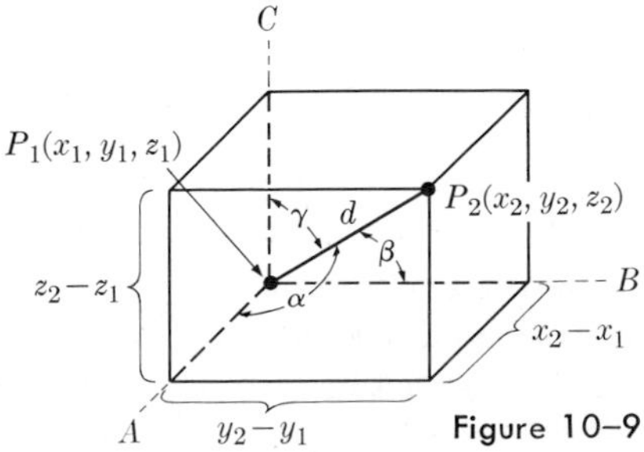

Figure 10–9

Corollary 1. *If $P_1(x_1, y_1, z_1)$ and $P_2(x_2, y_2, z_2)$ are two points on a line L, then*

$$x_2 - x_1, \qquad y_2 - y_1, \qquad z_2 - z_1$$

constitute a set of direction numbers for L.

Multiplying λ, μ, ν of Theorem 2 by the constant d, we obtain the result of the Corollary.

EXAMPLE 1. Find direction numbers and direction cosines for the line L passing through the points $P_1(1, 5, 2)$ and $P_2(3, 7, -4)$.

Solution. From the Corollary, 2, 2, -6 form a set of direction numbers. We compute

$$d = |P_1P_2| = \sqrt{4 + 4 + 36} = \sqrt{44} = 2\sqrt{11},$$

and so

$$\frac{1}{\sqrt{11}}, \qquad \frac{1}{\sqrt{11}}, \qquad -\frac{3}{\sqrt{11}}$$

form a set of direction cosines. Since L is undirected, it has two such sets, the other being

$$-\frac{1}{\sqrt{11}}, \qquad -\frac{1}{\sqrt{11}}, \qquad \frac{3}{\sqrt{11}}.$$

EXAMPLE 2. Do the three points $P_1(3, -1, 4)$, $P_2(1, 6, 8)$, and $P_3(9, -22, -8)$ lie on the same straight line?

Solution. A set of direction numbers for the line L_1 through P_1 and P_2 is $-2, 7, 4$. A set of direction numbers for the line L_2 through P_2 and P_3 is $8, -28, -16$. Since the second set is proportional to the first (with $k = -4$), we conclude that L_1 and L_2 have the same direction cosines. Therefore the two lines are parallel. However, they have the point P_2 in common and so must coincide.

From Theorem 2 and the statements in Example 2, we easily obtain the next result.

Corollary 2. *A line L_1 is parallel to a line L_2 if and only if a set of direction numbers of L_1 is proportional to a set of direction numbers of L_2.*

The angle between two intersecting lines in space is defined in the same way as the angle between two lines in the plane. It may happen that two lines L_1 and L_2 in space are neither parallel nor intersecting. Such lines are said to be **skew** to each other. Nevertheless, the angle between L_1 and L_2 can still be defined. Denote by L_1' and L_2' the lines passing through the origin and parallel to L_1 and L_2, respectively. The **angle between L_1 and L_2 is defined to be the angle between the intersecting lines L_1' and L_2'.**

Theorem 3. *If L_1 and L_2 have direction cosines λ_1, μ_1, ν_1 and λ_2, μ_2, ν_2, respectively, and if θ is the angle between L_1 and L_2, then*

$$\cos\theta = \lambda_1\lambda_2 + \mu_1\mu_2 + \nu_1\nu_2.$$

Proof. From the way we defined the angle between two lines we may consider L_1 and L_2 as lines passing through the origin. Let $P_1(x_1, y_1, z_1)$ be a point on L_1 and $P_2(x_2, y_2, z_2)$ a point on L_2, neither O (see Fig. 10–10). Denote by d_1 the distance of P_1 from O, by d_2 the distance of P_2 from O; let $d = |P_1P_2|$. We apply the Law of Cosines to triangle OP_1P_2, getting

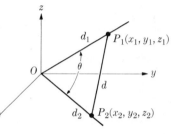

Figure 10–10

$$d^2 = d_1^2 + d_2^2 - 2d_1d_2\cos\theta$$

$$\Leftrightarrow \cos\theta = \frac{d_1^2 + d_2^2 - d^2}{2d_1d_2}$$

and

$$\cos\theta = \frac{x_1^2 + y_1^2 + z_1^2 + x_2^2 + y_2^2 + z_2^2 - (x_2 - x_1)^2 - (y_2 - y_1)^2 - (z_2 - z_1)^2}{2d_1d_2}.$$

After simplification we get

$$\cos\theta = \frac{x_1x_2 + y_1y_2 + z_1z_2}{d_1d_2} = \frac{x_1}{d_1}\cdot\frac{x_2}{d_2} + \frac{y_1}{d_1}\cdot\frac{y_2}{d_2} + \frac{z_1}{d_1}\cdot\frac{z_2}{d_2}$$

$$= \lambda_1\lambda_2 + \mu_1\mu_2 + \nu_1\nu_2.$$

Corollary. *Two lines L_1 and L_2 with direction numbers a_1, b_1, c_1 and a_2, b_2, c_2, respectively, are perpendicular if and only if*

$$a_1a_2 + b_1b_2 + c_1c_2 = 0.$$

EXAMPLE 3. Find the cosine of the angle between the line L_1, passing through the points $P_1(1, 4, 2)$ and $P_2(3, -1, 3)$, and the line L_2, passing through the points $Q_1(3, 1, 2)$ and $Q_2(2, 1, 3)$.

Solution. A set of direction numbers for L_1 is $2, -5, 1$. A set for L_2 is $-1, 0, 1$. Therefore direction cosines for the two lines are

$$L_1: \frac{2}{\sqrt{30}}, \frac{-5}{\sqrt{30}}, \frac{1}{\sqrt{30}}; \quad L_2: \frac{-1}{\sqrt{2}}, 0, \frac{1}{\sqrt{2}}.$$

We obtain

$$\cos\theta = -\frac{1}{\sqrt{15}} + 0 + \frac{1}{2\sqrt{15}} = -\frac{1}{2\sqrt{15}}.$$

We observe that two lines always have two possible supplementary angles of inter-section. If $\cos \theta$ is negative, we have obtained the obtuse angle and, if it is positive, the acute angle of intersection.

PROBLEMS

In problems 1 through 4, find a set of direction numbers and a set of direction cosines for the line passing through the given points.

1. $P_1(2, 3, 1)$, $P_2(0, 2, 6)$

2. $P_1(0, 4, 1)$, $P_2(2, 0, 1)$

3. $P_1(4, -5, -1)$, $P_2(6, 2, 0)$

4. $P_1(5, -1, 5)$, $P_2(5, 3, 5)$

In each of problems 5 through 8, a point P_1 and a set of direction numbers is given. Find another point which is on the line passing through P_1 and having the given set of direction numbers.

5. $P_1(1, 2, 5)$, direction numbers 5, 3, 2

6 $P_1(-5, 3, 0)$, direction numbers 0, 2, -1

7. $P_1(-3, 0, 4)$, direction numbers 5, 0, 0

8. $P_1(0, 0, 0)$, direction numbers 0, 2, -4

In each of problems 9 through 12, determine whether or not the three given points lie on a line.

9. $A(0, 3, 1)$, $B(2, 2, 2)$, $C(6, 0, 4)$ 10. $A(1, 2, -1)$, $B(-3, 4, 1)$, $C(-9, 7, 4)$

11. $A(-1, 4, 2)$, $B(1, 2, 1)$, $C(2, 0, 0)$ 12. $A(6, 5, 8)$, $B(1, -2, -3)$, $C(8, 4, 2)$

In each of problems 13 through 15, determine whether or not the line through the points P_1 and P_2 is parallel to the line through the points Q_1 and Q_2.

13. $P_1(0, 4, 8)$, $P_2(3, 1, 2)$; $Q_1(0, 0, 5)$, $Q_2(3, -3, -1)$

14. $P_1(1, 2, 1)$, $P_2(-1, 3, 2)$; $Q_1(4, 0, 1)$, $Q_2(0, 2, 3)$

15. $P_1(4, 3, 1)$, $P_2(5, -3, 2)$; $Q_1(1, 4, 6)$, $Q_2(8, 0, 5)$

In each of problems 16 through 18, determine whether or not the line through the points P_1 and P_2 is perpendicular to the line through the points Q_1 and Q_2.

16. $P_1(3, 2, 1)$, $P_2(5, 4, 0)$; $Q_1(2, 3, 1)$, $Q_2(6, 2, 1)$

17. $P_1(0, 2, -1)$, $P_2(2, 3, 1)$; $Q_1(4, 2, 1)$, $Q_2(4, 4, 0)$

18. $P_1(2, 0, -4)$, $P_2(0, 5, -1)$; $Q_1(2, 3, 0)$, $Q_2(1, 2, 1)$

In each of problems 19 through 21, find $\cos \theta$ where θ is the angle between the line L_1 passing through P_1 and P_2 and L_2 passing through Q_1 and Q_2.

19. $P_1(4, 2, 1)$, $P_2(1, -1, 4)$; $Q_1(1, 0, 5)$, $Q_2(-2, 3, -1)$

20. $P_1(5, 4, 0)$, $P_2(-2, -1, -3)$; $Q_1(4, 2, 1)$, $Q_2(1, 2, -5)$

21. $P_1(5, 0, 0)$, $P_2(0, 4, -2)$; $Q_1(6, 0, 0)$, $Q_2(1, 3, -2)$

22. A **regular tetrahedron** is a 4-sided figure each side of which is an equilateral triangle. Find 4 points in space which are the vertices of a regular tetrahedron with each edge of length 2 units.

23. A **regular pyramid** is a 5-sided figure with a square base and sides consisting of 4 congruent isosceles triangles. If the base has a side of length 4 and if the height of the pyramid is 6 units, find the area of each of the triangular faces.

24. The points $P_1(1, 2, 3)$, $P_2(2, 1, 2)$, $P_3(3, 0, 1)$, $P_4(5, 2, 7)$ are the vertices of a plane quadrilateral. Find the coordinates of the midpoints of the sides. What kind of quadrilateral do these four midpoints form?

25. Prove that the four interior diagonals of a parallelepiped bisect each other.

26. Let S be the set of points which satisfy $x^2 + y^2 + z^2 = 1$. Find the intersections of S with the line L passing through the origin and with direction numbers 2, 1, 3.

27. Let P, Q, R, S be the vertices of any quadrilateral in space. Prove that the lines joining the midpoints of the sides form a parallelogram.

28. The vertices of a tetrahedron are at $(3, 0, 0)$, $(6, 0, 0)$, $(0, 9, 0)$, and $(6, 12, 15)$. Show that the three lines joining the midpoints of opposite sides bisect each other. How general is this result?

▶3. EQUATIONS OF A LINE

In plane analytic geometry a single equation of the first degree,

$$Ax + By + C = 0,$$

is the equation of a line (so long as A and B are not both zero). *In the geometry of three dimensions such an equation represents a plane.* Although we shall postpone a systematic study of planes until the next section, we assert now that in three-dimensional geometry it is not possible to represent a line by a single first-degree equation.

A line in space is determined by two points. If $P_0(x_0, y_0, z_0)$ and $P_1(x_1, y_1, z_1)$ are given points, we seek an analytic method of representing the line L determined by these points. The result is obtained by solving a locus problem. A point $P(x, y, z)$, different from P_0, is on L if and only if the direction numbers determined by P and P_0 are proportional to those determined by P_1 and P_0 (Fig. 10–11). Calling the proportionality constant t,* we see that the conditions are

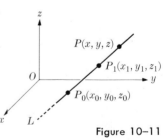

$P(x, y, z)$

$P_1(x_1, y_1, z_1)$

$P_0(x_0, y_0, z_0)$

L

Figure 10–11

$$x - x_0 = t(x_1 - x_0),$$
$$y - y_0 = t(y_1 - y_0),$$
$$z - z_0 = t(z_1 - z_0).$$

---

* In Eq. (3) following and in Chapter 12, the letter t has a special significance; otherwise it is better to use h as the parameter.

Thus we obtain the **two-point form of the parametric equations of a line**:

$$x = x_0 + (x_1 - x_0)t,$$
$$y = y_0 + (y_1 - y_0)t, \tag{1}$$
$$z = z_0 + (z_1 - z_0)t.$$

EXAMPLE 1. Find the parametric equations of the line through the points $A(3, 2, -1)$ and $B(4, 4, 6)$. Locate three additional points on the line.

Solution. Substituting in (1) we obtain

$$x = 3 + t, \qquad y = 2 + 2t, \qquad z = -1 + 7t.$$

To get an additional point on the line we let $t = 2$ and obtain $P_1(5, 6, 13)$; $t = -1$ yields $P_2(2, 0, -8)$ and $t = 3$ gives $P_3(6, 8, 20)$.

Theorem 4. *The parametric equations of a line L through the point $P_0(x_0, y_0, z_0)$ with direction numbers a, b, c are given by*

$$x = x_0 + at, \qquad y = y_0 + bt, \qquad z = z_0 + ct. \tag{2}$$

Proof. The point $P_1(x_0 + a, y_0 + b, z_0 + c)$ must be on L, since the direction numbers formed by P_0 and P_1 are just a, b, c. Using the two-point form (1) for the equations of a line through P_0 and P_1, we get (2) precisely.

If $\overrightarrow{L}$ is a directed line and α, β, γ are its direction angles, then

$$x = x_0 + t \cos \alpha, \qquad y = y_0 + t \cos \beta, \qquad z = z_0 + t \cos \gamma \tag{3}$$

are the equations of the directed line, where $P_0(x_0, y_0, z_0)$ is a given point on $\overrightarrow{L}$. Under these circumstances, the equations (3) set up a Cartesian coordinate system on $\overrightarrow{L}$ with origin at P_0. This coordinate system, in which the value of t determines the location of the point, agrees with the ordering on $\overrightarrow{L}$; that is, P_1 precedes P_2 if and only if $t_1 < t_2$. The directed distance $\overline{P_1P_2}$ between two points P_1, P_2 on $\overrightarrow{L}$ having t-coordinates t_1, t_2 is equal to $t_2 - t_1$. In contrast with equations (3), the parameter t in equations (2) has no special significance.

In Section 6 of Chapter 3, we derived a formula for finding the coordinates (x, y) of a point P which is h of the distance from a point $P_1(x_1, y_1)$ to a point

$P_2(x_2, y_2)$. The formulas we derived (p. 63) are

$$x = x_1 + h(x_2 - x_1), \qquad y = y_1 + h(y_2 - y_1)$$

and, in our derivation, we employed the parametric equations of a line as it appears in plane analytic geometry. In a *completely analogous way*, we have the formulas

$$x = x_1 + h(x_2 - x_1), \qquad y = y_1 + h(y_2 - y_1),$$
$$z = z_1 + h(z_2 - z_1)$$

(4)

for the coordinates of the point $P(x, y, z)$ which is h of the way from $P_1(x_1, y_1, z_1)$ to $P_2(x_2, y_2, z_2)$. No change is required in the method used to derive the formula in the plane. With the **point of division formula,** as (4) is called, we see that Eqs. (2) imply that $P(x, y, z)$ is t of the way from $P_0(x_0, y_0, z_0)$ to $P_1(x_0 + a, y_0 + b, z_0 + c)$.

EXAMPLE 2. Find the parametric equations of the line L through the point $A(3, -2, 5)$ with direction numbers 4, 0, -2. What is the relation of L to the coordinate planes?

Solution. Substituting in (2), we obtain

$$x = 3 + 4t, \qquad y = -2, \qquad z = 5 - 2t.$$

Since all points on the line must satisfy all three of the above equations, L must lie in the plane $y = -2$. This plane is parallel to the xz plane. Therefore L is parallel to the xz plane.

If none of the direction numbers is zero, the parameter t may be eliminated from the system of equations (2). We may write

$$\frac{x - x_0}{a} = \frac{y - y_0}{b} = \frac{z - z_0}{c}$$

(5)

for the equations of a line. For any value of t in (2) the ratios in (5) are equal. Conversely, if the ratios in (5) are all equal we may set the common value equal to t and (2) is satisfied.

If one of the direction numbers is zero, the form (5) may still be used if the zero in the denominator is interpreted properly. The equations

$$\frac{x - x_0}{a} = \frac{y - y_0}{b} = \frac{z - z_0}{0}$$

are understood to stand for the equations

$$\frac{x - x_0}{a} = \frac{y - y_0}{b} \qquad \text{and} \qquad z = z_0.$$

The system

$$\frac{x - x_0}{0} = \frac{y - y_0}{b} = \frac{z - z_0}{0}$$

stands for

$$x = x_0 \quad \text{and} \quad z = z_0.$$

We recognize these last two equations as those of planes parallel to coordinate planes. In other words, *a line is represented as the intersection of two planes.* This point will be discussed further in Section 5.

The two-point form for the equations of a line also may be written **symmetrically.** We have

$$\frac{x - x_0}{x_1 - x_0} = \frac{y - y_0}{y_1 - y_0} = \frac{z - z_0}{z_1 - z_0}.$$

PROBLEMS

In each of problems 1 through 4, find the equations, in both parametric and symmetric forms, of the line going through the given points. Find two additional points on each line.

1. $P_1(4, 2, 3)$, $P_2(2, -1, -3)$ 2. $P_1(3, 1, -1)$, $P_2(5, 2, 1)$
3. $P_1(-1, 1, 2)$, $P_2(2, 3, -1)$ 4. $P_1(1, -1, 2)$, $P_2(-1, 3, 1)$

In each of problems 5 through 9, find the equations of the line passing through the given point with the given direction numbers.

5. $P_1(-1, 0, 1)$, direction numbers $-3, 2, 1$
6. $P_1(3, -2, 1)$, direction numbers $-2, 3, -1$
7. $P_1(0, 4, 0)$, direction numbers $-2, 3, -1$
8. $P_1(0, 1, 2)$, direction numbers $3, 0, 1$
9. $P_1(-2, 3, -1)$, direction numbers $0, 2, 0$

In each of problems 10 through 14, decide whether or not L_1 and L_2 are perpendicular.

10. $L_1: \dfrac{x - 2}{4} = \dfrac{y + 1}{2} = \dfrac{z - 1}{-3}$; $L_2: \dfrac{x - 2}{3} = \dfrac{y + 1}{-3} = \dfrac{z - 1}{2}$

11. $L_1: \dfrac{x}{3} = \dfrac{y + 1}{1} = \dfrac{z + 1}{2}$; $L_2: \dfrac{x - 3}{2} = \dfrac{y + 1}{0} = \dfrac{z + 4}{-3}$

12. $L_1: \dfrac{x + 2}{3} = \dfrac{y - 2}{-1} = \dfrac{z + 3}{2}$; $L_2: \dfrac{x + 2}{-1} = \dfrac{y - 2}{1} = \dfrac{z + 3}{2}$

13. $L_1: \dfrac{x + 5}{5} = \dfrac{y - 1}{4} = \dfrac{z + 8}{3}$; $L_2: \dfrac{x - 4}{1} = \dfrac{y + 7}{3} = \dfrac{z + 4}{2}$

14. $L_1: \dfrac{x + 1}{0} = \dfrac{y - 2}{0} = \dfrac{z + 8}{1}$; $L_2: \dfrac{x - 3}{0} = \dfrac{y + 2}{1} = \dfrac{z - 1}{0}$

15. Find the equations of the medians of the triangle with vertices at $A(2, 4, 0)$, $B(4, 3, 1)$, $C(0, 2, 5)$.

16. Find the points of intersection of the line

$$x = 3 + t, \ y = 7 + 2t, \ z = 2 + 8t$$

with each of the coordinate planes.

17. Find the points of intersection of the line

$$\frac{x + 1}{7} = \frac{y + 1}{-2} = \frac{z - 2}{3}$$

with each of the coordinate planes.

18. Show that the following lines are coincident:

$$\frac{x - 1}{4} = \frac{y + 1}{2} = \frac{z}{-3}; \ \frac{x - 5}{4} = \frac{y - 1}{2} = \frac{z + 3}{-3}.$$

19. Find the equations of the line through $(3, 1, 5)$ which is parallel to the line

$$x = 4 - t, \qquad y = 2 + 3t, \qquad z = -4 + t.$$

20. Find the equations of the line through $(3, 1, -2)$ which is perpendicular to and intersects the line

$$\frac{x + 1}{1} = \frac{y + 2}{1} = \frac{z + 1}{1}.$$

[*Hint:* Let (x_0, y_0, z_0) be the point of intersection and determine its coordinates.]

21. Find the equations of the line through the point $(1, 2, -2)$ and perpendicular to the two lines with direction numbers $3, 1, -1$ and $2, -2, 1$.

22. A triangle has vertices at $A(2, 1, 6)$, $B(-3, 2, 4)$, and $C(5, 8, 7)$. Perpendiculars are drawn from these vertices to the xz plane. Locate the points A', B', and C' which are the intersections of the perpendiculars through A, B, C and the xz plane. Find the equations of the sides of the triangle $A'B'C'$.

▶4. THE PLANE

Any three points not on a straight line determine a plane. While this characterization of a plane is quite simple, it is not convenient for beginning the study of planes. Instead we use the fact that *there is exactly one plane which passes through a given point and is perpendicular to a given line.*

Let $P_0(x_0, y_0, z_0)$ be a given point, and suppose that a given line L goes through the point $P_1(x_1, y_1, z_1)$ and has direction numbers A, B, C.

Theorem 5. *The equation of the plane passing through P_0 and perpendicular to L is*

$$A(x - x_0) + B(y - y_0) + C(z - z_0) = 0.$$

Proof. We establish the result by solving a locus problem. Let $P(x, y, z)$ be a point (different from P_0) on the locus (Fig. 10-12). From Euclidean geometry we recall that if a line L_1 through P_0 and P is perpendicular to L, then P must be in the desired plane. A set of direction numbers for the line L_1 is

$$x - x_0, \qquad y - y_0, \qquad z - z_0.$$

Since L has direction numbers A, B, C, we conclude that the two lines L and L_1 are perpendicular if and only if their direction numbers satisfy the relation

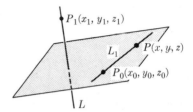

$$A(x - x_0) + B(y - y_0) + C(z - z_0) = 0,$$

which is the equation we seek.

Figure 10-12

Remark. Note that only the direction of L—and not the coordinates of P_1— enters the above equation. We obtain the same plane and the same equation if any line parallel to L is used in its stead.

EXAMPLE 1. Find the equation of the plane through the point $P_0(5, 2, -3)$ which is perpendicular to the line through the points $P_1(5, 4, 3)$ and $P_2(-6, 1, 7)$.

Solution. The line through the points P_1 and P_2 has direction numbers -11, -3, 4. The equation of the plane is

$$-11(x - 5) - 3(y - 2) + 4(z + 3) = 0$$

$$\Leftrightarrow 11x + 3y - 4z - 73 = 0.$$

All lines perpendicular to the same plane are parallel and therefore have proportional direction numbers.

DEFINITION. A set of **attitude numbers** of a plane is any set of direction numbers of a line perpendicular to the plane.

In Example 1 above, 11, 3, -4 form a set of attitude numbers of the plane.

EXAMPLE 2. What are sets of attitude numbers for planes parallel to the coordinate planes?

Solution. A plane parallel to the yz plane has an equation of the form $x - c = 0$, where c is a constant. A set of attitude numbers for this plane is 1, 0, 0. A plane parallel to the xz plane has attitude numbers 0, 1, 0, and any plane parallel to the xy plane has attitude numbers 0, 0, 1.

Since lines perpendicular to the same or parallel planes are themselves parallel, we get at once the next theorem.

Theorem 6. *Two planes are parallel if and only if their attitude numbers are proportional.*

Theorem 7. *If A, B, and C are not all zero, the locus of an equation of the form*

$$Ax + By + Cz + D = 0 \tag{1}$$

is a plane.

Proof. Suppose that $C \neq 0$, for example. Then the point $P_0(0, 0, -D/C)$ is on the locus, as its coordinates satisfy the above equation. Therefore we may write

$$A(x - 0) + B(y - 0) + C\left(z + \frac{D}{C}\right) = 0,$$

and the locus is the plane passing through P_0 perpendicular to any line with direction numbers A, B, C.

An equation of the plane through three points not on a line can be found by assuming that the plane has an equation of the form (1), substituting in turn the coordinates of the points, and solving simultaneously the three resulting equations. The fact that there are four constants, A, B, C, D, and only three equations is illusory, since we may divide through by one of them (say D) and obtain three equations in the unknowns A/D, B/D, C/D. This is equivalent to setting D (or one of the other constants) equal to some convenient value. An example illustrates the procedure.

EXAMPLE 3. Find an equation of the plane passing through the points $(2, 1, 3)$, $(1, 3, 2)$, $(-1, 2, 4)$.

Solution. Since the three points lie in the plane, each of them satisfies equation (1). We have

$$\begin{aligned}
(2, 1, 3): & \quad 2A + B + 3C + D = 0, \\
(1, 3, 2): & \quad A + 3B + 2C + D = 0, \\
(-1, 2, 4): & \quad -A + 2B + 4C + D = 0.
\end{aligned}$$

Solving for A, B, C in terms of D, we obtain

$$A = -\tfrac{3}{25}D, \qquad B = -\tfrac{4}{25}D, \qquad C = -\tfrac{5}{25}D.$$

Setting $D = -25$, we get the equation

$$3x + 4y + 5z - 25 = 0.$$

PROBLEMS

In each of problems 1 through 4, find the equation of the plane which passes through the given point P and has the given attitude numbers.

1. $P(6, 0, 5)$; $-1, 4, 2$ 2. $P(4, -3, 1)$; $5, 0, 2$

3. $P(-3, 1, -2)$; $-1, 4, 0$ 4. $P(6, 2, 0)$; $-5, 0, 0$

In each of problems 5 through 8, find the equation of the plane which passes through the three points.

5. $(1, 1, -2)$, $(3, 2, 0)$, $(-1, 0, 1)$ 6. $(1, 2, 2)$, $(3, -1, 2)$, $(-2, 3, -5)$

7. $(2, 3, -1)$, $(-1, 1, 2)$, $(1, 2, 3)$ 8. $(1, -1, 3)$, $(2, 2, 1)$, $(-1, 4, 2)$

In each of problems 9 through 12, find the equation of the plane passing through P_1 and perpendicular to the line L_1.

9. $P_1(3, 2, -1)$; $L_1: x = -1 - 4t, y = 1 + 2t, z = 3t$

10. $P_1(-3, 1, 2)$; $L_1: x = 3t, y = -2 + t, z = 1 - 2t$

11. $P_1(-2, 2, -1)$; $L_1: x = 2 - 2t, y = 3t, z = -1$

12. $P_1(-3, -1, 2)$; $L_1: x = -1 + 3t, y = 1 + 5t, z = -1 + 2t$

In each of problems 13 through 16, find the equations of the line through P_1 and perpendicular to the given plane M_1.

13. $P_1(1, -2, 3)$; $M_1: x + 2y + 3z - 1 = 0$

14. $P_1(-3, 1, -2)$; $M_1: -2x + 3y - z + 2 = 0$

15. $P_1(-2, -1, 0)$; $M_1: 2x + y + 1 = 0$

16. $P_1(-3, 2, -1)$; $M_1: y = 3$

In each of problems 17 through 20, find an equation of the plane through P_1 and parallel to the plane Φ.

17. $P_1(-1, 1, -2)$; $\Phi: -x + 3y + 2z - 1 = 0$

18. $P_1(2, -1, 3)$; $\Phi: -3x + 2y + z + 2 = 0$

19. $P_1(3, 2, -1)$; $\Phi: -3x + y - 2z + 3 = 0$

20. $P_1(2, 3, 0)$; $\Phi: y + 2z - 3 = 0$

In each of problems 21 through 23, find equations of the line through P_1 parallel to the given line L.

21. $P_1(3, 2, -1)$; $L: \dfrac{x + 1}{4} = \dfrac{y - 1}{3} = \dfrac{z - 2}{-2}$

22. $P_1(1, 0, 0)$; $L: \dfrac{x - 1}{-2} = \dfrac{y + 1}{1} = \dfrac{z - 2}{3}$

23. $P_1(0, 1, -2)$; $L: \dfrac{x + 1}{4} = \dfrac{y - 1}{2} = \dfrac{z + 1}{-1}$

In each of problems 24 through 28, find the equation of the plane containing L_1 and L_2.

24. $L_1: \dfrac{x-1}{1} = \dfrac{y+1}{2} = \dfrac{z-2}{3}$; $L_2: \dfrac{x-1}{2} = \dfrac{y+1}{1} = \dfrac{z-2}{-1}$

25. $L_1: \dfrac{x-2}{2} = \dfrac{y-1}{3} = \dfrac{z+2}{2}$; $L_2: \dfrac{x-2}{0} = \dfrac{y-1}{1} = \dfrac{z+2}{1}$

26. $L_1: \dfrac{x+1}{2} = \dfrac{y+2}{1} = \dfrac{z}{0}$; $L_2: \dfrac{x+1}{1} = \dfrac{y+2}{2} = \dfrac{z}{3}$

27. $L_1: \dfrac{x+2}{-1} = \dfrac{y}{2} = \dfrac{z-1}{3}$; $L_2: \dfrac{x}{-1} = \dfrac{y-2}{2} = \dfrac{z+1}{3}$ $(L_1 \parallel L_2)$

28. $L_1: \dfrac{x}{3} = \dfrac{y-2}{2} = \dfrac{z+1}{-1}$; $L_2: \dfrac{x+2}{3} = \dfrac{y+1}{2} = \dfrac{z}{-1}$ $(L_1 \parallel L_2)$

In problems 29 and 30, find the equation of the plane through P_1 and the given line L.

29. $P_1(2, 3, -1)$; $L: \dfrac{x}{-2} = \dfrac{y-2}{2} = \dfrac{z+1}{3}$

30. $P_1(3, 1, -2)$; $L: x = 2 - 2t,\ y = -1 + t,\ z = 2 + 2t$

31. Show that the plane $2x - 3y + z - 2 = 0$ is parallel to the line
$$\frac{x-2}{1} = \frac{y+2}{1} = \frac{z+1}{1}.$$

32. Show that the plane $5x - 3y - z - 6 = 0$ contains the line
$$x = 1 + 2t, \qquad y = -1 + 3t, \qquad z = 2 + t.$$

33. Find the equation of the line which passes through the point $(2, -1, 3)$ and is parallel to each of the planes $2x + 3y - z - 5 = 0$ and $x + 4y + z + 4 = 0$.

34. A plane has attitude numbers A, B, C, and a line has direction numbers a, b, c. What condition must be satisfied in order that the plane and line be parallel?

35. Show that the three planes $7x - 2y - 2z - 5 = 0$, $3x + 2y - 3z - 10 = 0$, and $7x + 2y - 5z - 16 = 0$ all contain a common line. Find the coordinates of two points on this line.

36. Find a condition that three planes $A_1x + B_1y + C_1z + D_1 = 0$, $A_2x + B_2y + C_2z + D_2 = 0$, and $A_3x + B_3y + C_3z + D_3 = 0$ either have a line in common or have no point in common.

▶5.   ANGLES.   DISTANCE FROM A POINT TO A PLANE

The angle between two lines was defined in Section 2. We recall that if line L_1 has direction cosines λ_1, μ_1, ν_1 and line L_2 has direction cosines λ_2, μ_2, ν_2, then

$$\cos \theta = \lambda_1\lambda_2 + \mu_1\mu_2 + \nu_1\nu_2,$$

where θ is the angle between L_1 and L_2.

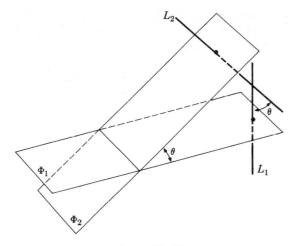

Figure 10–13

DEFINITION. Let Φ_1 and Φ_2 be two planes, and let L_1 and L_2 be two lines which are perpendicular to Φ_1 and Φ_2, respectively. Then the **angle between Φ_1 and Φ_2** is, by definition, the angle between L_1 and L_2. (See Fig. 10–13.) Furthermore we make the convention that we always select the acute angle between these lines as the angle between Φ_1 and Φ_2.

Theorem 8. *The angle θ between the planes $A_1x + B_1y + C_1z + D_1 = 0$ and $A_2x + B_2y + C_2z + D_2 = 0$ is given by*

$$\cos \theta = \frac{|A_1A_2 + B_1B_2 + C_1C_2|}{\sqrt{A_1^2 + B_1^2 + C_1^2}\,\sqrt{A_2^2 + B_2^2 + C_2^2}}.$$

Proof. From the definition of attitude numbers for a plane, we know that they are direction numbers of any line perpendicular to the plane. Converting to direction cosines, we get the above formula.

Corollary. *Two planes with attitude numbers A_1, B_1, C_1 and A_2, B_2, C_2 are perpendicular if and only if*

$$A_1A_2 + B_1B_2 + C_1C_2 = 0.$$

EXAMPLE 1. Find $\cos \theta$ where θ is the angle between the planes $3x - 2y + z = 4$ and $x + 4y - 3z - 2 = 0$.

Solution. Substituting in the formula of Theorem 8, we have

$$\cos \theta = \frac{|3 - 8 - 3|}{\sqrt{9 + 4 + 1}\,\sqrt{1 + 16 + 9}} = \frac{4}{\sqrt{91}}.$$

Two nonparallel planes intersect in a line. Every point on the line satisfies the equations of both planes and, conversely, every point which satisfies the equations of both planes must be on the line. *Therefore we may characterize any line in space by finding two planes which contain it.* Since every line has an unlimited number of planes which pass through it and since *any* two of them are sufficient to determine the line uniquely, we see that there is an unlimited number of ways of writing the equations of a line. The next example shows how to transform one representation into another.

EXAMPLE 2. The two planes

$$2x + 3y - 4z - 6 = 0 \qquad \text{and} \qquad 3x - y + 2z + 4 = 0$$

intersect in a line. (That is, the points which satisfy *both* equations constitute the line.) Find a set of parametric equations of the line of intersection.

Solution. We solve the above equations for x and y in terms of z, getting

$$x = -\tfrac{2}{11}z - \tfrac{6}{11}, \qquad y = \tfrac{16}{11}z + \tfrac{26}{11}$$

and

$$\frac{x + \tfrac{6}{11}}{-\tfrac{2}{11}} = \frac{y - \tfrac{26}{11}}{\tfrac{16}{11}} = \frac{z}{1}.$$

We can therefore write

$$x = -\tfrac{6}{11} - \tfrac{2}{11}t, \qquad y = \tfrac{26}{11} + \tfrac{16}{11}t, \qquad z = t,$$

which are the desired parametric equations.

Three planes may be parallel, may pass through a common line, may have no common points, or may have a unique point of intersection. If they have a unique point of intersection, the intersection point may be found by solving simultaneously the three equations of the planes. If they have no common point, an attempt to solve simultaneously will fail. A further examination will show whether or not two or more of the planes are parallel.

EXAMPLE 3. Determine whether or not the planes Φ_1: $3x - y + z - 2 = 0$; Φ_2: $x + 2y - z + 1 = 0$; Φ_3: $2x + 2y + z - 4 = 0$ intersect. If so, find the point of intersection.

Solution. Eliminating z between Φ_1 and Φ_2, we have

$$4x + y - 1 = 0. \tag{1}$$

Eliminating z between Φ_2 and Φ_3, we find

$$3x + 4y - 3 = 0. \tag{2}$$

We solve equations (1) and (2) simultaneously to get

$$x = \tfrac{1}{13}, \qquad y = \tfrac{9}{13}.$$

Substituting in the equation for Φ_1, we obtain $z = \tfrac{32}{13}$. Therefore the single point of intersection of the three planes is $(\tfrac{1}{13}, \tfrac{9}{13}, \tfrac{32}{13})$.

EXAMPLE 4. Find the point of intersection of the plane

$$3x - y + 2z - 3 = 0$$

and the line

$$\frac{x+1}{3} = \frac{y+1}{2} = \frac{z-1}{-2}.$$

Solution. We write the equations of the line in parametric form:

$$x = -1 + 3t, \qquad y = -1 + 2t, \qquad z = 1 - 2t.$$

The point of intersection is given by a value of t; call it t_0. This point must satisfy the equation of the plane. We have

$$3(-1 + 3t_0) - (-1 + 2t_0) + 2(1 - 2t_0) - 3 = 0$$
$$\Leftrightarrow t_0 = 1.$$

The desired point is $(2, 1, -1)$.

We now derive an important formula which tells us how to find the perpendicular distance from a point in space to a plane (Formula (3) below).

Theorem 9. *The distance d from the point $P_1(x_1, y_1, z_1)$ to the plane*

$$Ax + By + Cz + D = 0$$

is given by

$$d = \frac{|Ax_1 + By_1 + Cz_1 + D|}{\sqrt{A^2 + B^2 + C^2}}. \tag{3}$$

Proof. We write the equations of the line L through P_1 which is perpendicular to the plane. They are

$$L: x = x_1 + At, \qquad y = y_1 + Bt, \qquad z = z_1 + Ct.$$

Denote by (x_0, y_0, z_0) the intersection of L and the plane. Then

$$d^2 = (x_1 - x_0)^2 + (y_1 - y_0)^2 + (z_1 - z_0)^2. \tag{4}$$

Also (x_0, y_0, z_0) is on both the line and the plane. Therefore, we have for some value t_0

$$x_0 = x_1 + At_0, \qquad y_0 = y_1 + Bt_0, \qquad z_0 = z_1 + Ct_0 \tag{5}$$

and

$$Ax_0 + By_0 + Cz_0 + D$$
$$= 0 = A(x_1 + At_0) + B(y_1 + Bt_0) + C(z_1 + Ct_0) + D.$$

Thus, from (4) and (5), we write

$$d = \sqrt{A^2 + B^2 + C^2}\,|t_0|,$$

and now, inserting the relation

$$t_0 = \frac{-(Ax_1 + By_1 + Cz_1 + D)}{A^2 + B^2 + C^2}$$

in the preceding expression for d, we obtain the desired formula.

EXAMPLE 5. Find the distance from the point $(2, -1, 5)$ to the plane

$$3x + 2y - 2z - 7 = 0.$$

Solution.

$$d = \frac{|6 - 2 - 10 - 7|}{\sqrt{9 + 4 + 4}} = \frac{13}{\sqrt{17}}.$$

PROBLEMS

In each of problems 1 through 4, find $\cos \theta$ where θ is the angle between the given planes.

1. $2x + 2y - z - 2 = 0, -6x + 3y + 2z - 6 = 0$
2. $-3x + y + 2z - 1 = 0, x + y + z - 2 = 0$
3. $3x + 2y - z + 3 = 0, 2x + 3y - 2z - 3 = 0$
4. $4x + y - 1 = 0, 2x + z - 2 = 0$

In each of problems 5 through 8, find the equations in parametric form of the line of intersection of the given planes.

5. $-x + 3y + 2z + 5 = 0, 2x + 2y + z - 3 = 0$
6. $2x + y + 2z - 4 = 0, -3x + 2y + z + 5 = 0$
7. $-x + y + 2z + 4 = 0, 3x + 2y + 4z - 7 = 0$
8. $-4x + 2y + 3z + 7 = 0, 3x + 3y - 2z - 6 = 0$

In each of problems 9 through 12, find the point of intersection of the given line and the given plane.

9. $2x + 3y - z - 5 = 0, \dfrac{x - 1}{-2} = \dfrac{y - 1}{2} = \dfrac{z + 1}{3}$

10. $-4x + 2y + 3z + 15 = 0, \dfrac{x + 4}{3} = \dfrac{y + 3}{2} = \dfrac{z - 1}{-2}$

11. $2x + y + 3 = 0, \dfrac{x+2}{2} = \dfrac{y+1}{1} = \dfrac{z}{0}$

12. $x + 2y + 3z - 3 = 0, \dfrac{x-1}{1} = \dfrac{y+2}{2} = \dfrac{z-3}{3}$

In each of problems 13 through 16, find the distance from the given point to the given plane.

13. $(-1, 2, 1), \ 2x + y - 2z + 5 = 0$

14. $(2, 3, -1), \ -6x + 3y + 2z - 9 = 0$

15. $(2, -1, 3), \ 4x + 2y - 3z - 5 = 0$

16. $(-3, 0, 4), \ 2x + 3z - 7 = 0$

17. Find the equation of the plane through the line

$$\frac{x-2}{4} = \frac{y+1}{3} = \frac{z-1}{2}$$

which is perpendicular to the plane

$$-3x + 2y + z + 4 = 0.$$

18. Find the equation of the plane through the line

$$\frac{x-1}{-2} = \frac{y-2}{2} = \frac{z-2}{3}$$

which is parallel to the line

$$\frac{x+1}{1} = \frac{y+1}{3} = \frac{z-1}{2}.$$

19. Find the equation of the plane through the line

$$\frac{x+1}{2} = \frac{y+2}{3} = \frac{z}{-2}$$

which is parallel to the line

$$\frac{x-1}{4} = \frac{y-1}{2} = \frac{z+1}{3}.$$

20. Find the equations of every line through $(2, 1, 4)$ which is parallel to the plane

$$x + 2y + z - 4 = 0.$$

21. Find the equation of the plane through $(-1, 3, 2)$ and $(2, 1, -1)$ which is parallel to the line

$$\frac{x}{-2} = \frac{y-1}{3} = \frac{z+1}{2}.$$

In each of problems 22 through 26, find all the points of intersection of the three given planes. If the three planes pass through a line, find the equations of the line in parametric form.

22. $-2x + 2y + z - 1 = 0,\ x + 3y + 2z - 10 = 0,\ -3x + y + 2z + 2 = 0$

23. $3x + y + 2z - 4 = 0,\ x + 2y - 3z - 2 = 0,\ -2x + 3y + 2z - 5 = 0$

24. $2x + 3y - z - 4 = 0,\ -x + y + 2z - 3 = 0,\ 7x + 3y - 8z + 1 = 0$

25. $-2x + 2y + z - 3 = 0,\ x + y - z + 1 = 0,\ -7x + y + 5z - 3 = 0$

26. $3x + y + 2z - 5 = 0,\ -2x + 2y - z - 2 = 0,\ -13x + y - 8z + 11 = 0$

In each of problems 27 through 29, find the equations in parametric form of the line through the given point P_1 which intersects and is perpendicular to the given line L.

27. $P_1(2, 3, -1);$ $L: \dfrac{x}{3} = \dfrac{y - 1}{2} = \dfrac{z + 1}{-1}$

28. $P_1(3, -1, 2);$ $L: \dfrac{x + 3}{-3} = \dfrac{y}{2} = \dfrac{z - 2}{0}$

29. $P_1(4, 0, 2);$ $L: \dfrac{x - 3}{4} = \dfrac{y - 1}{3} = \dfrac{z - 2}{1}$

30. (a) If $A_1x + B_1y + C_1z + D_1 = 0$ and $A_2x + B_2y + C_2z + D_2 = 0$ are two intersecting planes, what is the locus of all points which satisfy

$$A_1x + B_1y + C_1z + D_1 + k(A_2x + B_2y + C_2z + D_2) = 0,$$

where k is a constant? (b) Find the equation of the plane passing through the point $(2, 1, -3)$ and the intersection of the planes $3x + y - z - 2 = 0$, $2x + y + 4z - 1 = 0$.

31. Given a regular tetrahedron with side of length a. Find the distance from a vertex to the opposite face.

32. Let P be a point inside a regular tetrahedron of side a, and denote by d_1, d_2, d_3, d_4 the distance from P to each of the faces. Show that $d_1 + d_2 + d_3 + d_4 =$ length of the altitude of the tetrahedron.

33. Let S be a regular pyramid (square base of side 2) and suppose the lateral edges are of length 4. (a) Find the distance from the top of the pyramid to the base. (b) Find the distance from one of the vertices of the base to a face opposite that vertex. (c) Find the angle between the base and one of the lateral faces.

34. A slice is made in a cube by cutting through a diagonal of the upper face and proceeding through one of the bottom vertices not directly below the diagonal. Find the angle between the base of the cube and the planar slice. See Fig. 10–14.

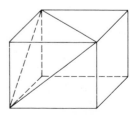

Figure 10–14

▶6. THE SPHERE. CYLINDERS

A **sphere** is the locus of all points at a given distance from a fixed point. The fixed point is called the **center** and the fixed distance is called the **radius.**

If the center is at the point (h, k, l), the radius is r, and (x, y, z) is any point on the sphere, then, from the formula for the distance between two points, we obtain the relation

$$(x - h)^2 + (y - k)^2 + (z - l)^2 = r^2. \tag{1}$$

Equation (1) is the **equation of a sphere.** If it is multiplied out and the terms collected we have the equivalent form

$$x^2 + y^2 + z^2 + Dx + Ey + Fz + G = 0. \tag{2}$$

EXAMPLE 1. Find the center and radius of the sphere with equation

$$x^2 + y^2 + z^2 + 4x - 6y + 9z - 6 = 0.$$

Solution. We complete the square by first writing

$$x^2 + 4x \quad\quad + y^2 - 6y \quad\quad + z^2 + 9z \quad\quad = 6;$$

then, adding the appropriate quantities to both sides, we have

$$x^2 + 4x + 4 + y^2 - 6y + 9 + z^2 + 9z + \tfrac{81}{4} = 6 + 4 + 9 + \tfrac{81}{4}$$
$$\Leftrightarrow (x + 2)^2 + (y - 3)^2 + (z + \tfrac{9}{2})^2 = \tfrac{157}{4}.$$

The center is at $(-2, 3, -\tfrac{9}{2})$ and the radius is $\tfrac{1}{2}\sqrt{157}$.

EXAMPLE 2. Find the equation of the sphere which passes through $(2, 1, 3)$, $(3, 2, 1)$, $(1, -2, -3)$, $(-1, 1, 2)$.

Solution. Substituting these points in the form (2) above for the equation of a sphere, we obtain

$$
\begin{array}{lll}
(2, 1, 3): & 2D + E + 3F + G = -14, \\
(3, 2, 1): & 3D + 2E + F + G = -14, \\
(1, -2, -3): & D - 2E - 3F + G = -14, \\
(-1, 1, 2): & -D + E + 2F + G = -6.
\end{array}
$$

Solving these by elimination (first G, then D, then F) we obtain, successively,

$$D + E - 2F = 0, \quad\quad D + 3E + 6F = 0, \quad\quad 3D + F = -8,$$

and

$$2E + 8F = 0, \quad\quad 3E - 7F = 8; \quad\quad \text{and so} \quad\quad -38E = -64.$$

Therefore

$$E = \tfrac{32}{19}, \qquad F = -\tfrac{8}{19}, \qquad D = -\tfrac{48}{19}, \qquad G = -\tfrac{178}{19}.$$

The desired equation is

$$x^2 + y^2 + z^2 - \tfrac{48}{19}x + \tfrac{32}{19}y - \tfrac{8}{19}z - \tfrac{178}{19} = 0.$$

A **cylindrical surface** is a surface which consists of a collection of parallel lines. Each of the parallel lines is called a **generator** of the **cylinder** or cylindrical surface.

The customary right circular cylinder of elementary geometry is clearly a special case of the type of cylinder we are considering. Figure 10–15 shows some examples of cylindrical surfaces. Note that a plane is a cylinder.

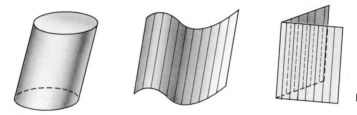

Figure 10–15

Theorem 10. *The locus of an equation of the form*

$$f(x, y) = 0$$

is a cylindrical surface with generators all parallel to the z axis. The surface intersects the xy plane in the curve

$$f(x, y) = 0, \qquad z = 0.$$

A similar result holds with axes interchanged.

Proof. Suppose that x_0, y_0 satisfies $f(x_0, y_0) = 0$. Then any point (x_0, y_0, z) for $-\infty < z < \infty$ satisfies the same equation, since z is absent. Therefore the line parallel to the z axis through $(x_0, y_0, 0)$ is a generator.

EXAMPLE 3. Describe and sketch the locus of the equation $x^2 + y^2 = 9$.

Solution. The locus is a right circular cylinder with generators parallel to the z axis (Theorem 10). It is sketched in **Fig. 10–16.**

EXAMPLE 4. Describe and sketch the locus of the equation $y^2 = 4z$.

Solution. According to Theorem 10 the locus is a cylinder with generators parallel to the x axis. The intersection with the yz plane is a parabola. The locus, called a parabolic cylinder, is sketched in **Fig. 10–17.**

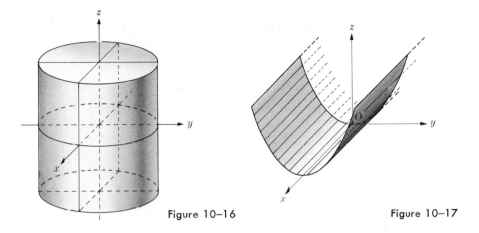

Figure 10–16   Figure 10–17

PROBLEMS

In each of problems 1 through 4, find the equation of the sphere with center C and radius r.

1. $C(1, 2, 0)$, $r = 4$ 2. $C(1, 3, -2)$, $r = 6$
3. $C(6, 3, -2)$, $r = 7$ 4. $C(-4, 2, 1)$, $r = 3$

In each of problems 5 through 9, determine the locus of the equation. If it is a sphere find its center and radius.

5. $x^2 + y^2 + z^2 - 4x + 2y + 1 = 0$
6. $x^2 + y^2 + z^2 + 6x - 4y + 2z - 2 = 0$
7. $x^2 + y^2 + z^2 - 2x - 2y + 4z + 7 = 0$
8. $x^2 + y^2 + z^2 + 4x + 4y - 2z + 9 = 0$
9. $x^2 + y^2 + z^2 + 2x - 6y + 4z + 10 = 0$

10. Find the equation of the locus of all points which are twice as far from $A(2, 3, -1)$ as from $B(-1, 0, 2)$.

11. Find the equation of the locus of all points which are three times as far from $A(-3, 2, 1)$ as from $B(5, -2, -3)$.

12. Find the equation of the locus of all points whose distances from the point $(4, 0, 0)$ are equal to their perpendicular distances from the yz plane.

In each of problems 13 through 24, describe and sketch the locus of the given equation.

13. $y = 3$ 14. $y^2 + z^2 = 16$ 15. $2y + z = 3$
16. $2x + y = 4$ 17. $y^2 = 4x$ 18. $x = 2 - z^2$
19. $4y^2 + z^2 = 16$ 20. $4y^2 - z^2 = 16$ 21. $y^2 + z^2 = 9$
22. $x^2 = 2 - 2y$ 23. $y^2 + z^2 - 2y = 0$ 24. $x^2 = z^2 + 4$

In each of problems 25 through 28, describe the curve of intersection, if any, of the given surface S and the given plane Φ.

25. $S: x^2 + y^2 + z^2 = 25$; $\Phi: x = 3$

26. $S: 4y = z^2 + x^2$; $\Phi: y = 4$

27. $S: 3x^2 + y^2 + 2z^2 = 12$; $\Phi: z = 4$

28. $S: y^2 + z^2 = x^2$; $\Phi: x + 2y = 4$

29. (a) Verify that the locus of the equation $(x - 2)^2 + (y - 1)^2 = 0$ is a straight line. Show that every straight line parallel to one of the coordinate axes can be represented by a *single* equation of the second degree. (b) Describe the locus of $(x + y + z - 1)^2 + (x - 2y - z - 3)^2 = 0$. Can all straight lines be represented by single equations of the second degree?

30. Let $x^2 + y^2 + z^2 + A_1x + B_1y + C_1z + D_1 = 0$ and $x^2 + y^2 + z^2 + A_2x + B_2y + C_2z + D_2 = 0$ be two spheres. The **radical plane**, given by

$$(A_1 - A_2)x + (B_1 - B_2)y + (C_1 - C_2)z + (D_1 - D_2) = 0$$

is obtained by subtracting the equations for the spheres. Show that the radical plane is perpendicular to the line joining the centers of the spheres.

31. Show that the three radical planes (see Problem 30) of three spheres intersect in a common line.

▶7.   QUADRIC SURFACES

In the plane an equation of the form

$$Ax^2 + Bxy + Cy^2 + Dx + Ey + F = 0$$

is the equation of a curve. More specifically, we have found that circles, parabolas, ellipses, and hyperbolas, i.e., all conic sections, are represented by such second-degree equations.

In three-space the most general equation of the second degree in x, y, and z has the form

$$ax^2 + by^2 + cz^2 + dxy + exz + fyz + gx + hy + kz + l = 0, \quad (1)$$

where the quantities $a, b, c, \ldots, l$ are positive or negative numbers or zero. The points in space satisfying such an equation all lie on a surface. Certain special cases, such as spheres and cylinders, were discussed in Section 6. Any second-degree equation which does not reduce to a cylinder, plane, line, or point corresponds to a surface which we call **quadric**. Quadric surfaces are classified into six types, and it can be shown that every second-degree equation which does not degenerate into a cylinder, a plane, etc., corresponds to one of these six types. The proof of this result involves the study of translation and rotation of coordinates in three-dimensional space, a topic beyond the scope of this book.

DEFINITIONS. The **x, y, and z intercepts** of a surface are, respectively, the x, y, and z coordinates of the points of intersection of the surface with the respective axes. When we are given an equation of a surface, we get the x intercept by setting y and z equal to zero and solving for x. We proceed analogously for the y and z intercepts.

The **traces** of a surface on the coordinate planes are the curves of intersections of the surface with the coordinate planes. When we are given a surface, we obtain the trace on the xz plane by first setting y equal to zero and then considering the resulting equation in x and z as the equation of a curve in the plane, as in plane analytic geometry. A **section of a surface by a plane** is the curve of intersection of the surface with the plane.

EXAMPLE 1. Find the x, y, and z intercepts of the surface

$$3x^2 + 2y^2 + 4z^2 = 12.$$

Describe the traces of this surface. Find the section of this surface by the plane $z = 1$ and by the plane $x = 3$.

Solution. We set $y = z = 0$, getting $3x^2 = 12$; the x intercepts are at 2 and -2. Similarly, the y intercepts are at $\pm\sqrt{6}$, the z intercepts at $\pm\sqrt{3}$. To find the trace on the xy plane, we set $z = 0$, getting

$$3x^2 + 2y^2 = 12 \quad \text{or} \quad \frac{x^2}{4} + \frac{y^2}{6} = 1, \quad z = 0.$$

We recognize this curve as an ellipse, with major semi-axis $\sqrt{6}$, minor semi-axis 2, foci at $(0, \sqrt{2}, 0)$, $(0, -\sqrt{2}, 0)$. Similarly, the trace on the xz plane is the ellipse

$$\frac{x^2}{4} + \frac{z^2}{3} = 1, \quad y = 0,$$

and the trace on the yz plane is the ellipse

$$\frac{y^2}{6} + \frac{z^2}{3} = 1, \quad x = 0.$$

The section of the surface by the plane $z = 1$ is the curve

$$\left\{ \begin{array}{l} 3x^2 + 2y^2 + 4 = 12 \\ z = 1 \end{array} \right. \Leftrightarrow \left\{ \begin{array}{l} \dfrac{x^2}{8/3} + \dfrac{y^2}{4} = 1, \\ z = 1 \end{array} \right.$$

which we recognize as an ellipse. The section by the plane $x = 3$ is the curve

$$\left\{ \begin{array}{l} 27 + 2y^2 + 4z^2 = 12 \\ x = 3 \end{array} \right. \Leftrightarrow \left\{ \begin{array}{l} 2y^2 + 4z^2 + 15 = 0 \\ x = 3 \end{array} \right. .$$

Since the sum of three positive quantities can never be zero, we conclude that the plane $x = 3$ does not intersect the surface. The section is empty.

DEFINITIONS. A surface is **symmetric with respect to the xy plane** if and only if the point $(x, y, -z)$ lies on the surface whenever (x, y, z) does; *it is* **symmetric with respect to the x axis** if and only if the point $(x, -y, -z)$ is on the locus whenever (x, y, z) is. Similar definitions are easily formulated for symmetry with respect to the remaining coordinate planes and axes.

The notions of intercepts, traces, and symmetry are useful in the following description of the six types of quadric surfaces.

(i) An **ellipsoid** is the locus of an equation of the form

$$\frac{x^2}{A^2} + \frac{y^2}{B^2} + \frac{z^2}{C^2} = 1.$$

The surface is sketched in Fig. 10–18. The x, y, and z intercepts are the numbers $\pm A$, $\pm B$, $\pm C$, respectively, and the traces on the xy, xz, and yz planes are, respectively, the ellipses

$$\frac{x^2}{A^2} + \frac{y^2}{B^2} = 1, \qquad \frac{x^2}{A^2} + \frac{z^2}{C^2} = 1, \qquad \frac{y^2}{B^2} + \frac{z^2}{C^2} = 1.$$

Sections made by the planes $y = k$ (k a constant) are the similar ellipses

$$\frac{x^2}{A^2(1 - k^2/B^2)} + \frac{z^2}{C^2(1 - k^2/B^2)} = 1, \qquad y = k, \quad -B < k < B.$$

Several such ellipses are drawn in Fig. 10–18.

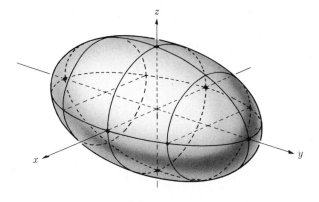

Figure 10–18

If $A = B = C$, we obtain a sphere while, if two of the three numbers are equal, the surface is an **ellipsoid of revolution,** also called a **spheroid.** If, for example, $A = B$ and $C > A$, the surface is called a **prolate spheroid,** exemplified by a football. On the other hand, if $A = B$ and $C < A$, we have an **oblate spheroid.** The earth is approximately the shape of an oblate spheroid, with the section at the equator being circular and the distance between the North and South poles being smaller than the diameter of the equatorial circle.

(ii) An **elliptic hyperboloid of one sheet** is the locus of an equation of the form

$$\frac{x^2}{A^2} + \frac{y^2}{B^2} - \frac{z^2}{C^2} = 1.$$

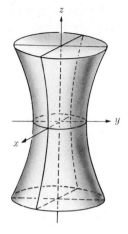

A locus of such a surface is sketched in Fig. 10–19. The x intercepts are at $\pm A$ and the y intercepts at $\pm B$. As for the z intercepts, we must solve the equation $-z^2/C^2 = 1$, which has no real solutions. Therefore the surface does not intersect the z axis. The trace on the xy plane is an ellipse, while the traces on the yz and xz planes are hyperbolas. The sections made by any plane $z = k$ are the ellipses

$$\frac{x^2}{A^2(1 + k^2/C^2)} + \frac{y^2}{B^2(1 + k^2/C^2)} = 1, \qquad z = k,$$

Figure 10–19

and the sections made by the planes $y = k$ are the hyperbolas

$$\frac{x^2}{A^2(1 - k^2/B^2)} - \frac{z^2}{C^2(1 - k^2/B^2)} = 1, \qquad y = k.$$

(iii) An **elliptic hyperboloid of two sheets** is the locus of an equation of the form

$$\frac{x^2}{A^2} - \frac{y^2}{B^2} - \frac{z^2}{C^2} = 1.$$

Such a locus is sketched in Fig. 10–20. We observe that we must have $|x| \geq A$, for otherwise the quantity $(x^2/A^2) < 1$ and the left side of the above equation will always be less than the right side. The x intercepts are at $x = \pm A$. There are no y and z intercepts. The traces on the xz and xy planes are hyperbolas; there is no trace on the yz plane. The sections made by the planes $x = k$ are the ellipses

$$\frac{y^2}{B^2(k^2/A^2 - 1)} + \frac{z^2}{C^2(k^2/A^2 - 1)} = 1, \quad x = k, \qquad \text{if } |k| > A,$$

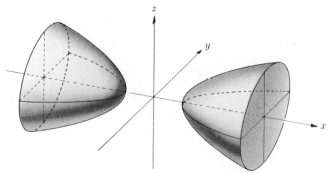

Figure 10–20

while the trace is empty if $|k| < A$. The sections by the planes $y = k$ and $z = k$ are hyperbolas.

(iv) An **elliptic paraboloid** is the locus of an equation of the form

$$\frac{x^2}{A^2} + \frac{y^2}{B^2} = z.$$

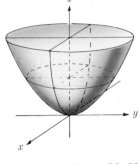

A typical elliptic paraboloid is sketched in Fig. 10–21.

All three intercepts are 0; the traces on the yz and xz planes are parabolas, while the trace on the xy plane consists of a single point, the origin. Sections made by planes $z = k$ are ellipses if $k > 0$, empty if $k < 0$. Sections made by planes $x = k$ and $y = k$ are parabolas.

Figure 10–21

If $A = B$, we have a **paraboloid of revolution,** and the sections made by the planes $z = k, k > 0$ are circles. The reflecting surfaces of telescopes, automobile headlights, etc., are always paraboloids of revolution. (See Chapter 6, Section 2.)

(v) A **hyperbolic paraboloid** is the locus of an equation of the form

$$\frac{x^2}{A^2} - \frac{y^2}{B^2} = z.$$

Such a locus is sketched in Fig. 10–22. As in the elliptic paraboloid, all intercepts are zero. The trace on the xz plane is a parabola opening upward; the trace on the yz plane is a parabola opening downward; and the trace on the xy plane is the pair of intersecting straight lines

$$y = \pm(B/A)x.$$

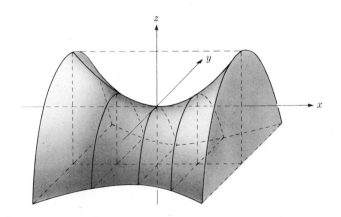

Figure 10–22

As Fig. 10–22 shows, the surface is "saddle-shaped"; sections made by planes $x = k$ are parabolas opening downward and those made by planes $y = k$ are parabolas opening upward. The sections made by planes $z = k$ are hyperbolas facing one way if $k < 0$ and the other way if $k > 0$. The trace on the xy plane corresponds to $k = 0$ and, as we saw, consists of two intersecting lines.

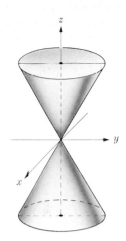

(vi) An **elliptic cone** is the locus of an equation of the form

$$\frac{x^2}{A^2} + \frac{y^2}{B^2} = \frac{z^2}{C^2}.$$

A typical cone of this type is shown in Fig. 10–23. Once again all intercepts are zero. The traces on the xz and yz planes are pairs of intersecting straight lines, while the trace on the xy plane is a single point, the origin. Planes parallel to the coordinate planes yield sections which are the familiar conic sections of plane analytic geometry.

Figure 10–23

EXAMPLE 2. Name and sketch the locus of $4x^2 - 9y^2 + 8z^2 = 72$. Indicate a few sections parallel to the coordinate planes.

Solution. We divide by 72, getting

$$\frac{x^2}{18} - \frac{y^2}{8} + \frac{z^2}{9} = 1,$$

which is an elliptic hyperboloid of one sheet. The x intercepts are $\pm 3\sqrt{2}$, the z intercepts are ± 3, and there are no y intercepts. The trace on the xy plane is the hyperbola

$$\frac{x^2}{18} - \frac{y^2}{8} = 1,$$

the trace on the xz plane is the ellipse

$$\frac{x^2}{18} + \frac{z^2}{9} = 1,$$

and the trace on the yz plane is the hyperbola

$$\frac{z^2}{9} - \frac{y^2}{8} = 1.$$

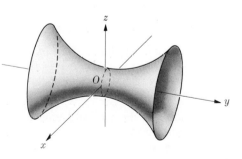

The surface is sketched in Fig. 10-24.

Figure 10–24

PROBLEMS

In problems 1 through 18, name and sketch the locus of each equation.

1. $\dfrac{x^2}{4} + \dfrac{y^2}{16} + \dfrac{z^2}{9} = 1$

2. $\dfrac{x^2}{9} + \dfrac{y^2}{9} + \dfrac{z^2}{12} = 1$

3. $\dfrac{x^2}{20} + \dfrac{y^2}{16} + \dfrac{z^2}{20} = 1$

4. $-\dfrac{x^2}{4} + \dfrac{y^2}{16} + \dfrac{z^2}{9} = 1$

5. $-\dfrac{x^2}{4} + \dfrac{y^2}{16} - \dfrac{z^2}{9} = 1$

6. $\dfrac{x^2}{4} + \dfrac{y^2}{16} - \dfrac{z^2}{9} = 1$

7. $\dfrac{x^2}{4} + \dfrac{z^2}{9} = 1 + \dfrac{y^2}{16}$

8. $\dfrac{x^2}{4} - \dfrac{y^2}{16} - \dfrac{z^2}{9} = 1$

9. $\dfrac{y}{4} = \dfrac{x^2}{9} + \dfrac{z^2}{4}$

10. $x = \dfrac{z^2}{4} - \dfrac{y^2}{9}$

11. $z = \dfrac{x^2}{8} + \dfrac{y^2}{8}$

12. $x^2 = 4y^2 + 4z^2$

13. $z^2 = x^2 + y^2$

14. $x^2 = y^2 - z^2$

15. $-3x^2 + 2y^2 + 6z^2 = 8$

16. $2x^2 + 4y^2 = 3z^2$

17. $3y = -5x^2 + 2z^2$

18. $\dfrac{z}{5} = 8x^2 - 2y^2$

19. A point moves so that the sum of its distances from $F_1(c, 0, 0)$ and $F_2(-c, 0, 0)$ is always $2a$. Show that the locus satisfies the equation

$$\frac{x^2}{a^2} + \frac{y^2}{a^2 - c^2} + \frac{z^2}{a^2 - c^2} = 1.$$

The surface is a **prolate spheroid**.

20. Show that the intersection of the hyperbolic paraboloid $x^2 - y^2 = z$ with the plane $z = x + y$ consists of two intersecting straight lines. Establish the same result for the intersection of this quadric with the plane $z = ax + ay$, where a is any number.

21. Show that the surface $z = xy$ has the property that each point on the surface is contained in a line which lies entirely on the surface. (Such surfaces are called **ruled surfaces**. Cylinders are special cases of ruled surfaces.)

22. Show that any hyperbolic paraboloid is a ruled surface. (See Problems 20 and 21.)

▶8.   TRANSLATION OF AXES*

Figure 10–25(a) shows a Cartesian coordinate system in three dimensions. Suppose we introduce a second Cartesian coordinate system, with axes x', y', and z' so located that the x' axis is parallel to the x axis and h units from it, the y' axis is parallel to the y axis and k units from it, and the z' axis is parallel to the z axis and l units from it (Fig. 10–25b). A point P in space will have coordinates in both systems. If its coordinates are (x, y, z) in the original system and (x', y', z') in the second system, the equations

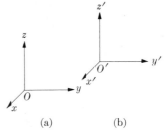

$$x' = x - h,$$
$$y' = y - k,$$
$$z' = z - l$$

hold.†   Two coordinate systems, xyz and $x'y'z'$, which satisfy these equations are said to be related by a **translation of axes.**

(a)   (b)

Figure 10–25

As in the case of plane analytic geometry (Chapter 6, Section 7), the method of translation of axes may be used to simplify second-degree equations, thereby making evident the nature of certain quadric surfaces. *The principal tool in this process is "completing the square."* We illustrate the method with three examples.

EXAMPLE 1. Use a translation of coordinates to identify the quadric surface

$$x^2 + 4y^2 + 3z^2 + 2x - 8y + 9z = 10.$$

Sketch.

Solution. We write

$$x^2 + 2x \qquad + 4(y^2 - 2y \qquad) + 3(z^2 + 3z \qquad) = 10.$$

Completing the square, we obtain

$$(x + 1)^2 + 4(y - 1)^2 + 3(z + \tfrac{3}{2})^2 = 10 + 1 + 4 + \tfrac{27}{4}.$$

Introducing the translation of coordinates

$$x' = x + 1,$$
$$y' = y - 1,$$
$$z' = z + \tfrac{3}{2},$$

---

* This section may be omitted without loss of continuity.

† A general proof that these relations hold for all positions of O, O', and P would involve vectors, as in Chapter 6, Section 7, and is omitted.

we find, for the equation of the surface,

$$x'^2 + 4y'^2 + 3z'^2 = \tfrac{87}{4},$$

which we recognize as an ellipsoid. The surface and both coordinate systems are sketched in Fig. 10–26.

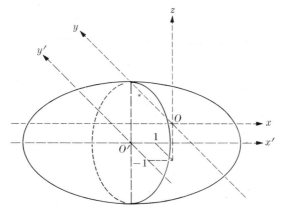

Figure 10–26

EXAMPLE 2. Use a translation of coordinates to identify the surface

$$2x^2 - 3y^2 + 6x - 12y - 4z = 0.$$

Sketch.

Solution. We write the equation in the form

$$2(x^2 + 3x \qquad) - 3(y^2 + 4y \qquad) = 4z$$

and complete the square, getting

$$2(x + \tfrac{3}{2})^2 - 3(y + 2)^2 = 4z + \tfrac{9}{2} - 12$$
$$= 4(z - \tfrac{15}{8}).$$

The translation of axes

$$x' = x + \tfrac{3}{2}, \qquad y' = y + 2, \qquad z' = z - \tfrac{15}{8},$$

yields the equation

$$2x'^2 - 3y'^2 = 4z'.$$

Dividing by 4, we get

$$\frac{x'^2}{2} - \frac{y'^2}{4/3} = z',$$

which we see is a hyperbolic paraboloid. The surface and both coordinate systems are sketched in Fig. 10-27.

EXAMPLE 3. Use a translation of coordinates to identify the surface

$$4x^2 - 3y^2 + 2z^2 + 8x + 18y - 8z = 15.$$

Sketch.

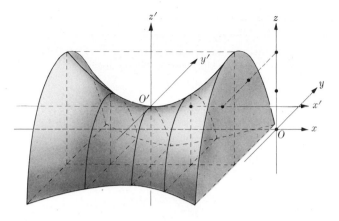

Figure 10–27

Solution. We write

$$4(x^2 + 2x \qquad) - 3(y^2 - 6y \qquad) + 2(z^2 - 4z \qquad) = 15.$$

Completing the square, we obtain

$$4(x + 1)^2 - 3(y - 3)^2 + 2(z - 2)^2 = 15 + 4 - 27 + 8 = 0.$$

The translation of axes,

$$x' = x + 1, \qquad y' = y - 3, \qquad z' = z - 2,$$

shows that the surface is an elliptic cone with equation

$$4x'^2 + 2z'^2 = 3y'^2.$$

The surface and coordinate systems are shown in Fig. 10–28.

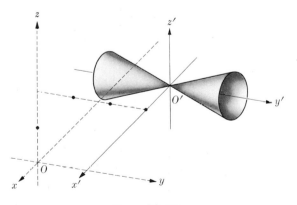

Figure 10–28

PROBLEMS

In each of problems 1 through 12, employ a translation of coordinates to identify the locus. Sketch the surface, showing both coordinate systems.

1. $x^2 + y^2 + z^2 - 3x + 4y - 8z = 0$

2. $2x^2 + 4y^2 + 6z^2 - 2x + 4y - 15z = 10$

3. $x^2 + 2y^2 - z^2 + 2x + 4y - 6z = 18$

4. $x^2 + 4y^2 + z^2 - 3x + 2y - 4z + 9 = 0$

5. $x^2 + 8z^2 + 2x - 3y + 16z = 0$

6. $2x^2 + 3y^2 + 4x + 2z = 5$

7. $2y^2 - 3z^2 + 4x - 3y + 2z = 0$

8. $x^2 + y^2 - z^2 + 2x + 4y - 2z = 0$

9. $x^2 + y^2 - z^2 + 4x + 8y + 6z + 11 = 0$

10. $2x^2 - 5z^2 + 3x - 4y + 10z = 4$

11. $x^2 - y^2 + 2x - 3y + 4z = 0$

12. $x^2 - 2y^2 - 3z^2 + 4x - 6y + 8z = 2$

13. Use the rotation of coordinates in the plane

$$x = x' \cos \theta - y' \sin \theta,$$
$$y = x' \sin \theta + y' \cos \theta,$$
$$z = z'$$

(see Chapter 6, Section 8) to eliminate the xy term in the following equation of a surface:

$$8x^2 - 4xy + 5y^2 + z^2 = 36.$$

Identify the surface and sketch.

14. Use the method of problem 13 to eliminate the xz term in the equation

$$2x^2 + y^2 - 2z^2 + 3xz = 25.$$

Identify the surface and sketch.

15. Use a rotation of coordinates, as in problem 13, and a translation of coordinates to simplify and identify the surface

$$x^2 - 4xz + 2y^2 + 4z^2 - 3x + 2y - 4z = 5.$$

Sketch.

16. Use a rotation of coordinates, as in problem 13, to show that $3x^2 = 2y + 2z$ is the equation of a parabolic cylinder.

17. Use a rotation of axes as in problem 13, and follow this by the rotation

$$x' = x'' \cos \phi - z'' \sin \phi, \quad z' = x'' \sin \phi + z'' \cos \phi, \quad y' = y''$$

to eliminate first the xy and then the $x'z'$ and $y'z'$ terms in the equation of the surface

$$8x^2 - 4xy + 5y^2 + \sqrt{5}\,xz + 2\sqrt{5}\,yz - 8z^2 = 153.$$

Identify the surface.

▶9. OTHER COORDINATE SYSTEMS

In plane analytic geometry we employed a Cartesian coordinate system for certain types of problems and a polar coordinate system for others. We saw that there are circumstances in which one system is more convenient than the other. A similar situation prevails in three-dimensional geometry, and we now take up systems of coordinates other than the Cartesian one which we have studied exclusively so far. One such system, known as **cylindrical coordinates,** is described in the following way. A point P in space with Cartesian coordinates (x, y, z) may also be located by replacing the x and y values with the corresponding polar coordinates r, θ and by allowing the z value to remain unchanged. In other words, to each ordered number triple of the form (r, θ, z), there is associated a point in space. The transformation from cylindrical to Cartesian coordinates is given by the equations

$$x = r\cos\theta, \qquad y = r\sin\theta, \qquad z = z.$$

The transformation from Cartesian to cylindrical coordinates is given by

$$r^2 = x^2 + y^2, \qquad \tan\theta = y/x, \qquad z = z.$$

If the coordinates of a point are given in one system, the above equations show how to get the coordinates in the other. Figure 10–29 exhibits the relation between the two systems. It is always assumed that the origins of the systems coincide and that $\theta = 0$ corresponds to the xz plane. We see that the locus $\theta = $ const consists of all points in a plane containing the z axis. The locus $r = $ const consists of all points on a right circular cylinder with the z axis as its central axis. (The term "cylindrical coordinates" comes from this fact.) The locus $z = $ const consists of all points in a plane parallel to the xy plane.

EXAMPLE 1. Find the cylindrical coordinates of the points whose Cartesian coordinates are $P(3, 3, 5)$, $Q(2, 0, -1)$, $R(0, 4, 4)$, $S(0, 0, 5)$, $T(2, 2\sqrt{3}, 1)$.

Solution. For the point P we have $r = \sqrt{9 + 9} = 3\sqrt{2}$, $\tan\theta = 1$, $\theta = \pi/4$, $z = 5$. Therefore the coordinates are $(3\sqrt{2}, \pi/4, 5)$. For Q we have $r = 2$, $\theta = 0$, $z = -1$. The coordinates are $(2, 0, -1)$. For R we get $r = 4$, $\theta = \pi/2$, $z = 4$. The result is $(4, \pi/2, 4)$. For S we see at once that the coordinates are $(0, \theta, 5)$ for any θ. For T we get $r = \sqrt{4 + 12} = 4$, $\tan\theta = \sqrt{3}$, $\theta = \pi/3$. The answer is $(4, \pi/3, 1)$.

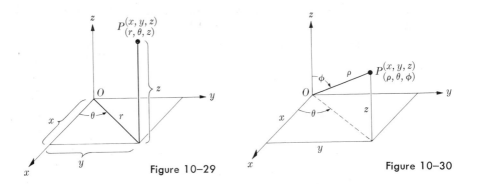

Figure 10–29   Figure 10–30

Remark. Just as polar coordinates do not give a one-to-one correspondence between ordered number pairs and points in the plane, so cylindrical coordinates do not give a one-to-one correspondence between ordered number triples and points in space.

A **spherical coordinate system** is defined in the following way. A point P with Cartesian coordinates (x, y, z) has spherical coordinates (ρ, θ, ϕ) where ρ is the distance of the point P from the origin, θ is the same quantity as in cylindrical coordinates, and ϕ is the angle that the directed line $\overrightarrow{OP}$ makes with the positive z direction. Figure 10–30 exhibits the relation between Cartesian and spherical coordinates. The transformation from spherical to Cartesian coordinates is given by the equations

$$x = \rho \sin \phi \cos \theta,$$

$$y = \rho \sin \phi \sin \theta,$$

$$z = \rho \cos \phi.$$

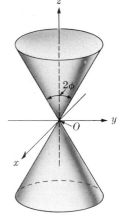

The transformation from Cartesian to spherical coordinates is given by

$$\rho^2 = x^2 + y^2 + z^2,$$

$$\tan \theta = \frac{y}{x},$$

and

$$\cos \phi = \frac{z}{\sqrt{x^2 + y^2 + z^2}}.$$

Figure 10–31

We note that the locus $\rho = $ const is a sphere with center at the origin (from which is derived the term "spherical coordinates"). The locus $\theta = $ const is a plane through the z axis, as in cylindrical coordinates. The locus $\phi = $ const is a cone with vertex at the origin and angle opening 2ϕ if $0 < \phi < \pi/2$. (See Fig. 10–31.) The lower nappe of the cone in Fig. 10–31 is or is not included according as negative values of ρ are or are not allowed.

EXAMPLE 2. Find an equation in spherical coordinates of the sphere

$$x^2 + y^2 + z^2 - 2z = 0.$$

Sketch the locus.

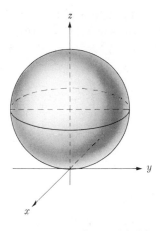

Solution. Since $\rho^2 = x^2 + y^2 + z^2$ and $z = \rho \cos \phi$, we have

$$\rho^2 - 2\rho \cos \phi = 0 \equiv \rho(\rho - 2 \cos \phi) = 0.$$

The locus of this equation is the locus of $\rho = 0$ and $\rho - 2 \cos \phi = 0$. The locus of $\rho = 0$ is on the locus of $\rho - 2 \cos \phi = 0$ (with $\phi = \pi/2$). Plotting the surface

$$\rho = 2 \cos \phi,$$

we get the surface shown in Fig. 10–32.

Figure 10–32

If ρ is constant, then the quantities (θ, ϕ) form a coordinate system on the surface of a sphere. Latitude and longitude on the surface of the earth also form a coordinate system. If we restrict θ so that $-\pi < \theta \leq \pi$, then θ is called the **longitude** of the point in spherical coordinates. If ϕ is restricted so that $0 \leq \phi \leq \pi$, then ϕ is called the **colatitude** of the point. That is, ϕ is $(\pi/2)$ − latitude, where latitude is taken in the ordinary sense—i.e., positive north of the equator and negative south of it.

PROBLEMS

---

1. Find a set of cylindrical coordinates for each of the points whose Cartesian coordinates are
   (a) $(3, 3, 7)$,   (b) $(4, 8, 2)$,   (c) $(-2, 3, 1)$.

2. Find the Cartesian coordinates of the points whose cylindrical coordinates are
   (a) $(2, \pi/3, 1)$,   (b) $(3, -\pi/4, 2)$,   (c) $(7, 2\pi/3, -4)$.

3. Find a set of spherical coordinates for each of the points whose Cartesian coordinates are
   (a) $(2, 2, 2)$,   (b) $(2, -2, -2)$,   (c) $(-1, \sqrt{3}, 2)$.

4. Find the Cartesian coordinates of the points whose spherical coordinates are
   (a) $(4, \pi/6, \pi/4)$,   (b) $(6, 2\pi/3, \pi/3)$,   (c) $(8, \pi/3, 2\pi/3)$.

5. Find a set of cylindrical coordinates for each of the points whose spherical coordinates are
   (a) $(4, \pi/3, \pi/2)$,   (b) $(2, 2\pi/3, 5\pi/6)$,   (c) $(7, \pi/2, \pi/6)$.

6. Find a set of spherical coordinates for each of the points whose cylindrical coordinates are
   (a) $(2, \pi/4, 1)$,   (b) $(3, \pi/2, 2)$,   (c) $(1, 5\pi/6, -2)$.

In each of problems 7 through 16, find an equation in cylindrical coordinates of the locus whose (x, y, z) equation is given. Sketch.

7. $x^2 + y^2 + z^2 = 9$ 8. $x^2 + y^2 + 2z^2 = 8$

9. $x^2 + y^2 = 4z$ 10. $x^2 + y^2 - 2x = 0$

11. $x^2 + y^2 = z^2$ 12. $x^2 + y^2 + 2z^2 + 2z = 0$

13. $x^2 - y^2 = 4$ 14. $xy + z^2 = 5$

15. $x^2 + y^2 - 4y = 0$ 16. $x^2 + y^2 + z^2 - 2x + 3y - 4z = 0$

In each of problems 17 through 22, find an equation in spherical coordinates of the locus whose (x, y, z) equation is given. Sketch.

17. $x^2 + y^2 + z^2 - 4z = 0$ 18. $x^2 + y^2 + z^2 + 2z = 0$

19. $x^2 + y^2 = z^2$ 20. $x^2 + y^2 = 4$

21. $x^2 + y^2 = 4z + 4$ (Solve for ρ in terms of ϕ.)

22. $x^2 + y^2 - z^2 + z - y = 0$

▶ 10.   LINEAR INEQUALITIES

Every plane divides three-dimensional space into two regions. If a plane is represented by the equation

$$Ax + By + Cz + D = 0,$$

then all points on one side of the plane satisfy the inequality

$$Ax + By + Cz + D > 0,$$

and all points on the other side satisfy the opposite inequality.

Two intersecting planes divide three-space into four regions, each of which may be described by a pair of linear inequalities. For example, the intersecting planes

$$\Phi_1: x - y + 2z - 4 = 0, \qquad \Phi_2: x + 2y - z + 2 = 0,$$

separate three-space into four regions which we label R_1, R_2, R_3, and R_4. Every point in the region R_1 satisfies the following pair of inequalities

$$x - y + 2z - 4 > 0,$$
$$x + 2y - z + 2 > 0;$$

every point in R_2 satisfies the pair of inequalities

$$x - y + 2z - 4 > 0,$$
$$x + 2y - z + 2 < 0,$$

and so forth. Figure 10–33 shows, in a schematic way, how Φ_1 and Φ_2 divide space into four regions.

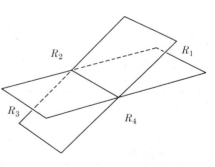

Figure 10–33

Two parallel planes divide three-dimensional space into three regions. For example, the parallel planes

$$2x + y - z + 1 = 0, \qquad 2x + y - z - 5 = 0,$$

separate space into three regions, R_1, R_2, and R_3, as shown in Fig. 10–34. Note that it is impossible for the two inequalities

$$2x + y - z < -1, \qquad 2x + y - z > 5,$$

to hold simultaneously, and so there is no region corresponding to this pair of inequalities.

Three planes divide three-dimensional space into at most eight regions. If the planes are parallel or have a line in common, then the division is into fewer than eight regions. In fact, it is not difficult to see that three distinct planes always decompose the entire space into 4, 6, 7, or 8 regions, depending on the circumstances.

An enumeration of all possible types of regions into which four planes may separate space is lengthy and tedious. The student can verify that there can be at *most* $2^4 = 16$ regions. Similarly, it is not difficult to show that n planes can divide space into at most 2^n regions.

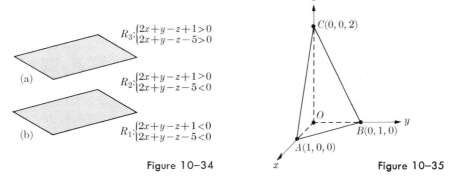

Figure 10–34 · Figure 10–35

Rather than attempt to describe in a general way the manner in which a system of inequalities determines a region, we shall consider several special cases which are of particular interest.

The system of inequalities

$$x > 0, \qquad y > 0, \qquad z > 0$$

determines a region which we call the first octant. We recognize that $x = 0$ is the equation of the yz plane. Similarly, $y = 0$ and $z = 0$ are the equations of the other two coordinate planes.

The four inequalities

$$x > 0, \qquad y > 0, \qquad z > 0, \qquad 2x + 2y + z - 2 < 0$$

enclose a region in the shape of a tetrahedron (Fig. 10–35).

A region determined by a system of linear inequalities has a boundary consisting of three parts. First, there are the parts we call **faces,** which are composed of portions of planes, such as triangles, quadrilaterals, etc. Two adjoining faces have in common a straight line segment which we call an **edge.** Finally, the points which are the intersections of edges are called **vertices.**

In the example of the tetrahedron in Fig. 10–35, the four vertices are at $A(1, 0, 0)$, $B(0, 1, 0)$, $C(0, 0, 2)$, and the origin O.

To determine the vertices of the boundary of a region defined by a system of linear inequalities, we may proceed by a method which is an extension of the one described in Chapter 3, Section 11. We do so by finding the vertices in a particular example.

EXAMPLE. Determine the region in three-space (if any) which satisfies all the inequalities

$$L_1: x \geq 0, \qquad L_2: y \geq 0, \qquad L_3: z \geq 0,$$
$$L_4: -x - y - 2z + 4 \geq 0, \qquad L_5: -x - y - 4z + 6 \geq 0.$$

Solution. We have a system of five linear inequalities, and we find all possible points of intersections of the five corresponding *equations.* Since three planes determine a point, there are $(5 \cdot 4 \cdot 3)/(1 \cdot 2 \cdot 3) = 10$ possible points of intersection. We find for these points of intersection:

$$P_1(0, 0, 0), \qquad P_2(0, 0, 2), \qquad P_3(0, 0, \tfrac{3}{2}), \qquad P_4(0, 4, 0),$$
$$P_5(0, 6, 0), \qquad P_6(0, 2, 1), \qquad P_7(4, 0, 0), \qquad P_8(6, 0, 0),$$
$$P_9(2, 0, 1), \qquad P_{10}(\text{no solution}).$$

The point P_1 was obtained by solving simultaneously L_1, L_2, and L_3; the point P_2 by solving simultaneously L_1, L_2, and L_4; the point P_3 by solving L_1, L_2, and L_5; and so forth.

The coordinates of P_1 satisfy all five inequalities and, therefore, P_1 is a vertex of the region to be determined. The coordinates of P_2 do not satisfy L_5 and this point is rejected. Continuing, we see that the vertices of the region are the points P_1, P_3, P_4, P_6, P_7, and P_9 (see Fig. 10–36). To determine an edge of the boundary, we check the vertices by pairs. The vertices P_1 and P_7 connect an edge, since P_1 was obtained from L_1, L_2, and L_3, while P_7 was obtained from L_2, L_3, and L_4. In general, if two vertices have in common two of the planes determining their intersection, then the line segment joining them is an edge. The edges of the region are shown in Fig. 10–36. We also see why P_{10} could not be obtained. The planes

$$z = 0, \qquad x + y + 2z - 4 = 0,$$

and

$$x + y + 4z - 6 = 0$$

intersect in parallel lines. (Note that the line through P_6 and P_9 is parallel to the line through P_4 and P_7.)

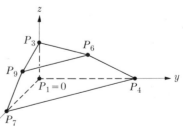

Figure 10–36

PROBLEMS

In each of the following problems, determine the region, if any, which satisfies the given system of inequalities.

1. $x \geq 0, \quad y \geq 0, \quad z \geq 0, \quad x + y + 2z - 4 \leq 0$

2. $x \geq 0, \quad y > 0, \quad z \geq 0, \quad 3x + 2y - 7 \leq 0$

3. $x > 0, \quad y > 0, \quad z < 0, \quad x + 2y - 2z - 4 \leq 0$

4. $x \geq 0, \quad y \geq 0, \quad z \geq 0, \quad 2x + y - 5 \leq 0, \quad z - 5 \leq 0$

5. $x \geq 0, \quad y \geq 0, \quad z \geq 0, \quad 3x + 2y - 9 \leq 0, \quad x \leq 2, \quad y \leq 2$

6. $x \geq 0, \quad y \geq 0, \quad z \geq 0, \quad x + y + 3z - 9 \leq 0, \quad x + y + 6z - 12 \leq 0,$
   $y \leq 15$

7. $x \leq 5, \quad z \geq -2, \quad y \leq 4, \quad x - y \leq 1, \quad z + x \geq 1, \quad y + z \geq 2$

8. $x \geq 0, \quad y \geq 0, \quad z \geq 0, \quad x + y + 2z - 1 \leq 0, \quad 3x + y \geq 12$

11 Vectors in Three Dimensions*

▶1. OPERATIONS WITH VECTORS

The development of vectors in three-dimensional space parallels the development in the plane, as given in Chapter 4. The student should review the material in that chapter since the same notation and terminology will be employed here.

A **directed line segment** $\overrightarrow{AB}$ is defined as before, except that now the **base** A and the **head** B may be situated anywhere in three-space. The **magnitude** of a directed line segment is its length. If $\overrightarrow{AB}$ is a directed line segment and $\vec{L}$ is a directed line in space, we define the **projection** of $\overrightarrow{AB}$ on $\vec{L}$ in the same way as in Chapter 3, Section 7; the projection is the number $\overline{A'B'}$ where A' and B' are the feet of the perpendicular dropped from A and B onto $\vec{L}$. As before, we denote this number by

$$\text{Proj}_{\vec{L}}\ \overrightarrow{AB}.$$

Theorem 1. *Suppose that A and B with coordinates (x_A, y_A, z_A) and (x_B, y_B, z_B), respectively, are given and that $\vec{L}$ is a directed line with direction cosines $\cos \alpha$, $\cos \beta$, $\cos \gamma$. Then we have the formula*

$$\text{Proj}_{\vec{L}}\ \overrightarrow{AB} = (x_B - x_A) \cos \alpha + (y_B - y_A) \cos \beta + (z_B - z_A) \cos \gamma. \qquad (1)$$

Proof. Let L_1 be the directed line through A and B directed so that A precedes B. Let $\cos \alpha_1$, $\cos \beta_1$, $\cos \gamma_1$ be the direction cosines of L_1. If φ is the angle between L and L_1, then

$$\text{Proj}_{\vec{L}}\ \overrightarrow{AB} = |AB| \cos \varphi$$
$$= |AB| (\cos \alpha_1 \cos \alpha + \cos \beta_1 \cos \beta + \cos \gamma_1 \cos \gamma).$$

---

* For an understanding of this chapter, we assume the reader is acquainted with determinants of the second and third order. For those unfamiliar with the subject a brief discussion is provided in Appendix 4.

Since

$$\cos \alpha_1 = \frac{x_B - x_A}{|AB|}, \qquad \cos \beta_1 = \frac{y_B - y_A}{|AB|}, \qquad \cos \gamma_1 = \frac{z_B - z_A}{|AB|},$$

formula (1) is obtained by substitution.

DEFINITION. Two directed line segments $\overrightarrow{AB}$ and $\overrightarrow{CD}$ are said to have the **same magnitude and direction** if and only if

$$\text{Proj}_{\vec{L}} \; \overrightarrow{AB} = \text{Proj}_{\vec{L}} \; \overrightarrow{CD}$$

for every directed line $\vec{L}$.

We recall that this definition is the same as that given for directed line segments in the plane in Chapter 4, Section 1. Whenever two directed line segments have the same magnitude and direction we say they are **equivalent** and we write

$$\overrightarrow{AB} \approx \overrightarrow{CD}.$$

In complete analogy with Theorem 1 of Chapter 4, we state the following result.

Theorem 2. *Suppose that A, B, C, and D are points in space with coordinates* (x_A, y_A, z_A), (x_B, y_B, z_B), (x_C, y_C, z_C), *and* (x_D, y_D, z_D), *respectively. If the equations*

$$x_B - x_A = x_D - x_C, \qquad y_B - y_A = y_D - y_C, \qquad z_B - z_A = z_D - z_C \quad (2)$$

hold, then $\overrightarrow{AB} \approx \overrightarrow{CD}$. *Conversely, if* $\overrightarrow{AB} \approx \overrightarrow{CD}$, *then equations (2) hold for any Cartesian coordinate system.*

The details of the proof are similar to the proof of Theorem 1, Chapter 4, and are left to the student.

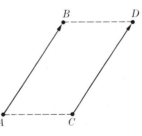

Remark. From Theorem 2, we see that two directed line segments $\overrightarrow{AB}$ and $\overrightarrow{CD}$ are equivalent if either one of the following two conditions holds:

(i) $\overrightarrow{AB}$ and $\overrightarrow{CD}$ are both on the same directed line $\vec{L}$ and their directed lengths are equal; or

(ii) the points A, C, D, and B are the vertices of a parallelogram as shown in Fig. 11–1.

Figure 11–1

If we are given a directed line segment $\overrightarrow{AB}$, it is clear that there is an unlimited number of equivalent ones. In fact, if C is any given point in three-space, we can use equations (1) of Theorem 1 to find the coordinates of the unique point D such that $\overrightarrow{CD} \approx \overrightarrow{AB}$.

DEFINITIONS.   A **vector** is the collection of all directed line segments having a given magnitude and direction. We shall use boldface letters to denote vectors. A particular directed line segment in a collection **v** is called a **representative** of the vector **v**. The **length** of a vector is the common length of all its representatives. A **unit vector** is a vector of length one. Two vectors are said to be **orthogonal** (or **perpendicular**) if any representative of one vector is perpendicular to any representative of the other. The **zero vector,** denoted by **0**, is the class of directed line "segments" of zero length (i.e., simply points). We make the convention that **0** is orthogonal to all vectors.

As in Chapter 4, Section 2, we can define the sum of two vectors. Given **u** and **v**, let $\overrightarrow{AB}$ be a representative of **u** and let $\overrightarrow{BC}$ be that representative of **v** which has its base at B. Then **u** + **v** is the vector which has representative $\overrightarrow{AC}$ as shown in Fig. 11–2. If $\overrightarrow{A'B'}$ and $\overrightarrow{B'C'}$ are other representatives of **u** and **v**, respectively, it follows from Theorem 1 that $\overrightarrow{A'C'} \approx \overrightarrow{AC}$. Therefore $\overrightarrow{A'C'}$ is also a representative of **u** + **v**. In other words, the rule for forming the sum of two vectors does not depend on the particular representatives we select in making the calculation.

Figure 11–2

Vectors may be multiplied by numbers (scalars). Given a vector **u** and a number c, let $\overrightarrow{AB}$ be a representative of **u** and let C be the point c of the way from A to B. Then $\overrightarrow{AC}$ is a representative of c**u**. It follows easily from Theorem 1 that if $\overrightarrow{A'B'}$ is another representative of **u** and C' is c of the way from A' to B', then $\overrightarrow{A'C'} \approx \overrightarrow{AC}$ and so is another representative of c**u**.

DEFINITIONS.   Suppose that a Cartesian coordinate system is given. Figure 11–3 shows such a system with the points $I(1, 0, 0), J(0, 1, 0)$, and $K(0, 0, 1)$ identified. The **unit vector i** is defined as the vector which has $\overrightarrow{OI}$ as one of its representatives. The **unit vector j** is defined as the vector which has $\overrightarrow{OJ}$ as one of its representatives. The **unit vector k** is defined as the vector which has $\overrightarrow{OK}$ as one of its representatives.

We now establish a direct extension of Theorem 2 in Chapter 4.

Theorem 3. *Suppose a vector* **w** *has* $\overrightarrow{AB}$ *as a representative. Denote the coordinates of A and B by* (x_A, y_A, z_A) *and* (x_B, y_B, z_B), *respectively. Then* **w** *may be expressed in the form*

$$\mathbf{w} = (x_B - x_A)\mathbf{i} + (y_B - y_A)\mathbf{j} + (z_B - z_A)\mathbf{k}.$$

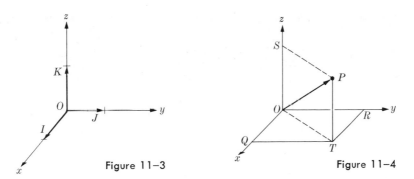

Figure 11–3   Figure 11–4

Proof. From equations (1) of Theorem 2, we know that **w** has the representative $\overrightarrow{OP}$ where P has coordinates $(x_B - x_A, y_B - y_A, z_B - z_A)$. Let $Q(x_B - x_A, 0, 0)$, $R(0, y_B - y_A, 0)$, $S(0, 0, z_B - z_A)$, and $T(x_B - x_A, y_B - y_A, 0)$ be as shown in Fig. 11–4. Then Q is $x_B - x_A$ of the way from O to $I(1, 0, 0)$, and similarly for R and S with regard to J and K. Therefore

$$\mathbf{v}(\overrightarrow{OQ}) = (x_B - x_A)\mathbf{i},$$
$$\mathbf{v}(\overrightarrow{QT}) = \mathbf{v}(\overrightarrow{OR}) = (y_B - y_A)\mathbf{j},$$
$$\mathbf{v}(\overrightarrow{TP}) = \mathbf{v}(\overrightarrow{OS}) = (z_B - z_A)\mathbf{k}.$$

Using the rule for addition of vectors, we find

$$\mathbf{v}(\overrightarrow{OP}) = \mathbf{v}(\overrightarrow{OQ}) + \mathbf{v}(\overrightarrow{QT}) + \mathbf{v}(\overrightarrow{TP}),$$

and the proof is complete.

EXAMPLE 1.   A vector **v** has $\overrightarrow{AB}$ as a representative. If A and B have coordinates $(3, -2, 4)$ and $(2, 1, 5)$, respectively, express **v** in terms of **i**, **j**, and **k**.

Solution. From Theorem 3, we obtain

$$\mathbf{v}(\overrightarrow{AB}) = (2 - 3)\mathbf{i} + (1 + 2)\mathbf{j} + (5 - 4)\mathbf{k} = -\mathbf{i} + 3\mathbf{j} + \mathbf{k}.$$

The next two theorems are direct extensions of Theorems 3 and 4 in Chapter 4. The proofs are left to the student. (See problem 12 at the end of the next section.) The **length** of a vector **v** is denoted by $|\mathbf{v}|$.

Theorem 4. *If* $\mathbf{v} = a\mathbf{i} + b\mathbf{j} + c\mathbf{k}$, *then*

$$|\mathbf{v}| = \sqrt{a^2 + b^2 + c^2}.$$

Theorem 5. *If* $\mathbf{v} = a_1\mathbf{i} + b_1\mathbf{j} + c_1\mathbf{k}$, $\mathbf{w} = a_2\mathbf{i} + b_2\mathbf{j} + c_2\mathbf{k}$, *then*

$$\mathbf{v} + \mathbf{w} = (a_1 + a_2)\mathbf{i} + (b_1 + b_2)\mathbf{j} + (c_1 + c_2)\mathbf{k}.$$

If h is any number, then

$$h\mathbf{v} = ha_1\mathbf{i} + hb_1\mathbf{j} + hc_1\mathbf{k}.$$

We conclude from the theorems above that the addition of vectors and their multiplication by numbers satisfy the following laws:

$$\left.\begin{array}{c} \mathbf{u} + (\mathbf{v} + \mathbf{w}) = (\mathbf{u} + \mathbf{v}) + \mathbf{w} \\ c(d\mathbf{v}) = (cd)\mathbf{v} \end{array}\right\} \text{Associative laws}$$

$$\mathbf{u} + \mathbf{v} = \mathbf{v} + \mathbf{u} \qquad \text{Commutative law}$$

$$\left.\begin{array}{c} (c + d)\mathbf{v} = c\mathbf{v} + d\mathbf{v} \\ c(\mathbf{u} + \mathbf{v}) = c\mathbf{u} + c\mathbf{v} \end{array}\right\} \text{Distributive laws}$$

$$1 \cdot \mathbf{u} = \mathbf{u}, \qquad 0 \cdot \mathbf{u} = \mathbf{0}, \qquad (-1)\mathbf{u} = -\mathbf{u}.$$

DEFINITION. Let $\mathbf{v}$ be any vector except $\mathbf{0}$. The **unit vector u in the direction of v** is defined by

$$\mathbf{u} = \frac{1}{|\mathbf{v}|}\,\mathbf{v}.$$

EXAMPLE 2. Given the vectors $\mathbf{u} = 3\mathbf{i} - 2\mathbf{j} + 4\mathbf{k}$ and $\mathbf{v} = 6\mathbf{i} - 4\mathbf{j} - 2\mathbf{k}$, express the vector $3\mathbf{u} - 2\mathbf{v}$ in terms of $\mathbf{i}$, $\mathbf{j}$, and $\mathbf{k}$.

Solution. $3\mathbf{u} = 9\mathbf{i} - 6\mathbf{j} + 12\mathbf{k}$ and $-2\mathbf{v} = -12\mathbf{i} + 8\mathbf{j} + 4\mathbf{k}$. Adding these vectors, we get $3\mathbf{u} - 2\mathbf{v} = -3\mathbf{i} + 2\mathbf{j} + 16\mathbf{k}$.

EXAMPLE 3. Given the vector $\mathbf{v} = 2\mathbf{i} - 3\mathbf{j} + \mathbf{k}$, find a unit vector in the direction of $\mathbf{v}$.

Solution. We have $|\mathbf{v}| = \sqrt{4 + 9 + 1} = \sqrt{14}$. The desired vector $\mathbf{u}$ is

$$\mathbf{u} = \frac{1}{\sqrt{14}}\,\mathbf{v} = \frac{2}{\sqrt{14}}\,\mathbf{i} - \frac{3}{\sqrt{14}}\,\mathbf{j} + \frac{1}{\sqrt{14}}\,\mathbf{k}.$$

EXAMPLE 4. Given the vector $\mathbf{v} = 2\mathbf{i} + 4\mathbf{j} - 3\mathbf{k}$, find the representative $\overrightarrow{AB}$ of $\mathbf{v}$ if the point A has coordinates $(2, 1, -5)$.

Solution. Denote the coordinates of B by x_B, y_B, z_B. Then we have

$$x_B - 2 = 2, \qquad y_B - 1 = 4, \qquad z_B + 5 = -3.$$

Therefore, $x_B = 4, y_B = 5, z_B = -8$.

PROBLEMS

In problems 1 through 6, express v in terms of i, j, and k, given that the endpoints A and B of the representative $\overrightarrow{AB}$ of v have the given coordinates:

1. $A(2, 1, 3), B(3, 2, 5)$
2. $A(0, 2, 4), B(3, 1, 5)$
3. $A(2, 1, -2), B(1, 2, 0)$
4. $A(-1, 2, -2), B(2, 1, -1)$
5. $A(0, 0, -3), B(2, -1, 3)$
6. $A(-1, 2, -3), B(2, -1, -2)$

In problems 7 through 10, in each case find a unit vector u in the direction of v. Express u in terms of i, j, and k.

7. $v = 2i - 2j - k$
8. $v = 7i + 4j - 4k$
9. $v = 3i - 2j - 4k$
10. $v = -3i - 2j + 6k$

In problems 11 through 17, find the representative $\overrightarrow{AB}$ of the vector v from the information given.

11. $v = i - j - 2k, A(2, -1, 1)$
12. $v = 2i - 3j + 4k, A(2, 1, -3)$
13. $v = -i + 2j + 3k, B(2, 0, 1)$
14. $v = 3i - 2j + k, B(1, 1, -2)$
15. $v = i - 2j + 4k$, the midpoint of segment AB has coordinates $(3, -2, 1)$
16. $v = 2i - 3j$, the midpoint of segment AB has coordinates $(-2, 1, 3)$
17. $v = i + 2k$, the point two-thirds of the way from A to B has coordinates $(3, -1, 2)$
18. Find a vector u in the direction $v = 2i - j + 2k$ and having two-thirds the length of v.
19. Given that $u = 2i - 3j - k, v = -i + 2j - 3k$, find $u + v$ in terms of $i, j,$ and k.
20. Given that $u = -i - 2j + k, v = 2i + j - k$, find $2u - 3v$ in terms of $i, j,$ and k.
21. Let a and b be any real numbers. Show that the vector i is orthogonal to $aj + bk$.

▶2. LINEAR DEPENDENCE AND INDEPENDENCE

Two vectors u and v, neither zero, are said to be **proportional** if and only if there is a number c such that $u = cv$; that is, each vector is a scalar multiple of the other. If $v_1, v_2, \ldots, v_k$ are any vectors and $c_1, c_2, \ldots, c_k$ are numbers, we call an expression of the form

$$c_1v_1 + c_2v_2 + \cdots + c_kv_k$$

a **linear combination** of the vectors $v_1, v_2, \ldots, v_k$. If two vectors u and v are proportional, the definition shows that a linear combination of them is the zero vector.

In fact, $\mathbf{u} - c\mathbf{v} = \mathbf{0}$. A set of vectors $\{\mathbf{v}_1, \mathbf{v}_2, \ldots, \mathbf{v}_k\}$ is **linearly dependent** if and only if there is a set of constants $\{c_1, c_2, \ldots, c_k\}$, *not all zero*, such that

$$c_1\mathbf{v}_1 + c_2\mathbf{v}_2 + \cdots + c_k\mathbf{v}_k = \mathbf{0}. \tag{1}$$

If no such set of constants exists, then the set $\{\mathbf{v}_1, \mathbf{v}_2, \ldots, \mathbf{v}_k\}$ is said to be **linearly independent.**

It is clear that any two proportional vectors are linearly dependent. As another example, the vectors $\mathbf{v}_1 = 2\mathbf{i} + 3\mathbf{j} - \mathbf{k}, \mathbf{v}_2 = -2\mathbf{i} - \mathbf{j} + \mathbf{k}, \mathbf{v}_3 = 2\mathbf{i} + 7\mathbf{j} - \mathbf{k}$ form a linearly dependent set since the selection $c_1 = 3, c_2 = 2, c_3 = -1$ shows that

$$c_1\mathbf{v}_1 + c_2\mathbf{v}_2 + c_3\mathbf{v}_3 = 3(2\mathbf{i} + 3\mathbf{j} - \mathbf{k}) + 2(-2\mathbf{i} - \mathbf{j} + \mathbf{k}) - (2\mathbf{i} + 7\mathbf{j} - \mathbf{k})$$
$$= \mathbf{0}.$$

A set $\{\mathbf{v}_1, \mathbf{v}_2, \ldots, \mathbf{v}_k\}$ is linearly dependent if and only if one member of the set can be expressed as a linear combination of the remaining members. To see this, we observe that in Eq. (1) one of the terms on the left-hand side, say $\mathbf{v}_i$, must have a nonzero coefficient and so may be transferred to the right-hand side. Dividing by the coefficient c_i, we express this particular $\mathbf{v}_i$ as a linear combination of the remaining $\mathbf{v}$'s. If some $\mathbf{v}_i$ is expressible in terms of the others, it follows by transposing $\mathbf{v}_i$ that $\mathbf{v}_1, \mathbf{v}_2, \ldots, \mathbf{v}_k$ are linearly dependent.

The following statement, a direct consequence of the definition of linear dependence, is often useful in proofs of theorems. If $\{\mathbf{v}_1, \mathbf{v}_2, \ldots, \mathbf{v}_k\}$ is a linearly independent set and if

$$c_1\mathbf{v}_1 + c_2\mathbf{v}_2 + \cdots + c_k\mathbf{v}_k = \mathbf{0},$$

then it follows that $c_1 = c_2 = \cdots = c_k = 0$.

The set $\{\mathbf{i}, \mathbf{j}, \mathbf{k}\}$ *is linearly independent.* To show this we observe that the equation

$$c_1\mathbf{i} + c_2\mathbf{j} + c_3\mathbf{k} = \mathbf{0} \tag{2}$$

holds if and only if $|c_1\mathbf{i} + c_2\mathbf{j} + c_3\mathbf{k}| = 0$. But

$$|c_1\mathbf{i} + c_2\mathbf{j} + c_3\mathbf{k}| = \sqrt{c_1^2 + c_2^2 + c_3^2},$$

and this last expression is zero if and only if $c_1 = c_2 = c_3 = 0$. Thus no nonzero constants satisfying (2) exist and $\{\mathbf{i}, \mathbf{j}, \mathbf{k}\}$ is a linearly independent set.

The proof of the next theorem is given in Appendix 5.

Theorem 6. *Let*
$$\mathbf{u} = a_{11}\mathbf{i} + a_{12}\mathbf{j} + a_{13}\mathbf{k},$$
$$\mathbf{v} = a_{21}\mathbf{i} + a_{22}\mathbf{j} + a_{23}\mathbf{k},$$
$$\mathbf{w} = a_{31}\mathbf{i} + a_{32}\mathbf{j} + a_{33}\mathbf{k},$$

and denote by D the determinant

$$D = \begin{vmatrix} a_{11} & a_{12} & a_{13} \\ a_{21} & a_{22} & a_{23} \\ a_{31} & a_{32} & a_{33} \end{vmatrix}.$$

Then the set $\{u, v, w\}$ *is linearly independent if and only if* $D \neq 0$.

EXAMPLE 1. Determine whether or not the vectors $u = 2i - j + k, v = i + 2j + k$, $w = -i + j + 3k$ form a linearly independent set.

Solution. Expanding D by its first row, we have

$$D = \begin{vmatrix} 2 & -1 & 1 \\ 1 & 2 & 1 \\ -1 & 1 & 3 \end{vmatrix} = 2 \begin{vmatrix} 2 & 1 \\ 1 & 3 \end{vmatrix} + \begin{vmatrix} 1 & 1 \\ -1 & 3 \end{vmatrix} + \begin{vmatrix} 1 & 2 \\ -1 & 1 \end{vmatrix}.$$

Therefore $D = 2(5) + 4 + 3 = 17 \neq 0$. The set is linearly independent.

Theorem 7.   *If* $\{u, v, w\}$ *is a linearly independent set and* r *is any vector, then there are constants* A_1, A_2, *and* A_3 *such that*

$$r = A_1 u + A_2 v + A_3 w. \tag{3}$$

Proof. According to Theorem 3, *every* vector can be expressed as a linear combination of i, j, and k. Therefore

$$u = a_{11}i + a_{12}j + a_{13}k,$$
$$v = a_{21}i + a_{22}j + a_{23}k,$$
$$w = a_{31}i + a_{32}j + a_{33}k,$$
$$r = b_1 i + b_2 j + b_3 k.$$

When we insert all these expressions in (3) and collect all terms on one side, we get a linear combination of i, j, and k equal to zero. Since $\{i, j, k\}$ is a linearly independent set, the coefficients of i, j, and k are equal to zero separately. Computing these coefficients, we get the equations

$$a_{11}A_1 + a_{21}A_2 + a_{31}A_3 = b_1,$$
$$a_{12}A_1 + a_{22}A_2 + a_{32}A_3 = b_2, \tag{4}$$
$$a_{13}A_1 + a_{23}A_2 + a_{33}A_3 = b_3.$$

We have here three equations in the three unknowns A_1, A_2, A_3. The determinant D' of the coefficients in (4) differs from the determinant D of Theorem 6 in that the rows and columns are interchanged. Since $\{u, v, w\}$ is an independent set, we know that $D \neq 0$; also Theorem 3 of Appendix 4 proves that $D = D'$, and so $D' \neq 0$. We now use Cramer's rule (Theorem 11, Appendix 4) to solve for A_1, A_2, A_3.

Note that the proof of Theorem 7 gives the method for finding A_1, A_2, A_3. We work an example.

EXAMPLE 2. Given the vectors $\mathbf{u} = 2\mathbf{i} + 3\mathbf{j} + \mathbf{k}$, $\mathbf{v} = -\mathbf{i} + \mathbf{j} + 2\mathbf{k}$, $\mathbf{w} = 3\mathbf{i} - \mathbf{j} + 3\mathbf{k}$, $\mathbf{r} = \mathbf{i} + 2\mathbf{j} - 6\mathbf{k}$, show that $\mathbf{u}$, $\mathbf{v}$, and $\mathbf{w}$ are linearly independent and express $\mathbf{r}$ as a linear combination of $\mathbf{u}$, $\mathbf{v}$, and $\mathbf{w}$.

Solution. Expanding D by its first row, we obtain

$$D = \begin{vmatrix} 2 & 3 & 1 \\ -1 & 1 & 2 \\ 3 & -1 & 3 \end{vmatrix} = 2 \begin{vmatrix} 1 & 2 \\ -1 & 3 \end{vmatrix} - 3 \begin{vmatrix} -1 & 2 \\ 3 & 3 \end{vmatrix} + \begin{vmatrix} -1 & 1 \\ 3 & -1 \end{vmatrix}$$

$$= 2(5) - 3(-9) + (-2) = 35.$$

Hence $D \neq 0$ and so $\{\mathbf{u}, \mathbf{v}, \mathbf{w}\}$ is linearly independent. Using equations (4), we now obtain the set of equations

$$2A_1 - A_2 + 3A_3 = 1,$$
$$3A_1 + A_2 - A_3 = 2,$$
$$A_1 + 2A_2 + 3A_3 = -6.$$

Solving these, we find that $A_1 = 1$, $A_2 = -2$, $A_3 = -1$. Finally, $\mathbf{r} = \mathbf{u} - 2\mathbf{v} - \mathbf{w}$.

PROBLEMS

In problems 1 through 5 state whether or not the given vectors are linearly independent.

1. $\mathbf{u} = -\mathbf{i} + \mathbf{j} + 4\mathbf{k}$, $\mathbf{v} = 5\mathbf{i} + 6\mathbf{j} - 3\mathbf{k}$, $\mathbf{w} = \mathbf{i} + 3\mathbf{j} - 12\mathbf{k}$

2. $\mathbf{u} = 3\mathbf{i} + 4\mathbf{j} + 5\mathbf{k}$, $\mathbf{v} = 4\mathbf{i} + 6\mathbf{j} + 4\mathbf{k}$, $\mathbf{w} = 2\mathbf{i} + 2\mathbf{j} + 6\mathbf{k}$

3. $\mathbf{u} = 2\mathbf{j} + 8\mathbf{k}$, $\mathbf{v} = \mathbf{j} - 7\mathbf{k}$, $\mathbf{w} = \mathbf{j} + 5\mathbf{k}$

4. $\mathbf{u} = -\mathbf{j} + 3\mathbf{k}$, $\mathbf{v} = \mathbf{i} + 2\mathbf{j} + 3\mathbf{k}$, $\mathbf{w} = 6\mathbf{i} + 6\mathbf{j} - 4\mathbf{k}$

5. $\mathbf{u} = \mathbf{j} + 3\mathbf{k}$, $\mathbf{v} = 3\mathbf{i} + 5\mathbf{j} - 10\mathbf{k}$, $\mathbf{w} = -\mathbf{j} + \mathbf{k}$, $\mathbf{r} = 4\mathbf{i} + 4\mathbf{j}$

In problems 6 through 11, show that $\mathbf{u}$, $\mathbf{v}$, and $\mathbf{w}$ are linearly independent and express $\mathbf{r}$ in terms of $\mathbf{u}$, $\mathbf{v}$, and $\mathbf{w}$.

6. $\mathbf{u} = -2\mathbf{i} - 3\mathbf{j} + \mathbf{k}$, $\mathbf{v} = 2\mathbf{i} + 4\mathbf{j}$, $\mathbf{w} = \mathbf{i} + 3\mathbf{j}$, $\mathbf{r} = 2\mathbf{i} + 5\mathbf{j} + \mathbf{k}$

7. $\mathbf{u} = -\mathbf{i} - 2\mathbf{j} + 3\mathbf{k}$, $\mathbf{v} = \mathbf{i} + 3\mathbf{j}$, $\mathbf{w} = \mathbf{i} + 2\mathbf{j} - \mathbf{k}$, $\mathbf{r} = 4\mathbf{i} + 6\mathbf{j} + 8\mathbf{k}$

8. $\mathbf{u} = -\mathbf{i} + \mathbf{j} + 2\mathbf{k}$, $\mathbf{v} = \mathbf{i} + 3\mathbf{k}$, $\mathbf{w} = -2\mathbf{i} + \mathbf{j} + 5\mathbf{k}$, $\mathbf{r} = -3\mathbf{i} - 2\mathbf{j} + 5\mathbf{k}$

9. $\mathbf{u} = \mathbf{j} + 3\mathbf{k}$, $\mathbf{v} = 2\mathbf{i} + \mathbf{j} + \mathbf{k}$, $\mathbf{w} = \mathbf{i} + 3\mathbf{j}$, $\mathbf{r} = -2\mathbf{i}$

10. $\mathbf{u} = -2\mathbf{i}$, $\mathbf{v} = \mathbf{i} + 2\mathbf{j} + \mathbf{k}$, $\mathbf{w} = -3\mathbf{i} - 2\mathbf{j} - 3\mathbf{k}$, $\mathbf{r} = 4\mathbf{i} + 6\mathbf{j} + 8\mathbf{k}$

11. $\mathbf{u} = -\mathbf{i} + 9\mathbf{k}$, $\mathbf{v} = 3\mathbf{i} + \mathbf{j} + 8\mathbf{k}$, $\mathbf{w} = -3\mathbf{i} - \mathbf{j} + 2\mathbf{k}$, $\mathbf{r} = 3\mathbf{i} + 4\mathbf{j} + 41\mathbf{k}$

12. Prove Theorem 5.

13. Show that any set of four vectors must be linearly dependent.

14. Show that if $\overrightarrow{OA}$, $\overrightarrow{OB}$, and $\overrightarrow{OC}$ are representatives of **u**, **v**, and **w**, respectively, and if $\{$**u**, **v**, **w**$\}$ is a linearly dependent set, then the three representatives lie in one plane.

▶ 3. THE INNER (SCALAR OR DOT) PRODUCT

Two vectors are said to be **parallel** or **proportional** when each is a scalar multiple of the other (and neither is zero). The representatives of parallel vectors are all parallel directed line segments.

By the **angle between two vectors v and w** (neither $= $ **0**), we mean the measure of the angle between any directed line containing a representative of **v** and an intersecting directed line containing a representative of **w** (Fig. 11–5). Two parallel vectors make an angle of 0 or π, depending on whether they are pointing in the same or opposite directions.

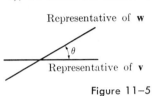

Figure 11–5

Theorem 8. *If θ is the angle between the vectors*

$$v = a_1 i + a_2 j + a_3 k$$

and

$$w = b_1 i + b_2 j + b_3 k,$$

then

$$\cos \theta = \frac{a_1 b_1 + a_2 b_2 + a_3 b_3}{|v| \cdot |w|}.$$

The proof is a straightforward extension of the proof of the analogous theorem in the plane (Theorem 5 of Chapter 4) and will therefore be omitted. (See problem 22 at the end of this section.)

EXAMPLE 1. Given the vectors $v = 2i + j - 3k$ and $w = -i + 4j - 2k$, find the cosine of the angle between **v** and **w**.

Solution.   We have $|v| = \sqrt{4 + 1 + 9} = \sqrt{14}$, $|w| = \sqrt{1 + 16 + 4} = \sqrt{21}$. Therefore

$$\cos \theta = \frac{-2 + 4 + 6}{\sqrt{14} \cdot \sqrt{21}} = \frac{8}{7\sqrt{6}}.$$

DEFINITIONS.   Given the vectors **u** and **v**, we define the **inner (scalar or dot) product**

$$u \cdot v$$

by the formula

$$\mathbf{u} \cdot \mathbf{v} = |\mathbf{u}| \, |\mathbf{v}| \cos \theta,$$

where θ is the angle between the vectors. If either $\mathbf{u}$ or $\mathbf{v}$ is $\mathbf{0}$, we define $\mathbf{u} \cdot \mathbf{v} = 0$. Two vectors $\mathbf{u}$ and $\mathbf{v}$ are **orthogonal** if and only if $\mathbf{u} \cdot \mathbf{v} = 0$.

Theorem 9. *The scalar product satisfies the laws*

(a) $\mathbf{u} \cdot \mathbf{v} = \mathbf{v} \cdot \mathbf{u}$;   (b) $\mathbf{u} \cdot \mathbf{u} = |\mathbf{u}|^2$.

(c) *If* $\mathbf{u} = a_1\mathbf{i} + b_1\mathbf{j} + c_1\mathbf{k}$ *and* $\mathbf{v} = a_2\mathbf{i} + b_2\mathbf{j} + c_2\mathbf{k}$,   *then*

$$\mathbf{u} \cdot \mathbf{v} = a_1a_2 + b_1b_2 + c_1c_2.$$

Proof. Parts (a) and (b) are direct consequences of the definition; part (c) follows from Theorem 8 since

$$\mathbf{u} \cdot \mathbf{v} = |\mathbf{u}| \cdot |\mathbf{v}| \cos \theta = |\mathbf{u}| \cdot |\mathbf{v}| \frac{a_1a_2 + b_1b_2 + c_1c_2}{|\mathbf{u}| \cdot |\mathbf{v}|}.$$

Corollary. (a) *If c and d are any numbers and if* $\mathbf{u}$, $\mathbf{v}$, $\mathbf{w}$ *are any vectors, then*

$$\mathbf{u} \cdot (c\mathbf{v} + d\mathbf{w}) = c(\mathbf{u} \cdot \mathbf{v}) + d(\mathbf{u} \cdot \mathbf{w}).$$

(b) *We have*

$$\mathbf{i} \cdot \mathbf{i} = \mathbf{j} \cdot \mathbf{j} = \mathbf{k} \cdot \mathbf{k} = 1, \qquad \mathbf{i} \cdot \mathbf{j} = \mathbf{i} \cdot \mathbf{k} = \mathbf{j} \cdot \mathbf{k} = 0.$$

EXAMPLE 2. Find the scalar product of the vectors

$$\mathbf{u} = 3\mathbf{i} + 2\mathbf{j} - 4\mathbf{k} \qquad \text{and} \qquad \mathbf{v} = -2\mathbf{i} + \mathbf{j} + 5\mathbf{k}.$$

Solution. $\mathbf{u} \cdot \mathbf{v} = 3(-2) + 2 \cdot 1 + (-4)(5) = -24.$

EXAMPLE 3. Express $|3\mathbf{u} + 5\mathbf{v}|^2$ in terms of $|\mathbf{u}|^2$, $|\mathbf{v}|^2$, and $\mathbf{u} \cdot \mathbf{v}$.

Solution. $\begin{aligned}[t] |3\mathbf{u} + 5\mathbf{v}|^2 &= (3\mathbf{u} + 5\mathbf{v}) \cdot (3\mathbf{u} + 5\mathbf{v}) \\ &= 9(\mathbf{u} \cdot \mathbf{u}) + 15(\mathbf{u} \cdot \mathbf{v}) + 15(\mathbf{v} \cdot \mathbf{u}) + 25(\mathbf{v} \cdot \mathbf{v}) \\ &= 9|\mathbf{u}|^2 + 30(\mathbf{u} \cdot \mathbf{v}) + 25|\mathbf{v}|^2. \end{aligned}$

Suppose that $\overrightarrow{AB}$ and $\overrightarrow{CD}$ are directed line segments as shown in Fig. 11–6. We denote by $\overrightarrow{L}$ the directed line through $\overrightarrow{CD}$, directed so that C precedes D. Drop perpendiculars from A and B to the line $\overrightarrow{L}$ and denote the feet of these perpendiculars by E and F respectively. We define the **projection of** $\overrightarrow{AB}$ **in the direction of** $\overrightarrow{CD}$ as the

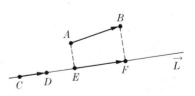

Figure 11–6

directed distance $\overrightarrow{EF}$ along $\vec{L}$. We observe that this projection is exactly the same as the projection of $\overrightarrow{AB}$ along $\vec{L}$.

Let $\mathbf{v}$ and $\mathbf{w}$ be nonzero vectors with $\overrightarrow{AB}$ a representative of $\mathbf{v}$ and $\overrightarrow{CD}$ a representative of $\mathbf{w}$. We define the **projection** of $\mathbf{v}$ along $\mathbf{w}$ to be the projection of $\overrightarrow{AB}$ in the direction of $\overrightarrow{CD}$. We write $\text{Proj}_\mathbf{w}\ \mathbf{v}$ for this quantity, and if θ is the angle between the vectors $\mathbf{v}$ and $\mathbf{w}$, it follows that

$$\text{Proj}_\mathbf{w}\ \mathbf{v} = |\mathbf{v}|\cos\theta.$$

From the formula for $\cos\theta$, we may also write

$$\text{Proj}_\mathbf{w}\ \mathbf{v} = |\mathbf{v}|\,\frac{\mathbf{v}\cdot\mathbf{w}}{|\mathbf{v}|\cdot|\mathbf{w}|} = \frac{\mathbf{v}\cdot\mathbf{w}}{|\mathbf{w}|}.$$

EXAMPLE 4. Find the projection of $\mathbf{v} = -\mathbf{i} + 2\mathbf{j} + 3\mathbf{k}$ along $\mathbf{w} = 2\mathbf{i} - \mathbf{j} - 4\mathbf{k}$.

Solution. $\mathbf{v}\cdot\mathbf{w} = (-1)(2) + (2)(-1) + (3)(-4) = -16$; $|\mathbf{w}| = \sqrt{21}$. Therefore, the projection of $\mathbf{v}$ along $\mathbf{w} = -16/\sqrt{21}$.

An application of scalar product to mechanics occurs in the calculation of work done by a constant force $\mathbf{F}$ when its point of application moves along a segment from A to B. The **work done** in this case is defined as the product of the distance from A to B and the projection of $\mathbf{F}$ along $\mathbf{v}$ $(\overrightarrow{AB})$. We have

$$\text{Projection of }\mathbf{F}\text{ along }\mathbf{v} = \frac{\mathbf{F}\cdot\mathbf{v}}{|\mathbf{v}|};$$

since the distance from A to B is exactly $|\mathbf{v}|$, we conclude that

$$\text{Work done by }\mathbf{F} = \mathbf{F}\cdot\mathbf{v}.$$

EXAMPLE 5. Find the work done by the force

$$\mathbf{F} = 5\mathbf{i} - 3\mathbf{j} + 2\mathbf{k}$$

as its point of application moves from the point $A(2, 1, 3)$ to $B(4, -1, 5)$.

Solution. We have $\mathbf{v}(\overrightarrow{AB}) = (4 - 2)\mathbf{i} + (-1 - 1)\mathbf{j} + (5 - 3)\mathbf{k} = 2\mathbf{i} - 2\mathbf{j} + 2\mathbf{k}$. Therefore, work done $= 5\cdot 2 + 3\cdot 2 + 2\cdot 2 = 20$.

Theorem 10. *If $\mathbf{u}$ and $\mathbf{v}$ are not $\mathbf{0}$, there is a unique number k such that $\mathbf{v} - k\mathbf{u}$ is orthogonal to $\mathbf{u}$. In fact, k can be found from the formula*

$$k = \frac{\mathbf{u}\cdot\mathbf{v}}{|\mathbf{u}|^2}.$$

Proof. $(\mathbf{v} - k\mathbf{u})$ is orthogonal to $\mathbf{u}$ if and only if $\mathbf{u} \cdot (\mathbf{v} - k\mathbf{u}) = 0$. But

$$\mathbf{u} \cdot (\mathbf{v} - k\mathbf{u}) = \mathbf{u} \cdot \mathbf{v} - k|\mathbf{u}|^2 = 0.$$

Therefore, selection of $k = \mathbf{u} \cdot \mathbf{v}/|\mathbf{u}|^2$ yields the result.

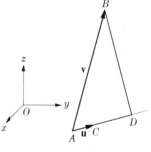

Figure 11–7 shows geometrically how k is to be selected. We drop a perpendicular from the head of $\mathbf{v}$ (point B) to the line containing $\mathbf{u}$ (point D). The directed segment $\overrightarrow{AD}$ gives the proper multiple of $\mathbf{u}(\overrightarrow{AC})$, and the directed segment $\overrightarrow{DB}$ represents the orthogonal vector.

Figure 11–7

EXAMPLE 6. Find a linear combination of $\mathbf{u} = 2\mathbf{i} + 3\mathbf{j} - \mathbf{k}$ and $\mathbf{v} = \mathbf{i} + 2\mathbf{j} + \mathbf{k}$ which is orthogonal to $\mathbf{u}$.

Solution. We select $k = (2 + 6 - 1)/14 = \frac{1}{2}$, and the desired vector is $\frac{1}{2}\mathbf{j} + \frac{3}{2}\mathbf{k}$.

PROBLEMS

In problems 1 through 5, find $\cos \theta$ where θ is the angle between the vectors $\mathbf{u}$ and $\mathbf{v}$.

1. $\mathbf{u} = \mathbf{i} - \mathbf{j} + 2\mathbf{k}$, $\mathbf{v} = 2\mathbf{i} + \mathbf{j} - \mathbf{k}$ 2. $\mathbf{u} = \mathbf{i} - 2\mathbf{j} + 2\mathbf{k}$, $\mathbf{v} = 2\mathbf{i} + \mathbf{j} - \mathbf{k}$

3. $\mathbf{u} = -\mathbf{i} - \mathbf{j} + 3\mathbf{k}$, $\mathbf{v} = \mathbf{i} + 2\mathbf{j} - \mathbf{k}$ 4. $\mathbf{u} = 3\mathbf{i} - 2\mathbf{j} + \mathbf{k}$, $\mathbf{v} = \mathbf{i} - 2\mathbf{j} + 3\mathbf{k}$

5. $\mathbf{u} = \mathbf{i} - \mathbf{j} + 3\mathbf{k}$, $\mathbf{v} = 2\mathbf{i} + \mathbf{j} - \mathbf{k}$

In each of problems 6 through 10, find the projection of the vector $\mathbf{v}$ along $\mathbf{u}$.

6. $\mathbf{u} = -2\mathbf{i} - \mathbf{j}$, $\mathbf{v} = \mathbf{i} + 2\mathbf{j} - 3\mathbf{k}$ 7. $\mathbf{u} = \mathbf{i} + 3\mathbf{j} - 2\mathbf{k}$, $\mathbf{v} = -\mathbf{i} - \mathbf{j} + \mathbf{k}$

8. $\mathbf{u} = \mathbf{i} - 2\mathbf{j} - 2\mathbf{k}$, $\mathbf{v} = 2\mathbf{i} - \mathbf{j} + 2\mathbf{k}$ 9. $\mathbf{u} = 3\mathbf{i} + \mathbf{j} - 2\mathbf{k}$, $\mathbf{v} = \mathbf{i} - \mathbf{j} + \mathbf{k}$

10. $\mathbf{u} = 2\mathbf{i} - \mathbf{j} + 2\mathbf{k}$, $\mathbf{v} = -\mathbf{i} + \mathbf{j} + \mathbf{k}$

In each of problems 11 through 14, find the work done by the force $\mathbf{F}$ when its point of application moves from A to B.

11. $\mathbf{F} = -32\mathbf{k}$, $A(2, 1, -3)$, $B(0, -1, 2)$

12. $\mathbf{F} = 2\mathbf{i} + 3\mathbf{j} - 4\mathbf{k}$, $A(2, -1, 3)$, $B(0, 2, -1)$

13. $\mathbf{F} = \mathbf{i} + 2\mathbf{j} - \mathbf{k}$, $A(2, 1, -1)$, $B(1, -1, 2)$

14. $\mathbf{F} = 3\mathbf{i} - 2\mathbf{j} - \mathbf{k}$, $A(2, 1, -3)$, $B(5, 2, 1)$

In each of problems 15 through 17, find a unit vector in the direction of $\mathbf{u}$.

15. $\mathbf{u} = 2\mathbf{i} + 2\mathbf{j} - \mathbf{k}$ 16. $\mathbf{u} = -2\mathbf{i} + 2\mathbf{j} - 3\mathbf{k}$

17. $\mathbf{u} = 4\mathbf{i} + 2\mathbf{j} - 3\mathbf{k}$

In each of problems 18 through 21, find the value of k so that $\mathbf{v} - k\mathbf{u}$ is orthogonal to $\mathbf{u}$. Also, find the value h so that $\mathbf{u} - h\mathbf{v}$ is orthogonal to $\mathbf{v}$.

18. $\mathbf{u} = \mathbf{i} + \mathbf{j}, \mathbf{v} = \mathbf{j} + \mathbf{k}$ 19. $\mathbf{u} = 2\mathbf{i} - \mathbf{j}, \mathbf{v} = \mathbf{j} + 2\mathbf{k}$

20. $\mathbf{u} = 2\mathbf{i} + \mathbf{j} - 2\mathbf{k}, \mathbf{v} = \mathbf{i} - \mathbf{j}$ 21. $\mathbf{u} = -\mathbf{i} - \mathbf{j} + 3\mathbf{k}, \mathbf{v} = \mathbf{i} + \mathbf{j} - \mathbf{k}$

22. Write a detailed proof of Theorem 8.

23. Show that if $\mathbf{u}$ and $\mathbf{v}$ are any vectors $\neq \mathbf{0}$, then $\mathbf{u}$ and $\mathbf{v}$ make equal angles with $\mathbf{w}$ if

$$\mathbf{w} = \left(\frac{|\mathbf{v}|}{|\mathbf{u}| + |\mathbf{v}|}\right)\mathbf{u} + \left(\frac{|\mathbf{u}|}{|\mathbf{u}| + |\mathbf{v}|}\right)\mathbf{v}.$$

24. Show that if $\mathbf{u}$ and $\mathbf{v}$ are any vectors, the vectors $|\mathbf{v}|\mathbf{u} + |\mathbf{u}|\mathbf{v}$ and $|\mathbf{v}|\mathbf{u} - |\mathbf{u}|\mathbf{v}$ are orthogonal.

In each of problems 25 through 27, determine the relation between g and h so that $g\mathbf{u} + h\mathbf{v}$ is orthogonal to $\mathbf{w}$.

25. $\mathbf{u} = 3\mathbf{i} - 2\mathbf{j} + \mathbf{k}, \quad \mathbf{v} = \mathbf{i} + 2\mathbf{j} - 3\mathbf{k}, \quad \mathbf{w} = -\mathbf{i} + \mathbf{j} + 2\mathbf{k}$

26. $\mathbf{u} = 2\mathbf{i} + \mathbf{j} - 2\mathbf{k}, \quad \mathbf{v} = \mathbf{i} - \mathbf{j} + \mathbf{k}, \quad \mathbf{w} = -\mathbf{i} + 2\mathbf{j} + 3\mathbf{k}$

27. $\mathbf{u} = \mathbf{i} + 2\mathbf{j} - 3\mathbf{k}, \quad \mathbf{v} = 3\mathbf{i} + \mathbf{j} - \mathbf{k}, \quad \mathbf{w} = 4\mathbf{i} - \mathbf{j} + 2\mathbf{k}$

In each of problems 28 through 30, determine g and h so that $\mathbf{w} - g\mathbf{u} - h\mathbf{v}$ is orthogonal to both $\mathbf{u}$ and $\mathbf{v}$.

28. $\mathbf{u} = 2\mathbf{i} - \mathbf{j} + \mathbf{k}, \quad \mathbf{v} = \mathbf{i} + \mathbf{j} + 2\mathbf{k}, \quad \mathbf{w} = 2\mathbf{i} - \mathbf{j} + 4\mathbf{k}$

29. $\mathbf{u} = \mathbf{i} + \mathbf{j} - 2\mathbf{k}, \quad \mathbf{v} = -\mathbf{i} + 2\mathbf{j} + 3\mathbf{k}, \quad \mathbf{w} = 5\mathbf{i} + 8\mathbf{k}$

30. $\mathbf{u} = 3\mathbf{i} - 2\mathbf{j}, \quad \mathbf{v} = 2\mathbf{i} - \mathbf{k}, \quad \mathbf{w} = 4\mathbf{i} - 2\mathbf{k}$

▶4.   THE VECTOR OR CROSS PRODUCT

We saw in Section 3 that the scalar product of two vectors $\mathbf{u}$ and $\mathbf{v}$ associates an ordinary number, i.e., a scalar, with each pair of vectors. The vector or cross product, on the other hand, associates a *vector* with each ordered pair of vectors. However, before defining the cross product, we shall discuss the notion of "right-handed" and "left-handed" triples of vectors.

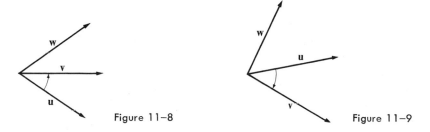

Figure 11-8   Figure 11-9

An *ordered* triple $\{\mathbf{u}, \mathbf{v}, \mathbf{w}\}$ of linearly independent vectors is said to be **right-handed** if the vectors are situated as in Fig. 11-8. If the ordered triple is situated as in Fig. 11–9, the vectors are said to form a **left-handed** triple. The notion of left-handed and right-handed triple is not defined if the vectors form a linearly dependent set.

DEFINITION. Two sets of triples of vectors are said to be **similarly oriented** if and only if both sets are right-handed or both are left-handed. Otherwise they are **oppositely oriented.**

Suppose $\{u_1, v_1, w_1\}$ and $\{u_2, v_2, w_2\}$ are ordered linearly independent sets of triples. From Theorem 7, it follows that we may express u_2, v_2, and w_2 in terms of u_1, v_1, and w_1 by equations of the form

$$u_2 = a_{11}u_1 + a_{12}v_1 + a_{13}w_1,$$
$$v_2 = a_{21}u_1 + a_{22}v_1 + a_{23}w_1,$$
$$w_2 = a_{31}u_1 + a_{32}v_1 + a_{33}w_1.$$

We denote by D the determinant

$$D = \begin{vmatrix} a_{11} & a_{12} & a_{13} \\ a_{21} & a_{22} & a_{23} \\ a_{31} & a_{32} & a_{33} \end{vmatrix}.$$

Although the proof is beyond the scope of this book, it is a fact that the two triples above are similarly oriented if and only if $D > 0$; they are oppositely oriented $\Leftrightarrow D < 0$. Note that the determinant cannot be zero, for then u_2, v_2, and w_2 would not be linearly independent. (See the proof of Theorem 6' in Appendix 5.)

It is also true that if $\{u_1, v_1, w_1\}$ and $\{u_2, v_2, w_2\}$ are similarly oriented and if $\{u_2, v_2, w_2\}$ and $\{u_3, v_3, w_3\}$ are similarly oriented, then $\{u_1, v_1, w_1\}$ and $\{u_3, v_3, w_3\}$ are similarly oriented.

The facts above lead to the following result.

Theorem 11. *If* $\{u, v, w\}$ *is a right-handed triple, then* (i) $\{v, u, -w\}$ *is a right-handed triple, and* (ii) $\{c_1u, c_2v, c_3w\}$ *is a right-handed triple provided that* $c_1c_2c_3 > 0$.

To prove (i) we apply the above determinant condition on similar orientation by regarding $\{u, v, w\}$ as $\{u_1, v_1, w_1\}$ and $\{v, u, -w\}$ as $\{u_2, v_2, w_2\}$. To establish (ii) we regard $\{c_1u, c_2v, c_3w\}$ as $\{u_2, v_2, w_2\}$. The details are left to the student.

DEFINITION. Given the vectors u and v, we define their **vector** or **cross product** $u \times v$ as follows:

(i)   if either u or v is 0,

$$u \times v = 0;$$

(ii) if **u** is proportional to **v**,

$$\mathbf{u} \times \mathbf{v} = \mathbf{0};$$

(iii) if otherwise,

$$\mathbf{u} \times \mathbf{v} = \mathbf{w}$$

where **w** has the three properties: (a) it is orthogonal to both **u** and **v**; (b) it has magnitude $|\mathbf{w}| = |\mathbf{u}| \cdot |\mathbf{v}| \sin \theta$, where θ is the angle between **u** and **v**, and (c) it is directed so that $\{\mathbf{u}, \mathbf{v}, \mathbf{w}\}$ is a right-handed triple.

Remark. We shall always assume that any coordinate triple $\{\mathbf{i}, \mathbf{j}, \mathbf{k}\}$ is right-handed. (We have assumed this up to now without pointing out this fact specifically.)

The proofs of the next two theorems are given in Appendix 5.

Theorem 12. *Suppose that* **u** *and* **v** *are any vectors, that* $\{\mathbf{i}, \mathbf{j}, \mathbf{k}\}$ *is a right-handed coordinate triple, and that* t *is any number.* Then

    (i) $\mathbf{v} \times \mathbf{u} = -(\mathbf{u} \times \mathbf{v})$,

    (ii) $(t\mathbf{u}) \times \mathbf{v} = t(\mathbf{u} \times \mathbf{v}) = \mathbf{u} \times (t\mathbf{v})$,

    (iii) $\mathbf{i} \times \mathbf{j} = -\mathbf{j} \times \mathbf{i} = \mathbf{k}$,

          $\mathbf{j} \times \mathbf{k} = -\mathbf{k} \times \mathbf{j} = \mathbf{i}$,

          $\mathbf{k} \times \mathbf{i} = -\mathbf{i} \times \mathbf{k} = \mathbf{j}$,

    (iv) $\mathbf{i} \times \mathbf{i} = \mathbf{j} \times \mathbf{j} = \mathbf{k} \times \mathbf{k} = \mathbf{0}$.

Theorem 13. *If* **u, v, w** *are any vectors, then*

    (i) $\mathbf{u} \times (\mathbf{v} + \mathbf{w}) = (\mathbf{u} \times \mathbf{v}) + (\mathbf{u} \times \mathbf{w})$ *and*

    (ii) $(\mathbf{v} + \mathbf{w}) \times \mathbf{u} = (\mathbf{v} \times \mathbf{u}) + (\mathbf{w} \times \mathbf{u})$.

With the aid of Theorems 12 and 13, the next theorem is easily established.

Theorem 14. *If*

$$\mathbf{u} = a_1\mathbf{i} + a_2\mathbf{j} + a_3\mathbf{k}$$

and

$$\mathbf{v} = b_1\mathbf{i} + b_2\mathbf{j} + b_3\mathbf{k},$$

then

$$\mathbf{u} \times \mathbf{v} = (a_2b_3 - a_3b_2)\mathbf{i} + (a_3b_1 - a_1b_3)\mathbf{j} + (a_1b_2 - a_2b_1)\mathbf{k}. \tag{1}$$

Proof. By using the laws in Theorems 12 and 13 we obtain (being careful to keep the order of the factors)

$$\mathbf{u} \times \mathbf{v} = a_1b_1(\mathbf{i} \times \mathbf{i}) + a_1b_2(\mathbf{i} \times \mathbf{j}) + a_1b_3(\mathbf{i} \times \mathbf{k})$$
$$+ a_2b_1(\mathbf{j} \times \mathbf{i}) + a_2b_2(\mathbf{j} \times \mathbf{j}) + a_2b_3(\mathbf{j} \times \mathbf{k})$$
$$+ a_3b_1(\mathbf{k} \times \mathbf{i}) + a_3b_2(\mathbf{k} \times \mathbf{j}) + a_3b_3(\mathbf{k} \times \mathbf{k}).$$

The result follows from Theorem 12, parts (iii) and (iv) by collecting terms.

The formula (1) above is useful in calculating the cross product. The following symbolic form is a great aid in remembering the formula. We write

$$\mathbf{u} \times \mathbf{v} = \begin{vmatrix} \mathbf{i} & \mathbf{j} & \mathbf{k} \\ a_1 & a_2 & a_3 \\ b_1 & b_2 & b_3 \end{vmatrix}$$

where it is understood that this "determinant" is to be expanded formally according to its first row. The student may easily verify that when the above expression is expanded, it is equal to (1).

EXAMPLE 1. Find $\mathbf{u} \times \mathbf{v}$ if $\mathbf{u} = 2\mathbf{i} - 3\mathbf{j} + \mathbf{k}$, $\mathbf{v} = \mathbf{i} + \mathbf{j} - 2\mathbf{k}$.

Solution. Carrying out the formal expansion, we obtain

$$\begin{vmatrix} \mathbf{i} & \mathbf{j} & \mathbf{k} \\ 2 & -3 & 1 \\ 1 & 1 & -2 \end{vmatrix} = \begin{vmatrix} -3 & 1 \\ 1 & -2 \end{vmatrix}\mathbf{i} - \begin{vmatrix} 2 & 1 \\ 1 & -2 \end{vmatrix}\mathbf{j} + \begin{vmatrix} 2 & -3 \\ 1 & 1 \end{vmatrix}\mathbf{k} = 5\mathbf{i} + 5\mathbf{j} + 5\mathbf{k}.$$

Remarks. In mechanics the cross product is used for the computation of the vector moment of a force $\mathbf{F}$ applied at a point B, about a point A. There are also applications of cross product to problems in electricity and magnetism. However, we shall confine our attention to applications to geometry.

Theorem 15. *The area of a parallelogram with adjacent sides AB and AC is given by*

$$|\mathbf{v}(\overrightarrow{AB}) \times \mathbf{v}(\overrightarrow{AC})|.$$

The area of $\triangle ABC$ is then $\frac{1}{2}|\mathbf{v}(\overrightarrow{AB}) \times \mathbf{v}(\overrightarrow{AC})|$.

Proof. From Fig. 11–10, we see that the area of the parallelogram is $|AB| \cdot h = |AB| \cdot |AC| \sin \theta$. The result then follows from the definition of cross product.

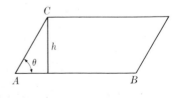

Figure 11–10

EXAMPLE 2. Find the area of $\triangle ABC$ with $A(-2, 1, 3)$, $B(1, -1, 1)$, $C(3, -2, 4)$.

Solution. We have $v(\overrightarrow{AB}) = 3i - 2j - 2k$, $v(\overrightarrow{AC}) = 5i - 3j + k$. From Theorem 14 we obtain

$$v(\overrightarrow{AB}) \times v(\overrightarrow{AC}) = -8i - 13j + k$$

and

$$\tfrac{1}{2}|-8i - 13j + k| = \tfrac{1}{2}\sqrt{64 + 169 + 1} = \tfrac{3}{2}\sqrt{26}.$$

The vector product may be used to find the equation of a plane through three points. The next example illustrates the technique.

EXAMPLE 3. Find the equation of the plane through the points $A(-1, 1, 2)$, $B(1, -2, 1)$, $C(2, 2, 4)$.

Solution. A vector normal to the plane will be perpendicular to both the vectors

$$v(\overrightarrow{AB}) = 2i - 3j - k$$

and

$$v(\overrightarrow{AC}) = 3i + j + 2k.$$

One such vector is the cross product

$$v(\overrightarrow{AB}) \times v(\overrightarrow{AC}) = -5i - 7j + 11k.$$

Therefore the numbers $-5, -7, 11$ form a set of *attitude numbers* (see Chapter 10, Section 4) of the desired plane. Using $A(-1, 1, 2)$ as a point on the plane, we get for the equation

$$-5(x + 1) - 7(y - 1) + 11(z - 2) = 0$$

or

$$5x + 7y - 11z + 20 = 0.$$

EXAMPLE 4. Find the perpendicular distance between the skew lines

$$L_1: \frac{x + 2}{2} = \frac{y - 1}{3} = \frac{z + 1}{-1}, \qquad L_2: \frac{x - 1}{-1} = \frac{y + 1}{2} = \frac{z - 2}{4}.$$

Solution. The vector $v_1 = 2i + 3j - k$ is a vector along L_1. The vector $v_2 = -i + 2j + 4k$ is a vector along L_2. A vector perpendicular to both v_1 and v_2 (i.e., to both L_1 and L_2) is

$$v_1 \times v_2 = 14i - 7j + 7k.$$

Call this common perpendicular w. The desired length may be obtained as a *projection*. Select any point on L_1 (call it P_1) and any point on L_2 (call it P_2). Then the desired length is the projection of the vector $v(\overrightarrow{P_1P_2})$ along w. To get this, we select $P_1(-2, 1, -1)$ on L_1 and $P_2(1, -1, 2)$ on L_2; and so

$$v(\overrightarrow{P_1P_2}) = 3i - 2j + 3k.$$

Therefore,

$$\text{Projection of } \mathbf{v}(\overrightarrow{P_1P_2}) \text{ along } \mathbf{w} = \frac{\mathbf{v}(\overrightarrow{P_1P_2}) \cdot \mathbf{w}}{|\mathbf{w}|}$$

$$= \frac{3 \cdot 14 + (-2)(-7) + 3(7)}{7\sqrt{6}} = \frac{11}{\sqrt{6}}.$$

PROBLEMS

In each of problems 1 through 6, find the cross product $\mathbf{u} \times \mathbf{v}$.

1. $\mathbf{u} = 2\mathbf{j}$, $\mathbf{v} = 2\mathbf{i} - 3\mathbf{k}$

2. $\mathbf{u} = \mathbf{i} + 2\mathbf{j}$, $\mathbf{v} = \mathbf{j} - 3\mathbf{k}$

3. $\mathbf{u} = 2\mathbf{i}$, $\mathbf{v} = \mathbf{i} - 2\mathbf{j} + 3\mathbf{k}$

4. $\mathbf{u} = \mathbf{i} - 2\mathbf{j} + 3\mathbf{k}$, $\mathbf{v} = -\mathbf{i} + \mathbf{k}$

5. $\mathbf{u} = 3\mathbf{i} - 2\mathbf{j} + \mathbf{k}$, $\mathbf{v} = 2\mathbf{i} + \mathbf{j} - 2\mathbf{k}$

6. $\mathbf{u} = -4\mathbf{i} + 2\mathbf{j} - 3\mathbf{k}$, $\mathbf{v} = \mathbf{i} - 2\mathbf{j} - \mathbf{k}$

In problems 7 through 11, find in each case the area of $\triangle ABC$ and the equation of the plane through A, B, and C.

7. $A(2, 1, -2)$, $B(1, 4, 1)$, $C(-1, 3, 2)$

8. $A(1, 2, 3)$, $B(2, 7, 4)$, $C(3, 3, 5)$

9. $A(4, 0, -1)$, $B(1, 4, 1)$, $C(2, 3, -1)$

10. $A(3, 1, 2)$, $B(1, 4, 4)$, $C(2, 3, 3)$

11. $A(1, 5, -3)$, $B(2, 3, 1)$, $C(3, 2, 0)$

In problems 12 through 14, find in each case the perpendicular distance between the given lines.

12. $\dfrac{x-2}{3} = \dfrac{y+1}{2} = \dfrac{z-1}{-2}$; $\quad \dfrac{x+1}{2} = \dfrac{y-1}{3} = \dfrac{z-2}{-3}$

13. $\dfrac{x}{2} = \dfrac{y-2}{-1} = \dfrac{z+1}{1}$; $\quad \dfrac{x-1}{1} = \dfrac{y+2}{-2} = \dfrac{z-1}{-1}$

14. $\dfrac{x+2}{2} = \dfrac{y-2}{-2} = \dfrac{z+2}{-1}$; $\quad \dfrac{x-3}{4} = \dfrac{y+3}{-3} = \dfrac{z-2}{2}$

In problems 15 through 19, use vector methods to find, in each case, the equations in symmetric form of the line through the given point P and parallel to the two given planes.

15. $P(2, 1, -2)$; $x + 2y - 3z + 2 = 0$, $3x - 2y + z - 4 = 0$

16. $P(1, -1, 3)$; $2x + y - z + 3 = 0$, $x - 2y - 2z + 3 = 0$

17. $P(-3, 2, 1)$; $3x - y + 2z - 4 = 0$, $x + y - 2z + 3 = 0$

18. $P(-1, 4, 2)$; $2x - 4y - 3z + 12 = 0$, $3x + 2y + z = 0$

19. $P(1, -3, -2)$; $x + y - 2z + 4 = 0$, $3x - y + 2z - 5 = 0$

In problems 20 and 21, find in each case equations in symmetric form of the line of intersection of the given planes. Use the method of vector products.

20. $3(x + 1) - 2(y - 1) - (z + 2) = 0$

$2(x + 1) + (y - 1) - 2(z + 2) = 0$

21. $2(x - 1) + 2(y + 1) - 3(z - 2) = 0$

$(x - 1) - 3(y + 1) + 2(z - 2) = 0$

In each of problems 22 through 26, find an equation of the plane through the given point or points and parallel to the given line or lines.

22. $(2, 1, -2)$; $\dfrac{x - 1}{3} = \dfrac{y + 1}{2} = \dfrac{z - 2}{-1}$, $\dfrac{x + 1}{2} = \dfrac{y - 1}{-1} = \dfrac{z + 1}{3}$

23. $(1, 2, -1)$; $\dfrac{x + 1}{2} = \dfrac{y - 2}{-2} = \dfrac{z + 1}{3}$, $\dfrac{x}{3} = \dfrac{y - 1}{2} = \dfrac{z - 1}{2}$

24. $(-1, 3, 2)$; $\dfrac{x - 2}{2} = \dfrac{y}{3} = \dfrac{z + 1}{-2}$, $\dfrac{x + 1}{1} = \dfrac{y - 2}{2} = \dfrac{z - 1}{-3}$

25. $(-3, 1, -2)$; $(2, -4, 3)$; $\dfrac{x - 3}{2} = \dfrac{y - 1}{-1} = \dfrac{z + 1}{3}$

26. $(2, 1, 0)$; $(1, 3, 2)$; $\dfrac{x - 1}{3} = \dfrac{y + 1}{-2} = \dfrac{z + 2}{2}$

In problems 27 through 29, find in each case the equation of the plane through the line L_1 which also satisfies the additional condition.

27. L_1: $\dfrac{x + 1}{3} = \dfrac{y - 1}{-2} = \dfrac{z - 1}{2}$;   through $(2, -1, 4)$

28. L_1: $\dfrac{x - 2}{2} = \dfrac{y + 1}{3} = \dfrac{z + 1}{1}$;   parallel to the line $\dfrac{x - 1}{4} = \dfrac{y}{1} = \dfrac{z + 2}{3}$

29. L_1: $\dfrac{x - 1}{1} = \dfrac{y + 2}{-2} = \dfrac{z - 2}{3}$;   perpendicular to the plane $3x - y + 2z - 3 = 0$

In problems 30 and 31, find the equation of the plane through the given points and perpendicular to the given planes.

30. $(2, -1, 1)$; $x + 2y - z + 2 = 0$, $2x - y + 2z - 3 = 0$

31. $(1, -2, 3)$; $(2, 3, 1)$; $2x + y - z + 7 = 0$

In problems 32 and 33, find equations in symmetric form of the line through the given point P, which is perpendicular to and intersects the given line. Use the cross product.

32. $P(2, 1, -1)$; $\dfrac{x - 1}{2} = \dfrac{y + 1}{-2} = \dfrac{z - 2}{3}$

33. $P(1, 2, 3)$; $\dfrac{x - 4}{3} = \dfrac{y - 2}{-2} = \dfrac{z + 3}{2}$

▶5. PRODUCTS OF THREE VECTORS

Since two types of multiplication, the scalar product and the cross product, may be performed on vectors, we can combine three vectors in several ways. For example, we can form the product

$$(\mathbf{u} \times \mathbf{v}) \cdot \mathbf{w}$$

and the product

$$(\mathbf{u} \times \mathbf{v}) \times \mathbf{w}.$$

Also, we can consider the combinations

$$\mathbf{u} \cdot (\mathbf{v} \times \mathbf{w})$$

and

$$\mathbf{u} \times (\mathbf{v} \times \mathbf{w}).$$

The next theorem gives a simple rule for computing $(\mathbf{u} \times \mathbf{v}) \cdot \mathbf{w}$ and also an elegant geometric interpretation of the quantity $|(\mathbf{u} \times \mathbf{v}) \cdot \mathbf{w}|$.

Theorem 16. *Suppose that* $\mathbf{u}_1$, $\mathbf{u}_2$, $\mathbf{u}_3$, *are vectors and that the points A, B, C, D are chosen so that*

$$\mathbf{v}(\overrightarrow{AB}) = \mathbf{u}_1,$$

$$\mathbf{v}(\overrightarrow{AC}) = \mathbf{u}_2,$$

$$\mathbf{v}(\overrightarrow{AD}) = \mathbf{u}_3.$$

Then

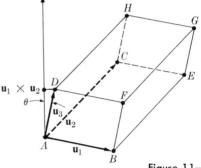

Figure 11–11

(i) *the quantity* $|(\mathbf{u}_1 \times \mathbf{u}_2) \cdot \mathbf{u}_3|$ *is the volume of the parallelepiped with one vertex at A and adjacent vertices at B, C, and D. (See Fig. 11–11.) This volume is zero if and only if the four points lie in a plane;*

(ii) *if* $\{\mathbf{i}, \mathbf{j}, \mathbf{k}\}$ *is a right-handed coordinate triple and if*

$$\mathbf{u}_1 = a_1\mathbf{i} + b_1\mathbf{j} + c_1\mathbf{k}, \qquad \mathbf{u}_2 = a_2\mathbf{i} + b_2\mathbf{j} + c_2\mathbf{k},$$
$$\mathbf{u}_3 = a_3\mathbf{i} + b_3\mathbf{j} + c_3\mathbf{k},$$

we have

$$(\mathbf{u}_1 \times \mathbf{u}_2) \cdot \mathbf{u}_3 = \begin{vmatrix} a_1 & b_1 & c_1 \\ a_2 & b_2 & c_2 \\ a_3 & b_3 & c_3 \end{vmatrix};$$

(iii) $(\mathbf{u}_1 \times \mathbf{u}_2) \cdot \mathbf{u}_3 = \mathbf{u}_1 \cdot (\mathbf{u}_2 \times \mathbf{u}_3).$

Proof. To prove (i), we note that $|\mathbf{u}_1 \times \mathbf{u}_2|$ is the area of the parallelogram $ABEC$ and that

$$|(\mathbf{u}_1 \times \mathbf{u}_2) \cdot \mathbf{u}_3| = |\mathbf{u}_1 \times \mathbf{u}_2| \cdot |\mathbf{u}_3| \cdot |\cos \theta|,$$

where θ is the angle between the two vectors $\mathbf{u}_3$ and $\mathbf{u}_1 \times \mathbf{u}_2$. The quantity $|\mathbf{u}_3| \, |\cos \theta|$ is the length of the projection of $\mathbf{u}_3$ on the normal to the plane of $ABEC$. Clearly, $(\mathbf{u}_1 \times \mathbf{u}_2) \cdot \mathbf{u}_3 = 0 \Leftrightarrow \mathbf{u}_1 \times \mathbf{u}_2 = \mathbf{0}$ or $\mathbf{u}_3 = \mathbf{0}$ or $\cos \theta = 0$. If $\cos \theta = 0$, then $\mathbf{u}_3$ is parallel to the plane of $\mathbf{u}_1$ and $\mathbf{u}_2$ and all four points lie in a plane. The proof of parts (ii) and (iii) follow from Theorems 9 and 14 and are left to the student. See problem 5 at the end of this section.

Theorem 17. *If* $\mathbf{u}$, $\mathbf{v}$, *and* $\mathbf{w}$ *are any vectors, then*

(i) $(\mathbf{u} \times \mathbf{v}) \times \mathbf{w} = (\mathbf{u} \cdot \mathbf{w})\mathbf{v} - (\mathbf{v} \cdot \mathbf{w})\mathbf{u}$,

(ii) $\mathbf{u} \times (\mathbf{v} \times \mathbf{w}) = (\mathbf{u} \cdot \mathbf{w})\mathbf{v} - (\mathbf{u} \cdot \mathbf{v})\mathbf{w}$.

Proof. If $\mathbf{u}$ and $\mathbf{v}$ are proportional or if $\mathbf{w}$ is orthogonal to both $\mathbf{u}$ and $\mathbf{v}$, then both sides of (i) are zero. Otherwise, we see that $(\mathbf{u} \times \mathbf{v}) \times \mathbf{w}$ is orthogonal to the perpendicular to the plane determined by $\mathbf{u}$ and $\mathbf{v}$. Hence $(\mathbf{u} \times \mathbf{v}) \times \mathbf{w}$ is in the plane of $\mathbf{u}$ and $\mathbf{v}$. We choose a right-handed coordinate triple $\{\mathbf{i}, \mathbf{j}, \mathbf{k}\}$ so that $\mathbf{i}$ is in the direction of $\mathbf{u}$ and $\mathbf{j}$ is in the plane of $\mathbf{u}$ and $\mathbf{v}$. Then there are numbers a_1, a_2, b_2, etc., so that

$$\mathbf{u} = a_1\mathbf{i}, \qquad \mathbf{v} = a_2\mathbf{i} + b_2\mathbf{j}, \qquad \mathbf{w} = a_3\mathbf{i} + b_3\mathbf{j} + c_3\mathbf{k}.$$

The student may now compute both sides of (i) to see that they are equal. The proof of (ii) is left to the student. (See problem 6 at the end of this section.)

EXAMPLE 1. Given $A(3, -1, 2)$, $B(1, 2, -2)$, $C(2, 1, -2)$, and $D(-1, 3, 2)$, find the volume of the parallelepiped having AB, AC, and AD as edges.

Solution. We have $\mathbf{u}_1 = \mathbf{v}(\overrightarrow{AB}) = -2\mathbf{i} + 3\mathbf{j} - 4\mathbf{k}$, $\mathbf{u}_2 = \mathbf{v}(\overrightarrow{AC}) = -\mathbf{i} + 2\mathbf{j} - 4\mathbf{k}$, $\mathbf{u}_3 = \mathbf{v}(\overrightarrow{AD}) = -4\mathbf{i} + 4\mathbf{j}$. We compute $\mathbf{u}_2 \times \mathbf{u}_3 = 16\mathbf{i} + 16\mathbf{j} + 4\mathbf{k}$. Therefore $|\mathbf{u}_1 \cdot (\mathbf{u}_2 \times \mathbf{u}_3)| = |-32 + 48 - 16| = 0$. Hence the four points are in a plane.

EXAMPLE 2. Find the equations of the line through the point $(3, -2, 1)$ perpendicular to the line L (and intersecting it) given by

$$L = \left\{(x, y, z) \,\middle|\, \frac{x - 2}{2} = \frac{y + 1}{-2} = \frac{z}{1}\right\}.$$

Solution. Let $P_0(3, -2, 1)$ and $P_1(2, -1, 0)$, $\mathbf{u} = 2\mathbf{i} - 2\mathbf{j} + \mathbf{k}$, $\mathbf{v} = \mathbf{v}(\overrightarrow{P_0P_1}) = -\mathbf{i} + \mathbf{j} - \mathbf{k}$. The plane containing L and P_0 has a normal perpendicular to $\mathbf{u}$ and $\mathbf{v}$. Hence this normal is proportional to $\mathbf{u} \times \mathbf{v}$. The desired line is in this plane and perpendicular to L. Therefore it has a direction w perpendicular to u and $u \times v$. Thus for some number c, we have

$$cw = u \times (u \times v) = (u \cdot v)u - (u \cdot u)v$$
$$= -5(2i - 2j + k) - 9(-i + j - k) = -i + j + 4k.$$

Consequently, the desired line has equations

$$\frac{x - 3}{-1} = \frac{y + 2}{1} = \frac{z - 1}{4}.$$

PROBLEMS

In problems 1 through 4, find the volume of the parallelepiped having edges AB, AC, and AD, or else show that A, B, C, and D lie on a plane or on a line. If they lie on a plane, find its equation; if they lie on a line, find its equations.

1. $A(1, 1, 2)$, $B(3, -1, 1)$, $C(-1, 2, 3)$, $D(2, -2, 1)$
2. $A(3, 1, -2)$, $B(1, 2, 1)$, $C(2, -1, 3)$, $D(4, 3, -7)$
3. $A(1, 2, -3)$, $B(3, 1, -2)$, $C(-1, 3, 1)$, $D(-3, 4, 3)$
4. $A(2, -1, -1)$, $B(2, 1, 2)$, $C(-1, 0, 3)$, $D(0, 2, 4)$
5. Prove Theorem 16, parts (ii) and (iii).
6. Complete the proof of Theorem 17.

In problems 7 through 10, compute $(u \times v) \times w$ directly and by using Theorem 17.

7. $u = i + j - 2k$, $v = 2i - j + k$, $w = 3i + 2j - k$
8. $u = -i + j + 2k$, $v = 3i + 2j - k$, $w = i - j + k$
9. $u = 2i - j - k$, $v = i + 2j + k$, $w = -i + j + 2k$
10. $u = 3i + 2j + k$, $v = 2i + j$, $w = i - k$
11. Show that every vector v satisfies the identity

$$i \times (v \times i) + j \times (v \times j) + k \times (v \times k) = 2v.$$

In problems 12 through 14, find equations of the lines through the given points and perpendicular to the given lines.

12. $(1, 3, -2)$; $\dfrac{x - 2}{3} = \dfrac{y + 1}{-2} = \dfrac{z - 1}{4}$

13. $(2, -1, 3)$; $\dfrac{x + 1}{2} = \dfrac{y - 2}{3} = \dfrac{z + 1}{-5}$

14. $(-1, 2, 4)$; $\dfrac{x - 1}{4} = \dfrac{y + 2}{-3} = \dfrac{z - 1}{2}$

In problems 15 and 16, express $(\mathbf{t} \times \mathbf{u}) \times (\mathbf{v} \times \mathbf{w})$ in terms of $\mathbf{v}$ and $\mathbf{w}$.

15. $\mathbf{t} = 2\mathbf{i} - \mathbf{j} + \mathbf{k}, \mathbf{u} = \mathbf{i} + \mathbf{j} - \mathbf{k}, \mathbf{v} = 3\mathbf{i} - 2\mathbf{j} + 2\mathbf{k}, \mathbf{w} = -\mathbf{i} + \mathbf{j} - 2\mathbf{k}$

16. $\mathbf{t} = \mathbf{i} + 2\mathbf{j} - \mathbf{k}, \mathbf{u} = 2\mathbf{i} - \mathbf{j} + \mathbf{k}, \mathbf{v} = -\mathbf{i} + \mathbf{j} + 2\mathbf{k}, \mathbf{w} = 3\mathbf{i} - 2\mathbf{j} + \mathbf{k}$

17. Derive a formula expressing $(\mathbf{t} \times \mathbf{u}) \times (\mathbf{v} \times \mathbf{w})$ in terms of $\mathbf{v}$ and $\mathbf{w}$.

18. Given $\mathbf{a} = \mathbf{i} - \mathbf{j} + \mathbf{k}, \mathbf{b} = 2\mathbf{i} + 3\mathbf{j} + \mathbf{k}, p = 1$. Solve $\mathbf{a} \cdot \mathbf{v} = p, \mathbf{a} \times \mathbf{v} = \mathbf{b}$.

19. Given that $\mathbf{a} \cdot \mathbf{b} = 0, \mathbf{a} \neq \mathbf{0}, \mathbf{b} \neq \mathbf{0}$, find a formula for the solution $\mathbf{v}$ of the equations

$$\mathbf{a} \cdot \mathbf{v} = p, \qquad \mathbf{a} \times \mathbf{v} = \mathbf{b}.$$

[*Hint:* Note that $\mathbf{a}, \mathbf{b}$, and $\mathbf{a} \times \mathbf{b}$ are mutually orthogonal.]

12 Analytic Geometry in Four Dimensions

▶ **1. THE SPACE R_4. THE DISTANCE FORMULA. STRAIGHT LINES**

The study of analytic geometry in two and three dimensions is useful in the solution of algebraic problems which involve two or three variables. Since we can easily visualize points, lines, planes, curves, and surfaces, we can use geometric concepts to gain insight into algebraic and analytic problems. In order to attack problems in four variables we shall develop a geometry of four dimensions. While we can no longer visualize the entities in such geometries, we can proceed by analogy with two and three dimensional geometry in defining various geometric quantities. By this means we obtain insights into many problems which otherwise would remain quite elusive.

DEFINITIONS. We define the **space** R_4 as the set of all ordered quadruples of real numbers. We use the symbol (x, y, z, v) to indicate such a quadruple which we call a **point** in R_4. The quadruple $(0, 0, 0, 0)$ is called the **origin** of the space. All points of the form $(x, 0, 0, 0)$ where x is any real number determine the x axis. Similarly, the points $(0, y, 0, 0)$ form the y axis, the points $(0, 0, z, 0)$ form the z axis, and the points $(0, 0, 0, v)$ form the v axis.

In any space the formula for the distance between two points is basic in determining the kind of geometry we shall have. In analogy with the formulas in two and three dimensions (see pages 41 and 204), we define the distance $|P_1P_2|$ between two points $P_1(x_1, y_1, z_1, v_1)$ and $P_2(x_2, y_2, z_2, v_2)$ by the formula

$$|P_1P_2| = \sqrt{(x_2 - x_1)^2 + (y_2 - y_1)^2 + (z_2 - z_1)^2 + (v_2 - v_1)^2}. \quad (1)$$

We call the space R_4 with the distance formula (1) the **Euclidean space of four dimensions**. By using ordered quintuples, sextuples, etc. of numbers and a distance formula analogous to (1), we can define Euclidean space in five, six, or any number of dimensions.

If we use a formula different from (1) for the distance between two points, we may obtain an entirely different kind of geometry. For example, the Theory of

Relativity employs R_4 with the distance formula

$$|P_1P_2| = \sqrt{(v_2 - v_1)^2 - (x_2 - x_1)^2 - (y_2 - y_1)^2 - (z_2 - z_1)^2}.$$

The quantity v is interpreted as time. In such a geometry two different points may be at zero distance from each other, a situation which cannot occur in Euclidean geometry.

In Euclidean spaces the straight line is of basic importance. We shall define it in analogy with the parametric equations for a straight line in two and three dimensions (pages 61 and 213).

DEFINITIONS. Let $P_1(x_1, y_1, z_1, v_1)$, $P_2(x_2, y_2, z_2, v_2)$ be two given points. The **straight line through P_1 and P_2** consists of all points $P(x, y, z, v)$ which satisfy the parametric equations

$$\left.\begin{array}{ll} x = x_1 + (x_2 - x_1)t, & y = y_1 + (y_2 - y_1)t, \\ z = z_1 + (z_2 - z_1)t, & v = v_1 + (v_2 - v_1)t, \end{array}\right\} \tag{2}$$

where the parameter t takes on all real values; we sometimes denote this line by $L(P_1P_2)$. The four numbers

$$x_2 - x_1, \qquad y_2 - y_1, \qquad z_2 - z_1, \qquad v_2 - v_1$$

are called **direction numbers** of the line through P_1 and P_2. Given a point P and and a quadruple a, b, c, d (not all zero) there is a unique line $L(P_1P_2)$ through P_1 which has direction numbers $a, b, c, d,$ The point P_2 has coordinates $P_2 = (x_1 + a, y_1 + b, z_1 + c, v_1 + d)$ and $L(P_1P_2)$ is the locus of the equations $x = x_1 + ta, y = y_1 + tb, z = z_1 + tc, v = v_1 + td.$ Denoting a given line by L, we can form the **directed line $\vec{L}$** by ordering the points on it according to increasing values of t. If Q_1 and Q_2 are two points which are on the line (2) with values of t_1 and t_2, respectively, we say that Q_1 **precedes** Q_2 if $t_1 < t_2$; we write $Q_1 < Q_2$. Of course we can give L the opposite ordering in terms of decreasing values of t, in which case we write $-\vec{L}$. If $Q_1, Q_2,$ and Q_3 are three points on $\vec{L}$, we say that Q_2 **is between Q_1 and Q_3** if either $Q_1 < Q_2$ and $Q_2 < Q_3$ or $Q_3 < Q_2$ and $Q_2 < Q_1$. In order that a straight line, defined according to (2), should satisfy our intuition that it is really "straight," we must have the formula

$$|Q_1Q_2| + |Q_2Q_3| = |Q_1Q_3| \tag{3}$$

whenever Q_2 is between Q_1 and Q_3 on the line $\vec{L}$. (See Fig. 12–1.)

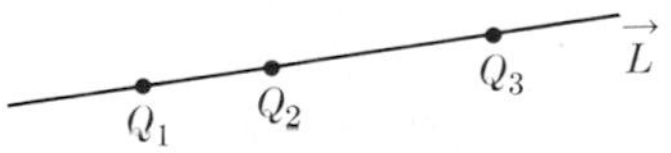

Figure 12–1

To establish (3), we first write L in the form

$$x = x_1 + at, \qquad y = y_1 + bt, \qquad z = z_1 + ct, \qquad v = v_1 + dt \tag{4}$$

where the direction numbers are $a = x_2 - x_1,$ and so forth for $b, c, d.$ Suppose

Q_1 corresponds to t_1 in (4), Q_2 corresponds to t_2, and Q_3 to t_3. If Q_2 is between Q_1 and Q_3, then $t_1 < t_2 < t_3$ or $t_3 < t_2 < t_1$. We write

$$Q_i(x_1 + at_i, y_1 + bt_i, z_1 + ct_i, v_1 + dt_i) \qquad \text{for } i = 1, 2, 3.$$

Then from (1), we have

$$|Q_1 Q_2| = \sqrt{(a^2 + b^2 + c^2 + d^2)(t_2 - t_1)^2}$$
$$|Q_1 Q_3| = \sqrt{(a^2 + b^2 + c^2 + d^2)(t_3 - t_1)^2}$$
$$|Q_2 Q_3| = \sqrt{(a^2 + b^2 + c^2 + d^2)(t_3 - t_2)^2}.$$

Consequently, the statement (3) becomes

$$|t_2 - t_1| + |t_3 - t_2| = |t_3 - t_1|.$$

However, this equation is easily verified whenever t_2 is between t_1 and t_3.

Theorem 1. *Let a, b, c, d and a_1, b_1, c_1, d_1 be proportional sets of direction numbers. Then the equations*

$$x = x_1 + at, \qquad y = y_1 + bt, \qquad z = z_1 + ct, \qquad v = v_1 + dt \quad \text{(5a)}$$

and

$$x = x_1 + a_1 t, \qquad y = y_1 + b_1 t, \qquad z = z_1 + c_1 t, \qquad v = v_1 + d_1 t \quad \text{(5b)}$$

are parametric equations of the same line L.

Proof. We set $x_2 = x_1 + a$, $y_2 = y_1 + b$, $z_2 = z_1 + c$, $v_2 = v_1 + d$, and then (5a) is the line L through (x_1, y_1, z_1, v_1) and (x_2, y_2, z_2, v_2) as given in (2). Since the direction numbers in (5a) and (5b) are proportional, there is a real number $\alpha \neq 0$ such that

$$a_1 = \alpha a, \qquad b_1 = \alpha b, \qquad c_1 = \alpha c, \qquad d_1 = \alpha d.$$

Hence we can write (5b) in the form

$$x = x_1 + a(\alpha t), \qquad y = y_1 + b(\alpha t), \qquad z = z_1 + c(\alpha t), \qquad v = v_1 + d(\alpha t).$$

Let P_0 be any point on the line (5a) corresponding to a value of t, say t_0. Then by choosing $t = t_0/\alpha$ in (5b) we get the same point P_0 satisfying the equations in (5b). Therefore every point on line (5a) is on (5b). Similarly, every point on the line (5b) is on (5a). The lines coincide.

Corollary. *Two lines L_1 and L_2 which have proportional direction numbers either do not intersect or coincide.*

Proof. If L_1 and L_2 have a point in common, then we label this point (x_1, y_1, z_1, v_1) and write the parametric equations of L_1 and L_2 in the form (5a) and (5b). Using Theorem 1, we see that the lines coincide.

It can be shown that any two sets of parametric equations for the same line L lead either to the same directed line $\vec{L}$ (for increasing values of the parameter) or to opposite directed lines $\vec{L}$ and $-\vec{L}$. Thus every line in Euclidean four space has only two possible orderings (as is the case for lines in the plane and in three space). Suppose that $\vec{L}$ is a given directed line. We define the **directed distance** $\overline{PQ}$ from P to Q along $\vec{L}$ by

$$\overline{PQ} = \begin{cases} |PQ| & \text{if } P = Q \text{ or } P < Q, \\ -|PQ| & \text{if } Q < P. \end{cases}$$

Theorem 2. *Suppose that $P(\bar{x}, \bar{y}, \bar{z}, \bar{v})$ is h of the way from $P_1(x_1, y_1, z_1, v_1)$ to $P_2(x_2, y_2, z_2, v_2)$ on the line $\vec{L}$ through P_1 and P_2. Then*

$$\begin{aligned}\bar{x} &= x_1 + h(x_2 - x_1), & \bar{y} &= y_1 + h(y_2 - y_1), \\ \bar{z} &= z_1 + h(z_2 - z_1), & \bar{v} &= v_1 + h(v_2 - v_1).\end{aligned} \tag{6}$$

Proof. We write $\vec{L}$ parametrically

$$x = x_0 + at, \quad y = y_0 + bt, \quad z = z_0 + ct, \quad v = v_0 + dt$$

and suppose P_1 corresponds to t_1, P_2 to t_2, and P to $\bar{t}$. Then

$$\bar{x} - x_1 = a(\bar{t} - t_1) \quad \text{and} \quad x_2 - x_1 = a(t_2 - t_1).$$

Also,

$$\overline{P_1P} = \sqrt{a_2 + b^2 + c^2 + d^2} \, (\bar{t} - t_1)$$

and

$$\overline{P_1P_2} = \sqrt{a^2 + b^2 + c^2 + d^2} \, (t_2 - t_1).$$

Since $\overline{P_1P} = h\overline{P_1P_2}$, we have $\bar{t} - t_1 = h(t_2 - t_1)$, and so

$$\bar{x} - x_1 = a(\bar{t} - t_1) = ah(t_2 - t_1) = h(x_2 - x_1), \quad \text{or} \quad \bar{x} = x_1 + h(x_2 - x_1).$$

The remaining formulas in (6) are obtained similarly.

In case $h = \frac{1}{2}$ in Theorem 2, the point P is called the **midpoint of the segment** P_1P_2. In this case, we observe that $\bar{t} = \frac{1}{2}(t_1 + t_2)$, $\bar{x} = \frac{1}{2}(x_1 + x_2)$, etc.

Corollary. *If P is the midpoint of P_1 and P_2, then $|PP_1| = |PP_2|$.*

EXAMPLE 1. Find the midpoint P of the segment AB, where $A = (3, -1, 2, 4)$, $B = (1, 1, 4, 2)$. Verify the corollary to Theorem 2 by showing that $|AP| = |BP|$.

Solution. We have $\bar{x} = \frac{1}{2}(3 + 1) = 2$, $\bar{y} = \frac{1}{2}(1 - 1) = 0$, $\bar{z} = \frac{1}{2}(2 + 4) = 3$, $\bar{v} = (\frac{1}{2}4 + 2) = 3$. Therefore $P = (2, 0, 3, 3)$. Also,

$$|AP| = \sqrt{(2 - 3)^2 + (0 + 1)^2 + (3 - 2)^2 + (3 - 4)^2} = 2,$$
$$|BP| = \sqrt{(2 - 1)^2 + (0 - 1)^2 + (3 - 4)^2 + (3 - 2)^2} = 2.$$

EXAMPLE 2. Find the point P which is $\frac{2}{3}$ of the way from $A(1, -2, 4, 2)$ to $B(-2, 4, 1, -1)$.

Solution. $\bar{x} = 1 + \frac{2}{3}(-2 - 1) = -1, \bar{y} = -2 + \frac{2}{3}(4 + 2) = 2, \bar{z} = 4 + \frac{2}{3}(1 - 4) = 2,$ $\bar{v} = 2 + \frac{2}{3}(-1 - 2) = 0.$ $P = (-1, 2, 2, 0).$

Let a, b, c, d be a set of direction numbers of a line L. We set

$$\alpha = \sqrt{a^2 + b^2 + c^2 + d^2}$$

and we form the set $(1/\alpha)a, (1/\alpha)b, (1/\alpha)c, (1/\alpha)d$ which we call a set of **direction cosines** of the line L. Another set of direction cosines consists of $-(1/\alpha)a, -(1/\alpha)b,$ $-(1/\alpha)c, -(1/\alpha)d$. These two sets determine the two possible orderings of L as a directed line.

We summarize the results in this section in the following theorem.

Theorem 3. (i) *There is a unique line in Euclidean four space which passes through a given point and has a given set of direction numbers. If (x_0, y_0, z_0, v_0) is the point and a, b, c, d are the direction numbers, then the line L has the parametric equations*

$$x = x_0 + at, \qquad y = y_0 + bt, \qquad z = z_0 + ct, \qquad v = v_0 + dt.$$

(ii) *There is a unique line in Euclidean four space which passes through two given points.*

EXAMPLE 3. Find a set of direction numbers and a set of direction cosines for the line passing through the two given points $A(2, 3, -1, 1)$ and $B(-1, 4, 2, 3)$.

Solution. A set of direction numbers is

$$-1 - 2, 4 - 3, 2 - (-1), 3 - 1 = -3, 1, 3, 2.$$

Also, $\sqrt{(-3)^2 + 1^2 + 3^2 + 2^2} = \sqrt{23}$ and a set of direction cosines is $-3/\sqrt{23},$ $1/\sqrt{23}, 3/\sqrt{23}, 2/\sqrt{23}.$

PROBLEMS

In each of problems 1 through 4, find the lengths of the sides of triangle ABC.

1. $A = (2, 1, -1, 3), B = (3, 3, 2, 1), C = (-1, 1, -1, 2)$
2. $A = (1, -2, 3, 2), B = (2, -1, 3, -1), C = (4, 2, -1, 3)$
3. $A = (-1, 4, 3, 2), B = (-2, 2, 4, 3), C = (1, 2, 3, -2)$
4. $A = (-3, 2, 4, -1), B = (2, 1, 2, 3), C = (-2, 1, 4, 3)$

In each of problems 5 through 8, find the midpoint P of the segment AB and verify the Corollary to Theorem 2.

5. $A = (3, 1, -2, 0)$, $B = (1, -1, 2, 2)$ 6. $A = (2, 1, 0, 3)$, $B = (-1, 2, 2, 1)$

7. $A = (1, -1, 2, 3)$, $B = (2, 1, 0, 2)$ 8. $A = (3, 2, 1, 0)$, $B = (2, -1, 3, 2)$

9. One endpoint of a segment is at $P = (2, 1, -1, 2)$ and the midpoint is at $Q = (3, -1, 0, 1)$. Find the other endpoint.

In each of problems 10 through 13, find the point h of the way from A to B.

10. $A = (2, -1, 1, 3)$, $B = (-1, 2, 4, 0)$, $h = \frac{1}{3}$

11. $A = (-2, 1, 4, 2)$, $B = (3, -4, -1, -3)$, $h = \frac{2}{5}$

12. $A = (2, -1, 3, -1)$, $B = (1, -2, 1, 2)$, $h = -1$

13. $A = (1, 2, -1, 1)$, $B = (2, -1, 2, 3)$, $h = 2$

In each of problems 14 through 17, find a set of direction numbers and a set of direction cosines for the line passing through the given points.

14. $A = (2, 1, -1, 3)$, $B = (1, 3, 2, 1)$ 15. $A = (-1, 2, -1, 1)$, $B = (2, 3, 1, -2)$

16. $A = (3, -1, 2, 1)$, $B = (2, 1, -1, 2)$

17. $A = (2, 1, -3, -2)$, $B = (-1, 2, 1, -3)$

In each of problems 18 through 21, a point P_1 and a set of direction numbers are given. Find another point which is on the line through P_1 and having the given set of direction numbers.

18. $P_1 = (2, -1, 3, 1)$, direction numbers $-1, 2, 1, -3$

19. $P_1 = (1, 2, 1, -1)$, direction numbers $1, -1, 3, 2$

20. $P_1 = (-1, 1, 2, 3)$, direction numbers $2, 1, -1, -2$

21. $P_1 = (1, 2, 1, -1)$, direction numbers $2, -1, 2, 3$

In each of problems 22 through 25, determine whether or not the three given points are on a line.

22. $A = (2, 1, -1, 3)$, $B = (0, -3, 1, 1)$, $C = (-3, -9, 4, -2)$

23. $A = (3, 2, -1, 1)$, $B = (1, 0, 1, 2)$, $C = (-1, -2, 3, 2)$

24. $A = (3, 1, 0, -1)$, $B = (2, 2, 2, 1)$, $C = (0, 4, 6, 4)$

25. $A = (2, -1, 1, 3)$, $B = (4, 1, -3, 1)$, $C = (7, 4, -9, -2)$

26. Let R_5 be all ordered quintuples of numbers of the form (x, y, z, v, w). Write the distance formula which yields Euclidean five space. Define the parametric equations of a line through two points.

27. (a) Establish the analogue of formula (3) in Euclidean five space. (b) State and prove the analogue of Theorem 1 in Euclidean five space.

28. Develop the midpoint formula for a line segment in Euclidean five space. Show how the corresponding formula may be established in Euclidean n space where n is any positive integer.

29. Show that the x axis is a straight line in Euclidean four space.

*30. Show that for any triangle in Euclidean four space the sum of the lengths of two sides of the triangle is greater than the length of the third side.

▶ 2. EQUATIONS OF A LINE

Suppose that L is a line in Euclidean four space and that it has parametric equations

$$x = x_0 + at, \qquad y = y_0 + bt, \qquad z = z_0 + ct, \qquad v = v_0 + dt. \qquad (1)$$

If none of the numbers a, b, c, or d is zero, we may divide each equation in (1) by the coefficient of t and obtain

$$\frac{x - x_0}{a} = \frac{y - y_0}{b} = \frac{z - z_0}{c} = \frac{v - v_0}{d} \qquad (2)$$

for all t. We call this system of equations the **equations of a line in symmetric form**. In case one or more of the denominators in (2) is zero, we still use the same form if we agree that a term such as $(z - z_0)/0$ is replaced by the equation $z = z_0$. We recall that this convention was employed for symmetric equations in three space (page 215). We note that the symmetric equations of a line through the two points $P_1(x_1, y_1, z_1, v_1)$, $P_2(x_2, y_2, z_2, v_2)$ are

$$\frac{x - x_1}{x_2 - x_1} = \frac{y - y_1}{y_2 - y_1} = \frac{z - z_1}{z_2 - z_1} = \frac{v - v_1}{v_2 - v_1}. \qquad (3)$$

EXAMPLE 1. Find the symmetric equations and the parametric equations of the coordinate axes.

Solution. The points $(0, 0, 0, 0)$ and $(1, 0, 0, 0)$ are on the x axis. Therefore by (3), we have for the line through these points

$$\frac{x - 0}{1 - 0} = \frac{y}{0} = \frac{z}{0} = \frac{v}{0} \qquad (4a)$$

or, parametrically,

$$x = 0 + t, \, y = 0, \, z = 0, \, v = 0. \qquad (4b)$$

We see that every point of the form $(x, 0, 0, 0)$ satisfies (4), and conversely. The statements for the y, z, and v axes are similar.

DEFINITIONS. Two lines L_1 and L_2 are **parallel** $\Leftrightarrow$ L_1 and L_2 have proportional direction numbers. Two lines L_1 and L_2 are **perpendicular** $\Leftrightarrow$ L_1 and L_2 intersect at some point V and for every point A on L_1 and B on L_2, we have (see Fig. 12–2)

$$|AB|^2 = |VA|^2 + |VB|^2. \qquad (5)$$

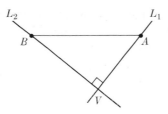

Figure 12–2

The next theorem is basic for parallel lines.

Theorem 4. *Let L_1 be a given line and $P_0(x_0, y_0, z_0, v_0)$ a given point not on L_1. Then there is a unique line L passing through P_0 which is parallel to L_1.*

Proof. Let a, b, c, d be direction numbers of L_1. Then the line L given by

$$x = x_0 + at, \qquad y = y_0 + bt, \qquad z = z_0 + ct, \qquad v = v_0 + dt$$

passes through P_0 and has direction numbers proportional to those of L_1. Hence L is parallel to L_1. If there were two lines parallel to L_1, then the Corollary to Theorem 1 would show that they coincide (since the point P_0 is common to both lines).

Theorem 5. *Let the line L_1 have direction numbers a_1, b_1, c_1, d_1 and L_2 the direction numbers a_2, b_2, c_2, d_2. Then L_1 is perpendicular to L_2 if and only if L_1 and L_2 intersect and*

$$a_1 a_2 + b_1 b_2 + c_1 c_2 + d_1 d_2 = 0. \tag{6}$$

Proof. Let $V(x_0, y_0, z_0, v_0)$ be the point of intersection of L_1 and L_2. We denote by $A(x_1, y_1, z_1, v_1)$ *any* point on L_1 and by $B(x_2, y_2, z_2, v_2)$ *any* point on L_2. We must show that (5) implies (6) and that (6) implies (5). We write the equations

$$L_1: x = x_0 + a_1 t, \qquad y = y_0 + b_1 t, \qquad z = z_0 + c_1 t, \qquad v = v_0 + d_1 t$$
$$L_2: x = x_0 + a_2 t, \qquad y = y_0 + b_2 t, \qquad z = z_0 + c_2 t, \qquad v = v_0 + d_2 t.$$

Then, letting t_1 correspond to A and t_2 correspond to B, we have

$$|AB|^2 = (a_2 t_2 - a_1 t_1)^2 + (b_2 t_2 - b_1 t_1)^2 + (c_2 t_2 - c_1 t_1)^2 + (d_2 t_2 - d_1 t_1)^2$$
$$|VA|^2 = (a_1^2 + b_1^2 + c_1^2 + d_1^2)t_1^2$$
$$|VB|^2 = (a_2^2 + b_2^2 + c_2^2 + d_2^2)t_2^2.$$

Therefore after multiplying out the right side of the first equation above, we find

$$|AB|^2 = |VA|^2 + |VB|^2 - 2(a_1 a_2 + b_1 b_2 + c_1 c_2 + d_1 d_2)t_1 t_2. \tag{7}$$

If L_1 and L_2 are perpendicular then (5) holds for all t_1 and t_2. Hence (7) can be valid only if $a_1 a_2 + b_1 b_2 + c_1 c_2 + d_1 d_2 = 0$. That is, (6) holds. Conversely, if

(6) holds, then equation (7) becomes identical with (5), and L_1 and L_2 must be perpendicular.

EXAMPLE 2. Given $P = (2, 3, -1, 1)$, $Q = (-1, 1, 0, -2)$, $R = (3, 5, 0, -1)$. Determine whether or not the line through the points P and Q is perpendicular to the line through P and R.

Solution. A set of direction numbers of PQ is $-3, -2, 1, -3$; direction numbers for PR are $1, 2, 1, -2$. We employ (6):

$$(-3) \cdot 1 + (-2) \cdot 2 + 1 \cdot 1 + (-3)(-2) = 0.$$

The lines are perpendicular.

Theorem 6. *Let L be a line with parametric equations*

$$x = x_0 + at, \qquad y = y_0 + bt, \qquad z = z_0 + ct, \qquad v = v_0 + dt \qquad (8)$$

where a, b, c, d are direction cosines of L. Let $P_1(x_1, y_1, z_1, v_1)$ be a point not on L. Then there is a unique line L_1 passing through P_1 which is perpendicular to L. The parametric equations of L_1 are

$$x = x_1 + At, \qquad y = y_1 + Bt, \qquad z = z_1 + Ct, \qquad v = v_1 + Dt \qquad (9)$$

where the direction numbers A, B, C, D are proportional to

$$x_0 - x_1 + a\bar{t}, \qquad y_0 - y_1 + b\bar{t}, \qquad z_0 - z_1 + c\bar{t}, \qquad v_0 - v_1 + d\bar{t} \qquad (10)$$

and

$$\bar{t} = a(x_1 - x_0) + b(y_1 - y_0) + c(z_1 - z_0) + d(v_1 - v_0). \qquad (11)$$

Proof. Suppose P_2 is the point of intersection of L and L_1 where L and L_1 are perpendicular. We denote the coordinates of P_2 by $(x_0 + a\bar{t}, y_0 + b\bar{t}, z_0 + c\bar{t}, v_0 + d\bar{t})$ for some value $\bar{t}$. Since P_2 is also on L_1, for some value t' the coordinates of P_2 are $(x_1 + At', y_1 + Bt', z_1 + Ct', v_1 + Dt')$. Thus we have

$$\begin{matrix} x_1 + At' = x_0 + a\bar{t}, & y_1 + Bt' = y_0 + b\bar{t}, \\ z_1 + Ct' = z_0 + c\bar{t}, & v_1 + Dt' = v_0 + d\bar{t} \end{matrix} \qquad (12)$$

and, since L_1 and L are perpendicular

$$aA + bB + cC + dD = 0. \qquad (13)$$

We solve the equations (12) for $A, B, C,$ and D, respectively, getting

$$A = \frac{1}{t'}(x_0 - x_1 + a\bar{t}), \qquad B = \frac{1}{t'}(y_0 - y_1 + b\bar{t}),$$

$$C = \frac{1}{t'}(z_0 - z_1 + c\bar{t}), \qquad D = \frac{1}{t'}(v_0 - v_1 + d\bar{t}).$$

We substitute these values in (13) and, taking the relation $a^2 + b^2 + c^2 + d^2 = 1$ into account, we get (11). We observe that A, B, C, D are proportional to the numbers (10).

EXAMPLE 3. Let L_1 have the parametric equations

$$x = -1 + 2t, \quad y = 2 - 2t, \quad z = 1 - 2t, \quad v = -1 + 2t.$$

Find the equations of the line L_2 through the point $(2, -1, 3, 2)$ which is perpendicular to L_1.

Solution. Direction numbers of L_1 are $2, -2, -2, 2$. Therefore direction cosines are $\frac{1}{2}, -\frac{1}{2}, -\frac{1}{2}, \frac{1}{2}$, obtained by dividing each direction number by

$$\sqrt{(2)^2 + (-2)^2 + (-2)^2 + (2)^2} = 4.$$

We write for L_1:

$$x = -1 + \tfrac{1}{2}t, \quad y = 2 - \tfrac{1}{2}t, \quad z = 1 - \tfrac{1}{2}t, \quad v = -1 + \tfrac{1}{2}t.$$

We can get the result by using $(x_0, y_0, z_0, v_0) = (-1, 2, 1, -1)$ and $a, b, c, d = \frac{1}{2}, -\frac{1}{2}, -\frac{1}{2}, \frac{1}{2}$ in equation (11) to find $\bar{t}$. Then we get the direction numbers of L_2 from (10). The result is

$$\bar{t} = \tfrac{1}{2}(2 + 1) - \tfrac{1}{2}(-1 - 2) - \tfrac{1}{2}(3 - 1) + \tfrac{1}{2}(2 + 1) = \tfrac{7}{2}$$

and

$$A = -1 - 2 + \tfrac{7}{4} = -\tfrac{5}{4}, \qquad B = 2 + 1 - \tfrac{7}{4} = \tfrac{5}{4},$$
$$C = 1 - 3 - \tfrac{7}{4} = -\tfrac{15}{4}, \qquad D = -1 - 2 + \tfrac{7}{4} = -\tfrac{5}{4}.$$

The equations of L_2 are (after multiplying the direction numbers by $-4/5$)

$$x = 2 + t, \quad y = -1 - t, \quad z = 3 + 3t, \quad v = 2 + t.$$

DEFINITION. Let $\vec{L}$ be a directed line and $\overrightarrow{AB}$ a directed line segment (that is, an ordered pair of points in R_4). We construct the line through A perpendicular to $\vec{L}$ and the line through B perpendicular to $\vec{L}$. The two points of intersection are denoted A', B', respectively (see Fig. 12–3). We define the **projection of $\overrightarrow{AB}$ along $\vec{L}$** to be the directed distance $\overline{A'B'}$. We denote this quantity by

$$\mathrm{Proj}_{\vec{L}} \ \overrightarrow{AB}.$$

We observe that the lines AA' and BB' while both perpendicular to $\vec{L}$ need not be parallel.

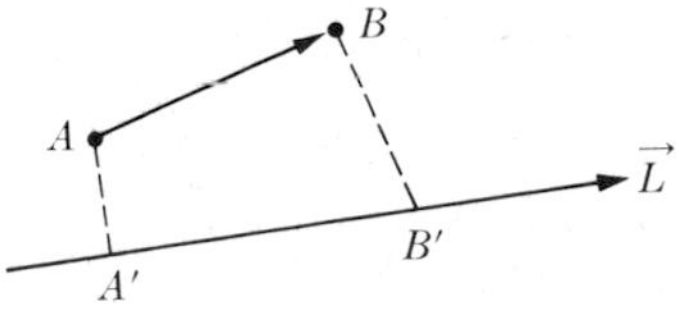

Figure 12–3

The next theorem gives a formula for finding the projection when $\vec{L}$ and the points A and B are known.

Theorem 7. *Let $\vec{L}$ be a directed line with direction cosines a, b, c, d. Let $\overrightarrow{AB}$ be a directed line segment. Then*

$$\text{Proj}_{\vec{L}} \; \overrightarrow{AB} = a(x_B - x_A) + b(y_B - y_A) + c(z_B - z_A) + d(v_B - v_A). \quad (14)$$

Proof. Let A' be the point of intersection of the line $\vec{L}$ and the perpendicular to $\vec{L}$ through the point A. Similarly, let B' be the point of intersection of $\vec{L}$ and the perpendicular to $\vec{L}$ through B. We write $A' = (x'_A, y'_A, z'_A, v'_A)$ and $B' = (x'_B, y'_B, z'_B, v'_B)$ for the corrdinates of these points. We seek $\overline{A'B'}$. The parametric equations of $\vec{L}$ may be written

$$x = x_0 + at, \quad y = y_0 + bt, \quad z = z_0 + ct, \quad v = v_0 + dt.$$

Then we let t_A correspond to A' and we let t_B correspond to B'. Hence, from Theorem 6, we get

$$B' = (x_0 + t_B a, \; y_0 + t_B b, \; z_0 + t_B c, \; v_0 + t_B d),$$
$$t_B = a(x_B - x_0) + b(y_B - y_0) + c(z_B - z_0) + d(v_B - v_0)$$
$$A' = (x_0 + t_A a, \; y_0 + t_A b, \; z_0 + t_A c, \; v_0 + t_A d),$$
$$t_A = a(x_A - x_0) + b(y_A - y_0) + c(z_A - z_0) + d(v_A - v_0).$$

Since the directed distance $\overline{A'B'} = t_B - t_A$, we get

$$\text{Proj}_{\vec{L}} \; \overrightarrow{AB} = \overline{A'B'} = t_B - t_A.$$

From formula (11) in Theorem 6, the result follows by subtraction.

EXAMPLE 4. Given the directed line $\vec{L}$

$$x = 2 + 3t, \quad y = -1 + t, \quad z = -2t, \quad v = 1 + 2t$$

and the points $A(2, 1, -1, 0)$, $B(-1, 0, 2, -2)$, find the projection of $\overrightarrow{AB}$ on $\vec{L}$.

Solution. The direction cosines of $\vec{L}$ are $1/\sqrt{2}, 1/3\sqrt{2}, -2/3\sqrt{2}, 2/3\sqrt{2}$. Employing (14), we obtain

$$\text{Proj}_{\vec{L}} \; \overrightarrow{AB} = \frac{1}{\sqrt{2}}(-3) + \frac{1}{3\sqrt{2}}(-1) - \frac{2}{3\sqrt{2}}(3) + \frac{2}{3\sqrt{2}}(-2) = -\frac{20}{3\sqrt{2}}.$$

PROBLEMS

In each of problems 1 through 4, find the equations in symmetric form of the line which passes through A and has the given direction numbers.

1. $A = (2, 1, -1, 3)$, direction numbers $1, -2, 2, 1$
2. $A = (1, 3, 2, -1)$, direction numbers $2, -1, 1, 2$
3. $A = (-1, 0, 3, 2)$, direction numbers $3, 2, -1, 4$
4. $A = (-2, 1, 2, -1)$, direction numbers $1, 3, 2, -1$

In each of problems 5 through 8, find equations in parametric form of the given line L.

5. $\dfrac{x-1}{1} = \dfrac{y+1}{-1} = \dfrac{z+2}{1} = \dfrac{v-1}{1}$

6. $\dfrac{x+2}{2} = \dfrac{y-1}{3} = \dfrac{z-2}{-1} = \dfrac{v+1}{2}$

7. $\dfrac{x-2}{4} = \dfrac{y-1}{-2} = \dfrac{z-1}{5} = \dfrac{v+2}{2}$

8. $\dfrac{x+1}{2} = \dfrac{y-2}{3} = \dfrac{z+1}{-2} = \dfrac{v-2}{4}$

In each of problems 9 through 12, find equations in symmetric form of the line L through A and parallel to the given line L_0.

9. $A = (2, 1, -2, -1)$; $L_0: \dfrac{x+1}{1} = \dfrac{y-2}{1} = \dfrac{z+1}{-1} = \dfrac{v+1}{-1}$

10. $A = (2, 3, -1, 2)$; $L_0: \dfrac{x-1}{2} = \dfrac{y+1}{-3} = \dfrac{z-1}{1} = \dfrac{v-1}{-2}$

11. $A = (1, -1, 2, 3)$; $L_0: \dfrac{x+2}{2} = \dfrac{y+2}{-4} = \dfrac{z-2}{5} = \dfrac{v+1}{-2}$

12. $A = (-1, 2, 3, -2)$; $L_0: \dfrac{x-2}{2} = \dfrac{y-1}{-1} = \dfrac{z+2}{2} = \dfrac{v+1}{0}$

In each of problems 13 through 16, find equations in symmetric form of the line L passing through the given points A and B.

13. $A = (2, 1, -1, 3)$, $B = (1, 2, 2, -1)$
14. $A = (1, -2, 1, 2)$, $B = (3, 1, -1, 2)$
15. $A = (1, 2, -1, -2)$, $B = (1, -1, -1, 3)$
16. $A = (-2, 3, 1, 2)$, $B = (2, 3, 1, 2)$

In each of problems 17 through 20, determine whether or not L_1 and L_2 are perpendicular.

17. $L_1: \dfrac{x-1}{2} = \dfrac{y+1}{-1} = \dfrac{z-2}{3} = \dfrac{v+1}{1}$;

$L_2: \dfrac{x-1}{1} = \dfrac{y+1}{2} = \dfrac{z-2}{-1} = \dfrac{v+1}{3}$

18. $L_1: \dfrac{x+1}{1} = \dfrac{y-2}{2} = \dfrac{z+1}{2} = \dfrac{v-1}{-3}$;

$L_2: \dfrac{x+1}{-1} = \dfrac{y-2}{1} = \dfrac{z+1}{-2} = \dfrac{v-1}{-1}$

19. $L_1: \dfrac{x-2}{0} = \dfrac{y+1}{-2} = \dfrac{z-1}{1} = \dfrac{v-2}{2}$;

$L_2: \dfrac{x-2}{2} = \dfrac{y+1}{-1} = \dfrac{z-1}{3} = \dfrac{v-2}{1}$

20. $L_1: \dfrac{x+2}{0} = \dfrac{y-1}{0} = \dfrac{z+2}{-1} = \dfrac{v+1}{3}$;

$L_2: \dfrac{x+2}{1} = \dfrac{y-1}{2} = \dfrac{z+2}{3} = \dfrac{v+1}{1}$

In each of problems 21 through 24, find $\mathrm{Proj}_{\vec{L}}\ \overrightarrow{AB}$ where $\vec{L}$ passes through C and D and is directed from C to D.

21. $A = (2, 1, -3, 2)$, $B = (4, 3, 1, -1)$, $C = (2, 1, -1, 2)$, $D = (3, 2, -2, 1)$
22. $A = (1, -1, 2, 3)$, $B = (2, 2, 1, 1)$, $C = (3, 2, -1, 1)$, $D = (5, 6, 4, -1)$
23. $A = (-2, 1, 4, 2)$, $B = (3, -1, 2, 3)$, $C = (2, 1, 1, -1)$, $D = (2, 2, 3, 1)$
24. $A = (3, -1, 2, 0)$, $B = (2, 1, -1, 1)$, $C = (1, 2, -1, -2)$, $D = (3, 2, 1, -2)$

In each of problems 25 through 28, find equations in symmetric form of the line L_2 which passes through the point P_1 and is perpendicular to Line L_1 (and intersects it).

25. $P_1 = (3, -1, 2, 1)$; $\quad L_1: \dfrac{x-1}{1} = \dfrac{y+2}{-1} = \dfrac{z-1}{1} = \dfrac{v+2}{-1}$

26. $P_1 = (2, 1, 1, -1)$; $\quad L_1: \dfrac{x+1}{4} = \dfrac{y-1}{-2} = \dfrac{z+1}{5} = \dfrac{v-1}{2}$

27. $P_1 = (-1, 2, -3, 1)$; $\quad L_1: \dfrac{x-2}{2} = \dfrac{y+1}{-1} = \dfrac{z-2}{0} = \dfrac{v+1}{2}$

28. $P_1 = (2, 3, 1, 2)$; $\quad L_1: \dfrac{x+2}{0} = \dfrac{y-2}{0} = \dfrac{z+1}{1} = \dfrac{v-2}{-2}$

29. (a) Suppose that the line through two points A and B is parallel to a given line $\vec{L}$. Show that

$$\mathrm{Proj}_{\vec{L}}\ \overrightarrow{AB} = \overline{AB}.$$

(b) Show that if the line through A and B is perpendicular to $\vec{L}$, then $\mathrm{Proj}_{\vec{L}}\ \overrightarrow{AB} = 0$.

30. Show that the four coordinate axes in Euclidean four dimensional space are perpendicular to each other.

*31. Let $\vec{L}$ be a given line and P_0 a point on $\vec{L}$. Show that there are three distinct lines through P_0 each of which is perpendicular to $\vec{L}$ and which are mutually perpendicular.

32. (a) (See Problems 26, 27, 28 in Section 1.) Derive the equations of a line in symmetric form in Euclidean five space. (b) Define parallel and perpendicular lines in Euclidean five space and state and prove the analogs of Theorems 4 and 5. (c) Show that the x, y, z, v, and w axes in five space are mutually perpendicular. (d) Define projection of a directed line segment on a line $\vec{L}$ in five space. State and prove the analog to Theorem 7.

33. Let R_n be ordered n-tuples of numbers for any positive integer n. Define Euclidean n-space. Develop the analogs of Problem 32(a) through (d) for n-space.

▶ 3. TWO DIMENSIONAL PLANES IN EUCLIDEAN FOUR SPACE

We recall that the set of all ordered pairs of real numbers (r, s) comprises R_2.

DEFINITIONS. A set of points S in Euclidean four space is called a **two-dimensional plane** or simply a **plane** $\Leftrightarrow$ the points of S can be put into one-to-one correspondence with the ordered pairs (r, s) of R_2 in such a way that for any two points P_1 and P_2 of S corresponding to (r_1, s_1) and (r_2, s_2), respectively, the distance formula

$$|P_1 P_2| = \sqrt{(r_2 - r_1)^2 + (s_2 - s_1)^2}$$

holds. (See Fig. 12–4.) The point $(0, 0)$ is called the **origin** of the (r, s) system and the points I and J which correspond to $(1, 0)$ and $(0, 1)$ are called the **first and second unit points** of the plane.

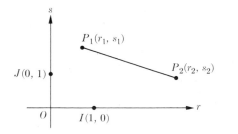

Figure 12–4

We must be careful to observe that every point P of S will have an ordered quadruple (x, y, z, v) representing its position in four space and an ordered pair (r, s) representing its position in the plane S.

With a plane defined in this way it can be shown that all the results of plane analytic geometry are valid in S as long as we restrict ourselves to the (r, s) system of coordinates. However, we wish to examine the relationship of all such planes to the four dimensional space containing them.

For convenience a point in a plane S will be designated by a double subscript to indicate its (r, s) coordinates. That is, P_{rs} corresponds to (r, s), P_{00} to $(0, 0)$, P_{10} to the point $(1, 0)$, and so forth. Of course, these points will also have ordered quadruples representing their positions in four space.

We now derive the parametric equations of a plane in Euclidean four space. Each such plane will be described by four equations involving the two parameters r and s.

Theorem 8. *Suppose that S is a plane in Euclidean four space and (r, s) is a coordinate system on S with origin at $P_{00} = (x_0, y_0, z_0, v_0)$ and with first and second unit points at $P_{10} = (x_0 + a_1, y_0 + b_1, z_0 + c_1, v_0 + d_1)$ and $P_{01} = (x_0 + a_2, y_0 + b_2, z_0 + c_2, v_0 + d_2)$, respectively. Then*

$$a_1^2 + b_1^2 + c_1^2 + d_1^2 = 1,$$
$$a_2^2 + b_2^2 + c_2^2 + d_2^2 = 1, \tag{1}$$
$$a_1 a_2 + b_1 b_2 + c_1 c_2 + d_1 d_2 = 0.$$

Also, every point P_{rs} in the plane S is given by the parametric equations

$$
\begin{aligned}
x &= x_0 + a_1 r + a_2 s, & y &= y_0 + b_1 r + b_2 s, \\
z &= z_0 + c_1 r + c_2 s, & v &= v_0 + d_1 r + d_2 s.
\end{aligned}
\tag{2}
$$

Proof. To establish (1), we observe that $|P_{00}P_{10}| = 1$ and $|P_{00}P_{01}| = 1$. Therefore from the distance formula between two points,

$$1 = \sqrt{a_1^2 + b_1^2 + c_1^2 + d_1^2}, \qquad 1 = \sqrt{a_2^2 + b_2^2 + c_2^2 + d_2^2}.$$

Referring to Fig. 12–4, we see that $|P_{10}P_{01}| = \sqrt{2}$, and so

$$\sqrt{2} = \sqrt{(a_2 - a_1)^2 + (b_2 - b_1)^2 + (c_2 - c_1)^2 + (d_2 - d_1)^2}.$$

Consequently,

$$
\begin{aligned}
2 &= a_1^2 + a_2^2 + b_1^2 + b_2^2 + c_1^2 + c_2^2 + d_1^2 + d_2^2 \\
&\quad - 2(a_1 a_2 + b_1 b_2 + c_1 c_2 + d_1 d_2) \\
&= 2 - 2(a_1 a_2 + b_1 b_2 + c_1 c_2 + d_1 d_2) \\
&\Leftrightarrow a_1 a_2 + b_1 b_2 + c_1 c_2 + d_1 d_2 = 0,
\end{aligned}
$$

and (1) is established.

To prove (2) we suppose that any point P_{rs} of S has coordinates (x, y, z, v) given by

$$
\begin{aligned}
&x_0 + a_1 r + a_2 s + e, & &y_0 + b_1 r + b_2 s + f, \\
&z_0 + c_1 r + c_2 s + g, & &v_0 + d_1 r + d_2 s + h.
\end{aligned}
$$

If we can show that $e = f = g = h = 0$, then (2) is established. In the (r, s)-system, we have

$$
\begin{aligned}
|P_{00}P_{rs}|^2 &= r^2 + s^2, \\
|P_{10}P_{rs}|^2 &= (r - 1)^2 + s^2, \\
|P_{01}P_{rs}|^2 &= r^2 + (s - 1)^2.
\end{aligned}
\tag{3}
$$

In the coordinate system of four space, we have

$$
\left.\begin{aligned}
|P_{00}P_{rs}|^2 &= (a_1r + a_2s + e)^2 + (b_1r + b_2s + f)^2 \\
&\quad + (c_1r + c_2s + g)^2 + (d_1r + d_2s + h)^2 \\
|P_{10}P_{rs}|^2 &= (a_1r + a_2s + e - a_1)^2 + (b_1r + b_2s + f - b_1)^2 \\
&\quad + (c_1r + c_2s + g - c_1)^2 + (d_1r + d_2s + h - d_1)^2 \\
|P_{01}P_{rs}|^2 &= (a_1r + a_2s + e - a_2)^2 + (b_1r + b_2s + f - b_2)^2 \\
&\quad + (c_1r + c_2s + g - c_2)^2 + (d_1r + d_2s + h - d_2)^2.
\end{aligned}\right\} \quad (4)
$$

Setting the corresponding formulas in (3) and (4) equal to each other and taking (1) into account implies that $e = f = g = h = 0$. The details are given in Appendix 6.

The next theorem shows that if the coordinates of the points P_{00}, P_{10}, and P_{01} are known, then the equations (2) determine the plane through these points.

Theorem 9. *Let (x_0, y_0, z_0, v_0) be given and suppose the quadruples a_1, b_1, c_1, d_1 and a_2, b_2, c_2, d_2 satisfy (1) of Theorem 8. Then the equations*

$$
\left.\begin{aligned}
x &= x_0 + a_1r + a_2s, & y &= y_0 + b_1r + b_2s, \\
z &= z_0 + c_1r + c_2s, & v &= v_0 + d_1r + d_2s
\end{aligned}\right\} \quad (5)
$$

determine a plane S as (r, s) ranges through all of R_2. The origin of S is at (x_0, y_0, z_0, v_0). The first unit is at $(x_0 + a_1, y_0 + b_1, z_0 + c_1, v_0 + d_1)$ and the second unit is at $(x_0 + a_2, y_0 + b_2, z_0 + c_2, v_0 + d_2)$.

Proof. Let P_{rs} and P_{tu} be two points which satisfy equations (5). From the distance formula, we have

$$
|P_{rs}P_{tu}|^2 = [(r - t)a_1 + (s - u)a_2]^2 + [(r - t)b_1 + (s - u)b_2]^2 \\
+ [(r - t)c_1 + (s - u)c_2]^2 + [(r - t)d_1 + (s - u)d_2]^2.
$$

We multiply out the right side and, taking (1) of Theorem 8 into account, we find

$$
|P_{rs}P_{tu}|^2 = (r - t)^2 + (s - u)^2.
$$

Therefore the equations (5) set up a one-to-one correspondence between S and R_2 with the proper distance formula. Hence S is a plane.

EXAMPLE 1. Given $P_{00} = (1, -1, 2, -3)$, and

$$
a_1, b_1, c_1, d_1 = -\frac{1}{2}, \frac{1}{2}, -\frac{1}{2}, -\frac{1}{2},
$$

$$
a_2, b_2, c_2, d_2 = \frac{2}{\sqrt{6}}, \frac{1}{\sqrt{6}}, -\frac{1}{\sqrt{6}}, 0.
$$

Find the first and second unit points and the parametric equations of the plane determined.

Solution. We have $P_{10} = (\frac{1}{2}, -\frac{1}{2}, \frac{5}{2}, -\frac{7}{2}), P_{01} = (1 + 2/\sqrt{6}, -1 + 1/\sqrt{6}, 2 - 1/\sqrt{6}, -3)$. Using equations (5), we find

$$x = 1 - \frac{1}{2}r + \frac{2}{\sqrt{6}}s, \qquad y = -1 + \frac{1}{2}r + \frac{1}{\sqrt{6}}s,$$

$$z = 2 - \frac{1}{2}r - \frac{1}{\sqrt{6}}s, \qquad v = -3 - \frac{1}{2}r.$$

Theorem 10. *Suppose that two distinct points lie in a plane S. Then the line L containing them is entirely in S.*

The proof is left to the reader. See Problem 5 at the end of this section and the hint given there.

Theorem 11. *Let L_1 and L_2 be two intersecting perpendicular lines in four space. Then there is a unique plane S containing these lines. Also, S consists of all points P lying on a line joining a point of L_1 with a point of L_2 together with P_0, the point of intersection of L_1 and L_2.*

Proof. We write the equations L_1 and L_2 in parametric form:

$$L_1: x = x_0 + a_1t, \qquad y = y_0 + b_1t, \qquad z = z_0 + c_1t, \qquad v = v_0 + d_1t$$
$$L_2: x = x_0 + a_2t, \qquad y = y_0 + b_2t, \qquad z = z_0 + c_2t, \qquad v = v_0 + d_2t$$

where $P_0 = (x_0, y_0, z_0, v_0)$ is the point of intersection of L_1 and L_2; a_1, b_1, c_1, d_1 are direction cosines of L_1 and a_2, b_2, c_2, d_2 are direction cosines of L_2. Since L_1 is perpendicular to L_2, we have

$$a_1a_2 + b_1b_2 + c_1c_2 + d_1d_2 = 0.$$

We use Theorem 9 to conclude that

$$x = x_0 + a_1r + a_2s, \qquad y = y_0 + b_1r + b_2s,\Big\}$$
$$z = z_0 + c_1r + c_2s, \qquad v = v_0 + d_1r + d_2s.\Big\} \qquad (6)$$

are the equations of a plane S containing L_1 and L_2. (L_1 corresponds to $s = 0$ and L_2 corresponds to $r = 0$.)

If P is a point of S and has coordinates (r, s), it lies on a line joining $P_{2r,0}$ and $P_{0,2s}$ (assuming $P \neq P_0$). See Fig. 12–5. Therefore if $P \in S$, then P is on a line joining a point of L_1 with a point of L_2. Conversely, suppose P is on a line joining L_1 to L_2, say $P_1 \in L_1$ and $P_2 \in L_2$. Then P is h of the way from P_1 to P_2 for some value of h. That is,

$$x = x_1 + h(x_2 - x_1), \qquad y = y_1 + h(y_2 - y_1),$$
$$z = z_1 + h(z_2 - z_1), \qquad v = v_1 + h(v_2 - v_1).$$

But since P_1 and P_2 satisfy (6), we see by substitution that (x, y, z, v), the coordinates of P, satisfy (6). Hence P is in S, and the theorem is established.

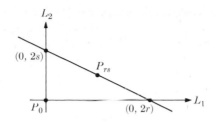

Figure 12–5

Corollary 1. *There is a unique plane in Euclidean four space which passes through a line L_1 and a point P_1 not on L_1.*

The proof is left to the reader.

Corollary 2. *There is a unique plane in Euclidean four space which passes through three points not on a line.*

Corollary 3. *There is a unique plane in Euclidean four space which contains two intersecting lines.*

EXAMPLE 2. Find the equations of the plane containing the line

$$L_1: x = 1 + \frac{1}{\sqrt{7}} t, \qquad y = -1 - \frac{2}{\sqrt{7}} t,$$

$$z = 2 + \frac{1}{\sqrt{7}} t, \qquad v = 3 - \frac{1}{\sqrt{7}} t$$

and passing through the point $P_1(2, 0, -1, 1)$.

Solution. We use the method of Theorem 6 to find the line L_2 which passes through P_1 and is perpendicular to L_1. We have, using formula (11) of Section 2:

$$\bar{t} = \frac{1}{\sqrt{7}} (2 - 1) - \frac{2}{\sqrt{7}} (0 + 1) + \frac{1}{\sqrt{7}} (-1 - 2) - \frac{1}{\sqrt{7}} (1 - 3) = -\frac{2}{\sqrt{7}}.$$

Therefore direction numbers of L_2 are

$$1 - 2 + \frac{1}{\sqrt{7}} \left(\frac{-2}{\sqrt{7}} \right), \qquad -1 - 0 - \frac{2}{\sqrt{7}} \left(\frac{-2}{\sqrt{7}} \right),$$

$$2 + 1 + \frac{1}{\sqrt{7}} \left(\frac{-2}{\sqrt{7}} \right), \qquad 3 - 1 - \frac{1}{\sqrt{7}} \left(\frac{-2}{\sqrt{7}} \right),$$

or

$$-\frac{9}{7}, \ -\frac{3}{7}, \ \frac{19}{7}, \ \frac{16}{7}.$$

The parametric equations of the desired plane are

$$x = 1 + \frac{1}{\sqrt{7}}r - \frac{9}{7}s, \qquad y = -1 - \frac{2}{\sqrt{7}}r - \frac{3}{7}s,$$

$$z = 2 + \frac{1}{\sqrt{7}}r + \frac{19}{7}s, \qquad v = 3 - \frac{1}{\sqrt{7}}r + \frac{16}{7}s.$$

We can "simplify" these equations by changing the parameters or, equivalently, using a proportional set of direction numbers. The same plane is represented by

$$x = 1 + r' - 9s', \qquad y = -1 - 2r' - 3s',$$
$$z = 2 + r' + 19s', \qquad v = 3 - r' + 16s'.$$

These equations still set up a one-to-one correspondence between the plane S and R_2, but the distance formula in terms of r', s' is not the same as that given in terms of r, s.

Proceeding by analogy with geometric quantities in two and three dimensions, we define an **angle** in four space as the union of two rays which are not parallel and have a common endpoint. From Corollary 3 of Theorem 11 above, we see that an angle always lies in a unique two-dimensional plane. We define the (radian) **measure of an angle** as its measure considered as an ordinary angle in the plane. We now wish to find a formula for determining the measure of angles.

Let $\overrightarrow{L_1}$ and $\overrightarrow{L_2}$ be directed lines which are not parallel. We let $\overrightarrow{L_1'}$ be the unique directed line parallel to $\overrightarrow{L_1}$ which passes through the origin; that is, through $(0, 0, 0, 0)$. Also, let $\overrightarrow{L_2'}$ be the line parallel to $\overrightarrow{L_2}$ which passes through the origin.

DEFINITION. The **angle between any two nonparallel lines** $\overrightarrow{L_1}$ and $\overrightarrow{L_2}$ is defined as the angle between $\overrightarrow{L_1'}$ and $\overrightarrow{L_2'}$. The **measure of this angle** is the measure of the angle between $\overrightarrow{L_1'}$ and $\overrightarrow{L_2'}$.

We note that the above definition determines an angle between lines which may not intersect.

Theorem 12. *Suppose that* $\overrightarrow{L_1}$ *and* $\overrightarrow{L_2}$ *have direction cosines* a_1, b_1, c_1, d_1 *and* a_2, b_2, c_2, d_2, *respectively, and that* $\overrightarrow{L_1}$ *is not parallel to* $\overrightarrow{L_2}$. *Then the measure of the angle* θ *between* $\overrightarrow{L_1}$ *and* $\overrightarrow{L_2}$ *is given by* $\cos\theta = a_1a_2 + b_1b_2 + c_1c_2 + d_1d_2$.

Proof. We construct the lines $\overrightarrow{L_1'}$, $\overrightarrow{L_2'}$ passing through the origin and parallel to $\overrightarrow{L_1}$ and $\overrightarrow{L_2}$. Let P_1 be one unit from O along $\overrightarrow{L_1'}$ and P_2 one unit from O along $\overrightarrow{L_2'}$. See Fig. 12–6. We apply the *Law of Cosines* to triangle OP_1P_2. We get

$$|P_1P_2|^2 = |OP_1|^2 + |OP_2|^2 - 2|OP_1|\,|OP_2|\cos\theta.$$

Since the coordinates of P_1 are (a_1, b_1, c_1, d_1) and that of P_2 are (a_2, b_2, c_2, d_2), we have

$$(a_2 - a_1)^2 + (b_2 - b_1)^2 + (c_2 - c_1)^2 + (d_2 - d_1)^2 = 2 - 2\cos\theta.$$

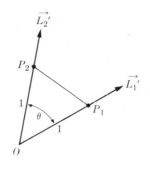

Figure 12-6

Multiplying out the left side, we obtain

$$\cos \theta = a_1 a_2 + b_1 b_2 + c_1 c_2 + d_1 d_2. \tag{7}$$

Remark. The formulas for $\cos \theta$ in the plane and in three space, given on pages 93 and 210, show the similarity with (7) in four dimensions. We can proceed by analogy in any number of dimensions.

EXAMPLE 3. Given $A = (2, -1, 0, 1)$, $V = (1, 0, 1, 2)$, $B = (6, -4, 3, 0)$, find $\cos \theta$ where $\theta = \text{meas} \angle AVB$.

Solution. The directed line $\overrightarrow{VA}$ has direction numbers $1, -1, -1, -1$ and therefore direction cosines $\frac{1}{2}, -\frac{1}{2}, -\frac{1}{2}, -\frac{1}{2}$. The directed line $\overrightarrow{VB}$ has direction numbers $5, -4, 2, -2$, and direction cosines $\frac{5}{7}, -\frac{4}{7}, \frac{2}{7}, -\frac{2}{7}$. Substituting in formula (7), we find

$$\cos \theta = \frac{5}{14} + \frac{4}{14} - \frac{2}{14} + \frac{2}{14} = \frac{9}{14}.$$

When we studied lines in three dimensions we discussed their parametric equations; three such equations determined a line. However, we also could eliminate the parameter and describe a line in three space by means of two equations involving x, y, z. Similarly, the equations of a plane in four space require four parametric equations with two parameters. We now show that the parameters can be eliminated and a plane in four space is determined by two simultaneous linear equations in the variables x, y, z, v. In order to establish this result we need the following Lemma.

Lemma 1. (i) *Two quadruples* a_1, a_2, a_3, a_4 *and* b_1, b_2, b_3, b_4 *are proportional if and only if all the two-by-two determinants*

$$\begin{vmatrix} a_i & a_j \\ b_i & b_j \end{vmatrix} = 0, \tag{8}$$

for all i, j *between* 1 *and* 4.

(ii) *If the two quadruples satisfy*

$$\left.\begin{array}{r} a_1^2 + a_2^2 + a_3^2 + a_4^2 = 1, \\ b_1^2 + b_2^2 + b_3^2 + b_4^2 = 1, \\ a_1b_1 + a_2b_2 + a_3b_3 + a_4b_4 = 0, \end{array}\right\} \qquad (9)$$

then they are not proportional.

Proof. Suppose the quadruples are proportional. Then there is a number h (not zero) such that $a_i = hb_i$, $i = 1, 2, 3, 4$. Since

$$\begin{vmatrix} a_i & a_j \\ b_i & b_j \end{vmatrix} = a_ib_j - a_jb_i,$$

we have $hb_ib_j - hb_jb_i = 0$ for all i, j, and so the determinants vanish. Now suppose all the determinants vanish. We wish to show the quadruples are proportional. Since not all the a_i are zero, we suppose $a_1 \neq 0$. Then we define $h = b_1/a_1$. Since $a_1b_2 - a_2b_1 = 0$, we find

$$0 = a_1b_2 - a_2 \cdot ha_1 = a_1(b_2 - ha_2).$$

Hence $b_2 \simeq ha_2$. Similarly $b_3 = ha_3$ and $b_4 = ha_4$ and the quadruples are proportional.

To prove (ii) we assume the quadruples are proportional. Then $b_1^2 + \cdots + b_4^2 = h^2(a_1^2 + \cdots + a_4^2) = h^2$. Therefore $h = \pm 1$. However

$$a_1b_1 + \cdots + a_4b_4 = h(a_1^2 + \cdots + a_4^2) = h \neq 0,$$

a contradiction. Therefore the quadruples are not proportional.

Theorem 13. (i) *A plane S in Euclidean four space is the solution set of a pair of linear equations of the form*

$$\left.\begin{array}{r} A_1x + B_1y + C_1z + D_1v + E_1 = 0 \\ A_2x + B_2y + C_2z + D_2v + E_2 = 0 \end{array}\right\} \qquad (10)$$

in which the quadruples A_1, B_1, C_1, D_1 and A_2, B_2, C_2, D_2 are not proportional.
   (ii) *Conversely, the solution set of such a pair of equations is a plane in four space.*

Proof. (i) Let S be a plane in four space. Then points of S satisfy the parametric equations

$$\left.\begin{array}{ll} x = x_0 + a_1r + a_2s, & y = y_0 + b_1r + b_2s, \\ z = z_0 + c_1r + c_2s, & v = v_0 + d_1r + d_2s. \end{array}\right\} \qquad (11)$$

where

$$a_1^2 + b_1^2 + c_1^2 + d_1^2 = 1, \qquad a_2^2 + b_2^2 + c_2^2 + d_2^2 = 1,$$

and

$$a_1a_2 + b_1b_2 + c_1c_2 + d_1d_2 = 0.$$

Using part (ii) of Lemma 1, we see that a_1, b_1, c_1, d_1 is not proportional to $a_2, b_2,$ c_2, d_2. Therefore at least one of the determinants in (8) is not zero. Suppose that

$$\begin{vmatrix} c_1 c_2 \\ d_1 d_2 \end{vmatrix} = c_1 d_2 - c_2 d_1 \neq 0.$$

If one of the other determinants is not zero the proof is the same. We write the last two equations in (11) in the form

$$c_1 r + c_2 s = z - z_0$$
$$d_1 r + d_2 s = v - v_0.$$

We set $D = c_1 d_2 - c_2 d_1$, and we solve these equations simultaneously for r and s. The result is

$$r = \frac{1}{D}[d_2(z - z_0) - c_2(v - v_0)], \qquad s = \frac{1}{D}[-d_1(z - z_0) + c_1(v - v_0)].$$

We substitute these values of r and s into the first two equations in (11) getting

$$\left. \begin{aligned} x - x_0 &= \frac{a_1}{D}[d_2(z - z_0) - c_2(v - v_0)] + \frac{a_2}{D}[-d_1(z - z_0) + c_1(v - v_0)] \\ y - y_0 &= \frac{b_1}{D}[d_2(z - z_0) - c_2(v - v_0)] + \frac{b_2}{D}[-d_1(z - z_0) + c_1(v - v_0)]. \end{aligned} \right\} \quad (12)$$

The equations (12) are in the form (10) with

$$A_1 = 1, \quad B_1 = 0, \quad C_1 = \frac{a_2 d_1 - a_1 d_2}{D}, \quad D_1 = \frac{a_1 c_2 - a_2 c_1}{D}$$

$$A_2 = 0, \quad B_2 = 1, \quad C_2 = \frac{b_2 d_1 - b_1 d_2}{D}, \quad D_2 = \frac{b_1 c_2 - b_2 c_1}{D}.$$

These quadruples are not proportional since $A_1 B_2 - A_2 B_1 = 1 \neq 0$.

Now, suppose (x, y, z, v) satisfies (12). If we define r and s as just above (12), we see that the equations (11) hold. Thus S is the solution set of (12).

(ii) The proof of this part is in Appendix 6. See, also, problems 18 through 21 at the end of this section.

EXAMPLE 4. Given a plane S having the parametric equations

$$\left. \begin{aligned} x &= 1 + \frac{1}{2}r + \frac{2}{\sqrt{10}}s, \qquad y = -2 - \frac{1}{2}r + \frac{2}{\sqrt{10}}s, \\ z &= -1 - \frac{1}{2}r + \frac{1}{\sqrt{10}}s, \qquad v = -\frac{1}{2}r - \frac{1}{\sqrt{10}}s. \end{aligned} \right\} \quad (13)$$

Eliminate the parameters, that is, find two linear equations in x, y, z, v which have S as a solution set.

Solution. Since the determinant

$$\begin{vmatrix} -\dfrac{1}{2} & \dfrac{1}{\sqrt{10}} \\[2ex] -\dfrac{1}{2} & -\dfrac{1}{\sqrt{10}} \end{vmatrix} = \dfrac{1}{\sqrt{10}} \neq 0,$$

we can solve the last two equations in (13) for r and s in terms of z and v. The result is

$$r = -(z + 1) - v \qquad s = \tfrac{1}{2}\sqrt{10}(z + 1) - \tfrac{1}{2}\sqrt{10}\, v.$$

Inserting these expressions in the first two equations in (13), we find

$$x = \frac{1}{2}z - \frac{3}{2}v + \frac{3}{2}, \qquad y = \frac{3}{2}z - \frac{1}{2}v - \frac{1}{2},$$

or

$$2x \quad - \quad z + 3v - 3 = 0$$
$$2y - 3z + \quad v + 1 = 0.$$

EXAMPLE 5. Given the lines

$$L_1: \frac{x-1}{1} = \frac{y+1}{1} = \frac{z-2}{-1} = \frac{v}{-1}; \qquad L_2: \frac{x-1}{2} = \frac{y+1}{2} = \frac{z-2}{1} = \frac{v}{2},$$

verify that L_1 and L_2 are perpendicular and find a set of parametric equations for the plane S containing both lines.

Solution. Since

$$1 \cdot 2 + 1 \cdot 1 + 1 \cdot (-1) + (-1) \cdot 2 = 0,$$

the lines are perpendicular. We choose $1, 1, -1, -1$ and $2, 1, 1, 2$ as the coefficients of r and s, respectively, in the parametric equations of a plane. Clearly, the point $P_0(1, -1, 2, 0)$ being on both L_1 and L_2 is in S. Hence we obtain

$$x = 1 + r + 2s, \qquad y = -1 + r + s,$$
$$z = 2 - r + s, \qquad v = -r + 2s.$$

PROBLEMS

In each of problems 1 through 4, find $\cos \theta$ where $\theta = \text{meas } \angle AVB$.

1. $A = (3, 1, -1, 2)$, $V = (2, 2, -2, 1)$, $B = (6, 4, 3, 3)$
2. $A = (2, 3, 1, 4)$, $V = (4, 1, 0, 2)$, $B = (3, -1, 2, 4)$
3. $A = (3, 1, -1, 2)$, $V = (2, 3, 1, 2)$, $B = (0, 3, 3, 1)$
4. $A = (0, 1, 2, -1)$, $V = (-2, -1, 3, 0)$, $B = (3, 2, -1, 3)$
5. Prove Theorem 10. [*Hint:* Use the fact that in a plane S there is a unique line in S which passes through two distinct points in S.]

In each of problems 6 through 9, find two linear equations in x, y, z, v of which S is the solution set, S being the plane having the given parametric equations.

6. $S: x = 1 + 2hr + ks, y = 2 - hr + 2ks, z = -1 - 2hr - ks,$
   $v = 1 - hr + \sqrt{2}ks, h = 1/\sqrt{10}, k = 1/\sqrt{10}$

7. $S: x = -1 + hr - ks, y = 1 + 2hr + ks, z = 2 - hr + ks,$
   $v = 1 - hr, h = 1/\sqrt{7}, k = 1/\sqrt{3}$

8. $S: x = 2 + 2hr - 2ks, y = -1 + 5hr + 2ks, z = -4hr + ks,$
   $v = 2 + 2hr - ks, h = 1/7, k = 1/\sqrt{10}$

9. $S: x = -2 + 2hr - 2ks, y = 1 - hr - ks, z = 1 + hr + ks,$
   $v = 1 + hr + 2ks, h = 1/\sqrt{7}, k = 1/\sqrt{10}$

In each of problems 10 through 13, verify that L_1 is perpendicular to L_2 and find a set of parametric equations for the plane S which contains L_1 and L_2.

10. $L_1: \dfrac{x-1}{1} = \dfrac{y+1}{1} = \dfrac{z-2}{-1} = \dfrac{v+1}{-1};$

   $L_2: \dfrac{x-1}{1} = \dfrac{y+1}{3} = \dfrac{z-2}{3} = \dfrac{v+1}{1}$

11. $L_1: \dfrac{x-2}{2} = \dfrac{y-1}{-3} = \dfrac{z+1}{-2} = \dfrac{v-1}{1};$

   $L_2: \dfrac{x-2}{3} = \dfrac{y-1}{1} = \dfrac{z+1}{2} = \dfrac{v-1}{1}$

12. $L_1: \dfrac{x+1}{1} = \dfrac{y-2}{-2} = \dfrac{z-1}{2} = \dfrac{v+2}{-2};$

   $L_2: \dfrac{x+1}{-2} = \dfrac{y-2}{1} = \dfrac{z-1}{-1} = \dfrac{v+2}{-3}$

13. $L_1: \dfrac{x+2}{3} = \dfrac{y+2}{-1} = \dfrac{z+2}{-2} = \dfrac{v-2}{2};$

   $L_2: \dfrac{x+2}{2} = \dfrac{y+2}{2} = \dfrac{z+2}{3} = \dfrac{v-2}{1}$

In each of problems 14 through 17, find parametric equations of the plane through the line L_1 and the point P_1; first verify that P_1 is not on L_1.

14. $L_1: \dfrac{x+2}{1} = \dfrac{y-1}{-1} = \dfrac{z+1}{1} = \dfrac{v-2}{-1};$ $P_1 = (2, 3, 1, -2)$

15. $L_1: \dfrac{x-1}{2} = \dfrac{y+1}{-3} = \dfrac{z-1}{2} = \dfrac{v+1}{1};$ $P_1 = (1, -2, 0, 1)$

16. $L_1: \dfrac{x+1}{3} = \dfrac{y-2}{-2} = \dfrac{z+2}{1} = \dfrac{v-1}{2}$; $P_1 = (2, 0, 1, 2)$

17. $L_1: \dfrac{x-2}{2} = \dfrac{y+2}{-1} = \dfrac{z-2}{-2} = \dfrac{v+2}{2}$; $P_1 = (1, 3, 2, -1)$

In each of problems 18 through 21, find parametric equations of the plane S which is the solution set of the given equations.

18. $x = -2v$, $y = 3z$

19. $x = -3z + 2v$, $y = 2z + 3v$

20. $x = -z + 2v$, $y = z + 2v$

21. $x = -z + 2v$, $y = 2z + v$

22. (a) Show that any two coordinate axes determine a plane. (b) Show that the plane determined by the x and y axes has only the point $(0, 0, 0, 0)$ in common with the plane determined by the z and v axes.

23. Let $P(x_0, y_0, z_0, v_0)$ be a fixed point in four space. Construct two planes (by writing their parametric equations) which have only the point P in common. [*Hint*: See Problem 22.]

24. Show that the following two planes S_1 and S_2 have no points in common. [*Hint*: Express each of S_1 and S_2 as the locus of a pair of linear equations in (x, y, z, v).]

$$S_1: x = 1 + 2r - s,\ y = -1 + r + 2s,\ z = 3r - s,\ v = 2 + r + 3s$$
$$S_2: x = 2 + r - \tfrac{1}{2}s,\ y = 2 + \tfrac{1}{2}r + s,\ z = \tfrac{3}{2}r - \tfrac{1}{2}s,\ v = -2 + \tfrac{1}{2}r + \tfrac{3}{2}s$$

25. (See Problems 26–28 in Section 1 and Problems 32 and 33 in Section 2.) (a) Derive the parametric equations of a plane in Euclidean five dimensional space. (b) Show that if two points lie in a plane in five space, then the line joining them is also in the plane. (c) Derive the formula for the angle between two lines in five space. (d) Show that any plane in five space is represented by three linear equations

$$A_1 x + B_1 y + C_1 z + D_1 v + E_1 w + F_1 = 0$$
$$A_2 x + B_2 y + C_2 z + D_2 v + E_2 w + F_2 = 0$$
$$A_3 x + B_3 y + C_3 z + D_3 v + E_3 w + F_3 = 0$$

where at least one 3×3 determinant of the coefficients is not zero.

26. Prove Corollary 1 of Theorem 11.

▶ 4. HYPERPLANES IN FOUR DIMENSIONAL SPACE

The set of all points of the form $(x, y, z, 0)$ in Euclidean four dimensional space makes up ordinary three dimensional space which we studied in the chapters on solid analytic geometry. This set of points is an example of a *hyperplane* in four space.

DEFINITION. Let H be a set of points which we can put into one-to-one correspondence with all ordered triples (r, s, t) of R_3. That is, points P of H not only have the usual quadruple of coordinates but also a triple of coordinates locating their position in H. Suppose $P_1 = (r_1, s_1, t_1)$ and $P_2 = (r_2, s_2, t_2)$

are any two points of H. If the formula

$$|P_1P_2| = \sqrt{(r_2 - r_1)^2 + (s_2 - s_1)^2 + (t_2 - t_1)^2}$$

holds for all such points, then H is called a **hyperplane** in four space. The point O of H which corresponds to $(0, 0, 0)$ in the (r, s, t) system is called the **origin** and the points I, J, and K which correspond respectively to $(1, 0, 0)$, $(0, 1, 0)$, and $(0, 0, 1)$ are called the **first**, **second**, and **third unit points**.

It can be shown that if H is a hyperplane defined this way, then all the theorems of plane and solid geometry and trigonometry can be deduced. In other words, a hyperplane behaves in all ways like a three-dimensional Euclidean space. We now wish to explore the relationship between hyperplanes and the four-dimensional space in which they are situated.

In plane analytic geometry we represent a *line* in two ways: (1) as a linear equation in x and y, and (2) as a system of two parametric equation involving a single parameter. In solid analytic geometry we represent a *plane* in two ways: (1) as a linear equation in x, y, and z, and (2) as a system of three parametric equations involving two parameters. We shall show that in four dimensional Euclidean geometry we represent a *hyperplane* in two ways: (1) as a linear equation in x, y, z, and v and (2) as a system of four parametric equations involving three parameters. In each of the three cases we just described, line, plane, and hyperplane, we represent the geometric entity of dimension one less than the given space either by means of one linear equation or by means of a system of parametric equations in which the number of parameters is one less than the dimension of the space. This statement holds for Euclidean space in any number of dimensions. We give a brief introduction below of how this result is obtained in four dimensions.

Theorem 14. *Suppose that H is a hyperplane in four space and (r, s, t) is a coordinate system on H with origin at $P_{000} = (x_0, y_0, z_0, v_0)$ and with unit points at $P_{100} = (x_0 + a_1, y_0 + b_1, z_0 + c_1, v_0 + d_1)$, $P_{010} = (x_0 + a_2, y_0 + b_2, z_0 + c_2, v_0 + d_2)$, $P_{001} = (x_0 + a_3, y_0 + b_3, z_0 + c_3, v_0 + d_3)$. (See Fig. 12–7.)*

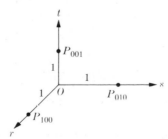

Figure 12–7

We define

$$g_{kl} = a_k a_l + b_k b_l + c_k c_l + d_k d_l,$$

where $k = 1, 2, 3$, $l = 1, 2, 3$. Then it follows that

$$g_{11} = g_{22} = g_{33} = 1 \quad \text{and} \quad g_{kl} = 0 \quad \text{if } k \neq l. \tag{1}$$

Furthermore, if P_{rst} is any point of H with coordinates (r, s, t), then P_{rst} is represented parametrically by the equations

$$\begin{aligned}
x &= x_0 + a_1 r + a_2 s + a_3 t, & y &= y_0 + b_1 r + b_2 s + b_3 t, \\
z &= z_0 + c_1 r + c_2 s + c_3 t, & v &= v_0 + d_1 r + d_2 s + d_3 t.
\end{aligned} \tag{2}$$

Proof. The proof is analogous to the proof of Theorem 8 which begins on page 286 and is completed in Appendix 6. We start the proof and leave most of the details to the reader. We have

$$|P_{000}P_{100}|^2 = g_{11} = 1, \quad |P_{000}P_{010}|^2 = g_{22} = 1, \quad |P_{000}P_{001}|^2 = g_{33} = 1,$$

and the first part of (1) is established. Also,

$$\begin{aligned}
|P_{100}P_{010}|^2 = 2 &= (a_2 - a_1)^2 + (b_2 - b_1)^2 + (c_2 - c_1)^2 + (d_2 - d_1)^2 \\
&= 2 - 2(a_1 a_2 + b_1 b_2 + c_1 c_2 + d_1 d_2)
\end{aligned}$$

which shows that $g_{12} = 0$. The remaining formulas in (1) are established similarly. For the remainder of the proof, we write the coordinates of a point P_{rst} of H in the form

$$\begin{aligned}
x &= x_0 + a_1 r + a_2 s + a_3 t + e, & y &= y_0 + b_1 r + b_2 s + b_3 t + f \\
z &= z_0 + c_1 r + c_2 s + c_3 t + g, & v &= v_0 + d_1 r + d_2 s + d_3 t + h.
\end{aligned}$$

We wish to show that $e = f = g = h = 0$. We know that

$$\begin{aligned}
|P_{000}P_{rst}|^2 &= r^2 + s^2 + t^2, & |P_{100}P_{rst}|^2 &= (r - 1)^2 + s^2 + t^2, \\
|P_{010}P_{rst}|^2 &= r^2 + (s - 1)^2 + t^2, & |P_{001}P_{rst}|^2 &= r^2 + s^2 + (t - 1)^2.
\end{aligned}$$

In the coordinates of four space, we have

$$\begin{aligned}
|P_{000}P_{rst}|^2 = &(a_1 r + a_2 s + a_3 t + e)^2 + (b_1 r + b_2 s + b_3 t + f)^2 \\
&+ (c_1 r + c_2 s + c_3 t + g)^2 + (d_1 r + d_2 s + d_3 t + h)^2
\end{aligned}$$

and similar formulas for the distance of P_{rst} from the unit points. The remainder of the proof follows the proof of Theorem 8(b) in Appendix 6.

We state the analog of Theorem 9 for hyperplanes.

Theorem 15. *Let (x_0, y_0, z_0, v_0) be given and suppose the three quadruples $a_k, b_k, c_k, d_k, k = 1, 2, 3$ satisfy conditions (1) of Theorem 14. Then the parametric* *equations*

$$x = x_0 + a_1 r + a_2 s + a_3 t, \qquad y = y_0 + b_1 r + b_2 s + b_3 t,$$
$$z = z_0 + c_1 r + c_2 s + c_3 t, \qquad v = v_0 + d_1 r + d_2 s + d_3 t, \tag{3}$$

determine a hyperplane H as (r, s, t) ranges through all of R_3. The origin of H is at (x_0, y_0, z_0, v_0). The three unit points are at $(x_0 + a_k, y_0 + b_k, z_0 + c_k, v_0 + d_k)$, $k = 1, 2, 3$, respectively.

The proof is analogous to the proof of Theorem 9 and is omitted.

EXAMPLE 1. Given the point $P_{000} = (2, 1, 0, -1)$ and the three quadruples

$$a_1, b_1, c_1, d_1 = -\tfrac{1}{2}, -\tfrac{1}{2}, \tfrac{1}{2}, -\tfrac{1}{2};$$

$$a_2, b_2, c_2, d_2 = \frac{2}{\sqrt{10}}, \frac{-1}{\sqrt{10}}, \frac{-1}{\sqrt{10}}, \frac{-2}{\sqrt{10}};$$

$$a_3, b_3, c_3, d_3 = \frac{3}{\sqrt{26}}, 0, \frac{4}{\sqrt{26}}, \frac{1}{\sqrt{26}}.$$

Find the first, second, and third unit points and the parametric equations of the hyperplane determined.

Solution. We have

$$P_{100} = \left(\frac{3}{2}, \frac{1}{2}, \frac{1}{2}, -\frac{3}{2} \right),$$

$$P_{010} = \left(2 + \frac{2}{\sqrt{10}}, 1 - \frac{1}{\sqrt{10}}, -\frac{1}{\sqrt{10}}, -1 - \frac{2}{\sqrt{10}} \right),$$

$$P_{001} = \left(2 + \frac{3}{\sqrt{26}}, 1, \frac{4}{\sqrt{26}}, -1 + \frac{1}{\sqrt{26}} \right).$$

Using equations (3), we obtain

$$x = 2 - \frac{1}{2}r + \frac{2}{\sqrt{10}}s + \frac{3}{\sqrt{26}}t, \qquad y = 1 - \frac{1}{2}r - \frac{1}{\sqrt{10}}s,$$

$$z = \frac{1}{2}r - \frac{1}{\sqrt{10}}s + \frac{4}{\sqrt{26}}t, \qquad v = -1 - \frac{1}{2}r - \frac{2}{\sqrt{10}}s + \frac{1}{\sqrt{26}}t.$$

The following simple facts are stated without proof. See Problems 28, 29, and 30 at the end of this section.

Theorem 16. (i) *If two points of a line L lie in a hyperplane H, then L is in H.* (ii) *If three points not on a line lie in a hyperplane H, then the plane S determined by these three points is in H.* (iii) *If a line L_1 and a point P not on L_1 lie in a hyperplane H, then the plane S determined by L_1 and P is in H.*

To show that hyperplanes are represented by linear equations in x, y, z, and v we need the following lemma.

Lemma 2. (i) *Suppose that the quadruples* a_k, b_k, c_k, d_k, $k = 1, 2, 3$ *satisfy the conditions*

$$g_{kl} = a_k a_l + b_k b_l + c_k c_l + d_k d_l = \begin{cases} 1 & \text{if } k = l \\ 0 & \text{if } k \neq l \end{cases}. \tag{4}$$

Then there is a quadruple a, b, c, d *which satisfies the conditions*

$$a^2 + b^2 + c^2 + d^2 = 1 = \begin{vmatrix} a & b & c & d \\ a_1 & b_1 & c_1 & d_1 \\ a_2 & b_2 & c_2 & d_2 \\ a_3 & b_3 & c_3 & d_3 \end{vmatrix} \tag{5}$$

$$a_k a + b_k b + c_k c + d_k d = 0 \quad \text{for } k = 1, 2, 3. \tag{6}$$

(ii) *Let* a, b, c, d *be any quadruple such that* $a^2 + b^2 + c^2 + d^2 = 1$. *Then there are three quadruples* a_k, b_k, c_k, d_k, $k = 1, 2, 3$ *such that* (4) *and* (6) *hold.*

The proof of this Lemma is in Appendix 6.

Theorem 17. (i) *Suppose that H is a hyperplane in Euclidean four space. Let* (r, s, t) *be a coordinate system with origin at* (x_0, y_0, z_0, v_0) *and with unit points at* $(x_0 + a_k, y_0 + b_k, z_0 + c_k, v_0 + d_k)$, $k = 1, 2, 3$. *Suppose a, b, c, d is a quadruple satisfying* (5) *and* (6) *of Lemma* 2. *Then H is the solution set of the linear equation*

$$a(x - x_0) + b(y - y_0) + c(z - z_0) + d(v - v_0) = 0. \tag{7}$$

(ii) *Suppose that a, b, c, d is any quadruple not all zeros. Then the solution set of* (7) *is a hyperplane through the point* (x_0, y_0, z_0, v_0).

Proof. (i) Suppose that (x, y, z, v) is a point of H. Then there is a triple (r, s, t) such that the parametric equations (3) in Theorem 15 hold. Then, using (6) we conclude that (7) holds. Hence every point of H satisfies (7). Now suppose (x, y, z, v) satisfies (7). We define r, s, t by

$$\left. \begin{array}{l} r = a_1(x - x_0) + b_1(y - y_0) + c_1(z - z_0) + d_1(v - v_0) \\ s = a_2(x - x_0) + b_2(y - y_0) + c_2(z - z_0) + d_2(v - v_0) \\ t = a_3(x - x_0) + b_3(y - y_0) + c_3(z - z_0) + d_3(v - v_0) \end{array} \right\} \tag{8}$$

We consider (8) and (7) as four equations in the four unknowns $x - x_0$, $y - y_0$, $z - z_0$, $v - v_0$. Since the determinant of the coefficients is not zero (by Lemma 2), we can solve in terms of r, s, and t, and the solution is unique. However, the parametric equations (3) represent *a* solution and hence they are *the* solution. Therefore every point satisfying (7) is in H.

(ii) If we multiply (7) by any constant the solution set is unchanged. Therefore given a, b, c, d, we multiply by $(a^2 + b^2 + c^2 + d^2)^{-1/2}$ and obtain a quadruple which satisfies (5). Hence we may suppose (5) holds for a, b, c, d. From Lemma 2 there exist three quadruples $a_k, b_k, c_k, d_k, k = 1, 2, 3$ which satisfy (4) and (6). Let H be the solution set of the corresponding parametric equations (3). Then by part (i), H is the solution set of (7).

Corollary. *Any hyperplane is the locus of a linear equation of the form*

$$Ax + By + Cz + Dv + E = 0 \tag{9}$$

with $A^2 + B^2 + C^2 + D^2 > 0$, and the locus of any such equation is a hyperplane.

DEFINITION. If a hyperplane has the equation (9), then the quadruple A, B, C, D is called a **set of attitude members** of the hyperplane. Any set proportional to A, B, C, D is also a set of attitude numbers for the same hyperplane. See page 217 for the corresponding definition for planes in three space.

EXAMPLE 2. Given the hyperplane H in parametric form:

$$x = 1 + hr + ks - lt, \ y = -1 + 2hr - ks - lt, \ z = 2 - hr + ks - 2lt,$$
$$v = 1 + hr + 2ks + lt,$$

with $h = k = 1 = 1/\sqrt{7}$, find a linear equation in x, y, z, v, for H.

Solution. We write

$$
\begin{array}{rrrl}
2hr - & ks - & lt & = y + 1 \\
-hr + & ks - & 2lt & = z - 2 \\
hr + & 2ks + & lt & = v - 1
\end{array}
$$

and we solve these three equations in three unknowns for hr, ks, and lt. The result is

$$hr = \frac{1}{14}(5y - z + 3v - 4), \ ks \frac{1}{14}(-y + 3z + 5v - 12),$$

$$lt = \frac{1}{14}(-3y - 5z + v + 6).$$

By substituting in the equation $x = 1 + hr + ks - lt$, we find

$$x = 1 + \frac{1}{14}(5y - z + 3v - 4) + \frac{1}{14}(-y + 3z + 5v - 12) - \frac{1}{14}(-3y - 5z + v - 6)$$

or

$$14x - 7y - 7z - 7v + 8 = 0.$$

In three dimensional analytic geometry, we established results such as: through a given point there is exactly one plane perpendicular to a given line. The next theorems establish corresponding results in four dimensions.

Theorem 18. *Suppose that L is a line in four space and $P_0 = (x_0, y_0, z_0, v_0)$ is a point on L. Then the set H consisting of P_0 and all points P such that the line*

P_0P is perpendicular to L is a hyperplane. If L has direction numbers A, B, C, D, then H has the equation

$$A(x - x_0) + B(y - y_0) + C(z - z_0) + D(v - v_0) = 0. \qquad (10)$$

Proof. Clearly, P_0 satisfies (10). Now if $P \neq P_0$, then the line through $P(x, y, z, v)$ and P_0 has direction numbers $x - x_0$, $y - y_0$, $z - z_0$, $v - v_0$. The condition that the line P_0P is perpendicular to L is precisely (10). The locus of (10) is a hyperplane by Theorem 17.

DEFINITION. We say that a **line L and a hyperplane H** are **perpendicular** $\Leftrightarrow$ L and H intersect at some point P_0 and every line in H through P_0 is perpendicular to L.

Theorem 19. (i) *There is a unique hyperplane H in four space which passes through a given point $P_0(x_0, y_0, z_0, v_0)$ and is perpendicular to a given line L. If L has direction numbers A, B, C, D, then H is given by*

$$A(x - x_0) + B(y - y_0) + C(z - z_0) + D(v - v_0) = 0. \qquad (11)$$

(ii) *If H is a given hyperplane, then there is a unique line L which passes through a given point $P_0(x_0, y_0, z_0, v_0)$ and is perpendicular to H. If H has attitude numbers A, B, C, D, then the equations of L are*

$$\frac{x - x_0}{A} = \frac{y - y_0}{B} = \frac{z - z_0}{C} = \frac{v - v_0}{D}.$$

Proof. (i) Suppose that H is perpendicular to L. By definition, there is a point $P_1(x_1, y_1, z_1, v_1)$ where they intersect. Hence H has the equation

$$A(x - x_1) + B(y - y_1) + C(z - z_1) + D(v - v_1) = 0.$$

Since P_0 is in H, we have

$$A(x_0 - x_1) + B(y_0 - y_1) + C(z_0 - z_1) + D(v_0 - v_1) = 0.$$

Subtracting this last equation from the one above it, we get (11).
The proof of part (ii) is left to the reader.

EXAMPLE 3. Given the hyperplane $2x - 3y + z - v - 5 = 0$, find the equations of the line perpendicular to the hyperplane which passes through the midpoint of the line segment joining $A(2, -1, 4, 0)$ and $B(-1, 3, 6, -4)$.

Solution. The midpoint of the line segment AB is $(\frac{1}{2}, 1, 5, -2)$. A set of attitude numbers of the plane is $2, -3, 1, -1$. The equations of the line in symmetric form are

$$\frac{x - \frac{1}{2}}{2} = \frac{y - 1}{-3} = \frac{z - 5}{1} = \frac{v + 2}{-1}.$$

DEFINITIONS. Two hyperplanes H_1 and H_2 are parallel $\Leftrightarrow H_1$ and H_2 do not intersect. A hyperplane H and a line L are parallel $\Leftrightarrow H$ and L do not intersect. A hyperplane H and a plane S are parallel $\Leftrightarrow H$ and S do not intersect.

Theorem 20. *Suppose that H_1 and H_2 are hyperplanes in four space given by the equations*

$$\left.\begin{array}{l} A_1x + B_1y + C_1z + D_1v + E_1 = 0 \\ A_2x + B_2y + C_2z + D_2v + E_2 = 0. \end{array}\right\} \qquad (12)$$

Then, (i) if the quadruples A_1, B_1, C_1, D_1 and A_2, B_2, C_2, D_2 are not proportional, the intersection of H_1 and H_2 is a plane. (ii) If the quadruples are proportional, the hyperplanes are either parallel or coincident.

Proof. (i) According to Theorem 13, a plane is characterized by points which satisfy two simultaneous linear equations. Since the points of intersection of H_1 and H_2 satisfy (12), the result follows. (ii) If the quadruples are proportional, there is a number h such that $A_1 = hA_2$, $B_1 = hB_2$, $C_1 = hC_2$, $D_1 = hD_2$. If it happens that $E_1 = hE_2$, then both equations in (12) have the same solution set and H_1 and H_2 coincide. If $E_1 \neq hE_2$, then no point of H_1 can be in H_2. The hyperplanes are parallel.

Corollary. *Let H_1 be a given hyperplane and $P_1(x_1, y_1, z_1, v_1)$ a point not in H_1. Then there is a unique hyperplane H_2 which passes through P_1 and is parallel to H_1. If the equation of H_1 is $A_1x + B_1y + C_1z + D_1v + E_1 = 0$, then the equation of H_2 is $A_1(x - x_1) + B_1(y - y_1) + C_1(z - z_1) + D_1(v - v_1) = 0$.*

EXAMPLE 4. Find the equation of the hyperplane parallel to H_1: $2x - y - 2z - 3v - 1 = 0$ and passing through the point $P_1(1, 2, -1, -1)$.

Solution. We observe that since $2(1) - (2) - 2(-1) - 3(-1) - 1 \neq 0$, the point P_1 is not in H_1. Since H_1 has attitude numbers $2, -1, -2, -3$, the desired hyperplane has the equation

$$2(x - 1) - (y - 2) - 2(z + 1) - 3(v + 1) = 0.$$

The next result shows that three mutually perpendicular lines which pass through a point cannot lie in the same plane in four space. This fact, which is obvious in three dimensions, requires proof in higher dimensional space.

Theorem 21. *Let L_1, L_2, and L_3 be three lines through the point $P_0(x_0, y_0, z_0, v_0)$. Suppose that L_1 and L_2 are perpendicular, L_1 and L_3 are perpendicular, and L_2 and L_3 are perpendicular. Then there is no plane which contains L_1, L_2, and L_3.*

Proof. Let a_k, b_k, c_k, d_k be direction cosines for L_k, $k = 1, 2, 3$. According to Theorem 11, there is a unique plane S which is determined by the perpendicular lines L_1 and L_2. In fact, the parametric equations of S are

$$x = x_0 + a_1 r + a_2 s, \qquad y = y_0 + b_1 r + b_2 s, \atop z = z_0 + c_1 r + c_2 s, \qquad v = v_0 + d_1 r + d_2 s. \Big\} \tag{13}$$

The point $P_1 = (x_0 + a_3, y_0 + b_3, z_0 + c_3, v_0 + d_3)$ is on L_3, and we shall show that it is not in S. If P_1 were in S, then its coordinates would satisfy (13) for some specific values of r and s. That is, for some r_1 and s_1 we would have

$$x_0 + a_3 = x_0 + a_1 r_1 + a_2 s_1, \qquad y_0 + b_3 = y_0 + b_1 r_1 + b_2 s_1,$$
$$z_0 + c_3 = z_0 + c_1 r_1 + c_2 s_1, \qquad v_0 + d_3 = v_0 + d_1 r_1 + d_2 s_1.$$

Therefore,

$$a_3 = a_1 r_1 + a_2 s_1, \qquad b_3 = b_1 r_1 + b_2 s_1, \atop c_3 = c_1 r_1 + c_2 s_1, \qquad d_3 = d_1 r_1 + d_2 s_1. \Big\} \tag{14}$$

Since the three lines L_1, L_2, L_3 are mutually perpendicular, we have

$$a_1 a_3 + b_1 b_3 + c_1 c_3 + d_1 d_3 = 0 \quad \text{and} \quad a_2 a_3 + b_2 b_3 + c_2 c_3 + d_2 d_3 = 0.$$

Substituting from (14) into these equations, we get $r_1 = s_1 = 0$. Hence $a_3 = b_3 = c_3 = d_3 = 0$, a contradiction since $a_3^2 + b_3^2 + c_3^2 + d_3^2 = 1$.

Theorem 22. *Let L_1, L_2, and L_3 be three mutually perpendicular lines which pass through the point $P_0(x_0, y_0, z_0, v_0)$. Then there is a unique hyperplane H which contains the three lines.*

Proof. Let a_k, b_k, c_k, d_k be the direction cosines of L_k, $k = 1, 2, 3$. Then the hyperplane H with parametric equations

$$x = x_0 + a_1 r + a_2 s + a_3 t, \qquad y = y_0 + b_1 r + b_2 s + b_3 t,$$
$$z = z_0 + c_1 r + c_2 s + c_3 t, \qquad v = v_0 + d_1 r + d_2 s + d_3 t$$

contains the lines L_1, L_2, and L_3. Note that L_1 corresponds to $s = t = 0$, L_2 to $r = t = 0$, and L_3 to $r = s = 0$. To show that H is unique, suppose there is another hyperplane H' containing L_1, L_2, and L_3. Then the intersection of H and H' would also contain L_1, L_2 and L_3. From Theorem 20, the intersection of H and H' is a plane. But by Theorem 21 this is impossible. Hence $H = H'$.

Since the equation $x = 0$ is of the first degree it determines a hyperplane. It is the hyperplane containing the y, z, and v axes and therefore contains the origin. Similarly, each of the equations $y = 0$, $z = 0$, and $v = 0$ determines a hyperplane. We call these the **coordinate hyperplanes**. Through any point in space there are four hyperplanes which are parallel to the coordinate hyperplanes. For example, through the point $P_0 = (1, 2, -3, -1)$, the four hyperplanes $x = 1$, $y = 2$, $z = -3$, $v = -1$ are parallel to the corresponding coordinate hyperplanes given by $x = 0$, $y = 0$, $z = 0$, $v = 0$, respectively.

The next theorem describes the manner in which lines and hyperplanes may intersect.

Theorem 23. *Let $H: Ax + By + Cz + Dv + E = 0$ be a given hyperplane and*

$$L: \frac{x - x_0}{a} = \frac{y - y_0}{b} = \frac{z - z_0}{c} = \frac{v - v_0}{d}$$

a given line. (i) *If $Aa + Bb + Cc + Dd \neq 0$, then H and L intersect in a single point.* (ii) *If $Aa + Bb + Cc + Dd = 0$, then L is in H if $P_0(x_0, y_0, z_0, v_0)$ is in H. Otherwise L and H are parallel.*

Proof. (i) If $P(x, y, z, v)$ is on L, then there is a number h such that $x = x_0 + ah$, $y = y_0 + bh$, $z = z_0 + ch$, $v = v_0 + dh$. If P is also in H, then

$$A(x_0 + ah) + B(y_0 + bh) + C(z_0 + ch) + D(v_0 + dh) + E = 0.$$

Therefore,

$$Ax_0 + By_0 + Cz_0 + Dv_0 + E + h(Aa + Bb + Cc + Dd) = 0. \quad (15)$$

Since $Aa + Bb + Cc + Dd \neq 0$, there is a unique value of h for which (15) holds. This value of h yields the point of intersection.

(ii) If $Aa + Bb + Cc + Dd = 0$ and $Ax_0 + By_0 + Cz_0 + Dv_0 + E \neq 0$, there is no value of h for which (15) is true. Hence H and L are parallel. On the other hand, if $Ax_0 + By_0 + Cz_0 + Dv_0 + E = 0$, then (15) holds for all values of h, and L is in H.

EXAMPLE 5. Given the hyperplane $H: 3x - 2y + 2z - 3v + 4 = 0$ and the line

$$L: \frac{x - 1}{2} = \frac{y + 1}{-3} = \frac{z - 2}{1} = \frac{v + 1}{-2}.$$

Determine whether L is parallel to H, contained in H, or intersects H in a single point. If the last case is valid, find the point of intersection.

Solution. $Aa + Bb + Cc + Dd = 3(2) - 2(-3) + 2(1) - 3(-2) = 20 \neq 0$. Hence L and H intersect in a point. Now the point is given by $x = 1 + 2h$, $y = -1 - 3h$, $z = 2 + h$, $v = -1 - 2h$. We substitute these equations into the equation for H, getting

$$3(1 + 2h) - 2(-1 - 3h) + 2(2 + h) - 3(-1 - 2h) + 4 = 0 \Leftrightarrow h = -\tfrac{3}{5}.$$

The point of intersection is $(-\tfrac{1}{5}, \tfrac{4}{5}, \tfrac{7}{5}, \tfrac{1}{5})$.

DEFINITION. If H is a hyperplane we know from Theorem 19 that all lines perpendicular to H are parallel. We define **the angle between two hyperplanes** (which are not parallel) to be the angle between two lines each perpendicular to one of the hyperplanes.

EXAMPLE 6. Given the hyperplanes

$$H_1: 2x - y + z - 2v + 4 = 0$$
$$H_2: x + y - 3z + v - 5 = 0,$$

find the angle between them.

Solution. $2, -1, 1, -2$, is a set of direction numbers of all lines perpendicular to H_1. Similarly, $1, 1, -3, 1$ is a set of direction numbers of all lines perpendicular to H_2. The corresponding direction cosines are

$$\frac{2}{\sqrt{10}}, -\frac{1}{\sqrt{10}}, \frac{1}{\sqrt{10}}, -\frac{2}{\sqrt{10}} \quad \text{and} \quad \frac{1}{\sqrt{12}}, \frac{1}{\sqrt{12}}, \frac{-3}{\sqrt{12}}, \frac{1}{\sqrt{12}}.$$

Hence

$$\cos \theta = \frac{1}{\sqrt{120}} (2 - 1 - 3 - 2) = \frac{-2}{\sqrt{30}}.$$

PROBLEMS

In each of problems 1 through 4, find the equations of the line L perpendicular to the given hyperplane H and passing through the given point P_1.

1. $P_1(2, -1, 3, 2)$; $H: 2x + 2y - z - 3v + 4 = 0$
2. $P_1(-1, 2, -1, 4)$; $H: x + y - 2z + 3v + 4 = 0$
3. $P_1(1, -2, 2, 1)$; $H: 2x - y + 2z + v - 3 = 0$
4. $P_1(0, 1, 0, -2)$; $H: 3x - 2y + z - v + 7 = 0$

In each of problems 5 through 8, find the equation of the hyperplane perpendicular to the given line L_1 and passing through the given point P_1.

5. $P_1(2, 0, 1, -1)$; $L_1: \dfrac{x+1}{2} = \dfrac{y-1}{3} = \dfrac{z+2}{-2} = \dfrac{v+1}{-4}$

6. $P_1(1, 2, -1, -2)$; $L_1: \dfrac{x-2}{1} = \dfrac{y+1}{2} = \dfrac{z+1}{-3} = \dfrac{v-1}{2}$

7. $P_1(-2, 1, 0, 2)$; $L_1: \dfrac{x+2}{3} = \dfrac{y-2}{2} = \dfrac{z}{-1} = \dfrac{v+2}{-2}$

8. $P_1(-1, 2, -1, 1)$; $L_1: \dfrac{x+2}{3} = \dfrac{y-1}{0} = \dfrac{z+1}{-2} = \dfrac{v-2}{-3}$

In each of problems 9 through 12, find the equation of the hyperplane H parallel to the given hyperplane H_1 and passing through the given point P_1.

9. $P_1(-2, 3, 1, 2)$; $H_1: 2x + y - 3z - 2v + 5 = 0$
10. $P_1(1, -2, 0, 3)$; $H_1: x - 2y + 3z + v - 6 = 0$
11. $P_1(0, 1, 2, -1)$; $H_1: 2x - y + 2z - v + 10 = 0$
12. $P_1(1, -1, 1, 2)$; $H_1: x + 3y - 2z + 2v + 7 = 0$

In each of problems 13 through 18, determine whether the hyperplane H is parallel to L, whether L is contained in H, or whether L and H intersect in a single point. In the latter case, find the point of intersection.

13. $L: \dfrac{x+1}{2} = \dfrac{y-2}{3} = \dfrac{z+1}{-2} = \dfrac{v+2}{1}$; $H: x + 2y + 3z + 2v - 5 = 0$

14. $L: \dfrac{x-2}{1} = \dfrac{y+1}{2} = \dfrac{z-1}{3} = \dfrac{v+1}{-2}$; $H: x + 2y - 3z - 2v + 3 = 0$

15. $L: \dfrac{x+2}{3} = \dfrac{y-1}{2} = \dfrac{z+2}{-2} = \dfrac{v-1}{1}$; $H: 2x - 3y - 2z - 4v + 7 = 0$

16. $L: \dfrac{x-1}{2} = \dfrac{y+2}{-3} = \dfrac{z-2}{-2} = \dfrac{v}{3}$; $H: x - 2y + 3z + 2v + 1 = 0$

17. $L: \dfrac{x}{3} = \dfrac{y-2}{-2} = \dfrac{z+1}{1} = \dfrac{v-2}{1}$; $H: 2x + y - 2z - 3v + 5 = 0$

18. $L: \dfrac{x-3}{1} = \dfrac{y+3}{-2} = \dfrac{z+3}{-3} = \dfrac{v-2}{2}$; $H: x + 2y - 3z - 3v + 3 = 0$

19. Prove that the distance d from a point $P_1(x_1, y_1, z_1, v_1)$ to the hyperplane $H: Ax + By + Cz + Dv + E = 0$ is given by

$$d = \frac{|Ax_1 + By_1 + Cz_1 + Dv_1 + E|}{\sqrt{A^2 + B^2 + C^2 + D^2}}.$$

[*Hint*: See the corresponding formulas in two and three dimensions on pages 65 and 223.]

In each of problems 20 through 23, find an equation in x, y, z, and v of the hyperplane H which has the parametric equations given.

20 $H: x = 2 + hr + 2ks + lt, \ y = -1 - hr + ks + lt, \ z = 3 - hr - ks + lt,$
     $v = 2 + hr - 2ks + lt; \ h = \frac{1}{2}, \ k = 1/\sqrt{10}, \ l = \frac{1}{2}$

21. $H: x = -1 + hr + 3ks + lt, \ y = 2 + 2hr - ks + 2lt, \ z = 3 + 2ks - 3lt,$
     $v = -2 - hr + ks + 5lt; \ h = 1/\sqrt{6}, \ k = 1/\sqrt{15}, \ l = 1/\sqrt{39}$

22. $H: x = 3 + 2hr + ks + 2lt, \ y = -1 + hr + 3lt, \ z = 2 - 3hr + 2ks + lt,$
     $v = 1 + 2hr + 2ks - 2lt; \ h = 1/3\sqrt{2}, \ k = 1/3, \ l = 1/3\sqrt{2}$

23. $H: x = -2 + 2hr + 3ks, \ y = 3 + hr + lt, \ z = 1 - hr + 2ks - lt,$
     $v = -1 - 2hr + 2ks + lt; \ h = 1/\sqrt{10}, \ k = 1/\sqrt{17}, \ l = 1/\sqrt{3}$

In each of problems 24 through 26, find the cosine of the angle of intersection of the given hyperplanes H_1 and H_2.

24. $H_1: 2x - y + z - 3v - 2 = 0;$ $H_2: x + y - z - 4v - 2 = 0$

25. $H_1: x + y + z + 5v - 4 = 0;$ $H_2: 2x - z + v - 3 = 0$

26. $H_1: x - y = 0;$ $H_2: 2x + y + z + 6v - 2 = 0$

27. Prove Theorem 15.

28. Prove Theorem 16(i). [*Hint*: Use the fact that there is a unique line L through two given points.]

29. Prove Theorem 16(ii).

30. Prove Theorem 16(iii).

31. Write out a proof of the Corollary to Theorem 17.

32. Prove Theorem 19(ii).

33. Write out a proof of the Corollary to Theorem 20.

34. Show that the following four hyperplanes have a unique point in common and find the coordinates of that point.

$$H_1: 2x + 3y - z + v + 8 = 0; \qquad H_2: x - 2y + 3z + 2v - 12 = 0$$
$$H_3: x + 2y + 2z + 3v = 0; \qquad H_4: 2x - 4y + z - 5v - 18 = 0$$

*35. A four dimensional Euclidean space in five dimensional space is called a **hyperplane** in five space. Develop the parametric equations (5 equations in 4 parameters) of such a hyperplane. Show that such a hyperplane is represented by a linear equation of the form $Ax + By + Cz + Dv + Ew + F = 0$.

The Axioms of Algebra

In Chapter 1 we stated that the customary "laws of algebra" could be derived from the axioms given there. In this appendix we carry out this derivation.

We assume that we are given a set of objects which we call **real numbers** which satisfy the axioms stated in Chapter 1, Section 1. For convenience we restate these axioms.

Axioms of addition and subtraction

A–1. Closure property. *If a and b are numbers, there is one and only one number, denoted by a + b, called their* **sum.**

A–2. Commutative law. *For every two numbers a and b, we have*

$$b + a = a + b.$$

A–3. Associative law. *For all numbers a, b, and c, we have*

$$(a + b) + c = a + (b + c).$$

A–4. Existence of a zero. *There is one and only one number 0, called* **zero,** *such that a + 0 = a for any number a.*

We remark that it is not necessary to assume in Axiom A–4 that there is *only one* number 0 with the given property. The uniqueness of this number may be established as follows: suppose 0 and $0'$ are two numbers such that $a + 0 = a$ and $a + 0' = a$ for every number a. Then $0 + 0' = 0$ and $0' + 0 = 0'$. By Axiom A–2, $0 + 0' = 0' + 0$ and so $0 = 0'$; the two numbers are the same.

A–5. Existence of a negative. *If a is any number, there is one and only one number x such that a + x = 0. This number is called the* **negative** *of a and is denoted by* **−a.**

As in Axiom A–4, it is not necessary to assume there is *only one* such number with the given property. The argument is similar to the one given after Axiom A–4.

Theorem 1. *If a and b are any numbers, then there is one and only one number x such that a + x = b. This number x is given by x = b + (−a).*

Proof. We must establish two results: (i) that $b + (-a)$ satisfies $a + x = b$, and (ii) that no other number satisfies $a + x = b$. To prove (i), we suppose that $x = b + (-a)$. Then, using the commutative and associative laws, we have

$$a + x = a + [b + (-a)] = a + [(-a) + b] = [a + (-a)] + b$$
$$= 0 + b = b.$$

Therefore (i) holds. To prove (ii), we suppose that x is some number such that $a + x = b$. Adding $(-a)$ to both sides of this equation, we find that

$$(a + x) + (-a) = b + (-a).$$

Now,

$$(a + x) + (-a) = a + [x + (-a)] = a + [(-a) + x]$$
$$= [a + (-a)] + x = 0 + x = x.$$

We conclude that $x = b + (-a)$, and the uniqueness of the solution is established.

Notation. We denote the number $b + (-a)$ by $b - a$.

Thus far addition has been defined *only for two numbers.* By means of the associative law we can define addition for three, four, and in fact, any finite number of elements. Since $(a + b) + c$ and $a + (b + c)$ are the same, we define $a + b + c$ as this common value. The following lemma is an easy consequence of the associative and commutative laws of addition.

Lemma 1. *If a, b, and c are any numbers, then*

$$a + b + c = a + c + b = b + a + c = b + c + a$$
$$= c + a + b = c + b + a.$$

The formal details of writing out a proof are left to the student.

The next lemma is useful in the proof of Theorem 2 below.

Lemma 2. *If a, b, c, and d are numbers, then*

$$(a + c) + (b + d) = (a + b) + (c + d).$$

Proof. Using Lemma 1 and the axioms, we have

$$(a + c) + (b + d) = [(a + c) + b] + d$$
$$= (a + c + b) + d = (a + b + c) + d$$
$$= [(a + b) + c] + d = (a + b) + (c + d).$$

Theorem 2. (i) *If a is a number, then* $-(-a) = a$. (ii) *If a and b are numbers, then*

$$-(a + b) = (-a) + (-b).$$

Proof. (i) From the definition of negative, we have

$$(-a) + [-(-a)] = 0,$$
$$(-a) + a = a + (-a) = 0.$$

Axiom A–5 states that the negative of $(-a)$ is *unique*. Therefore, $a = -(-a)$.

(ii) From the definition of negative, we know that

$$(a + b) + [-(a + b)] = 0.$$

Furthermore, using Lemma 2, we have

$$(a + b) + [(-a) + (-b)] = [a + (-a)] + [b + (-b)]$$
$$= 0 + 0 = 0.$$

The result follows from the "only one" part of Axiom A–5.

Theorem 2 can be stated in words in the familiar form: (i) *the negative of* $(-a)$ *is a* and (ii) *the negative of a sum is the sum of the negatives.*

Axioms of multiplication and division

M–1. *Closure property.* *If a and b are numbers, there is one and only one number, denoted by ab (or a* $\times$ *b or a* $\cdot$ *b), called their* **product.**

M–2. *Commutative law.* *For every two numbers a and b,*

$$ba = ab.$$

M–3. *Associative law.* *For all numbers a, b, and c, we have*

$$(ab) \cdot c = a \cdot (bc).$$

M–4. *Existence of a unit.* *There is one and only one number u, different from zero, such that au* $=$ *a for every number a. This number u is called the* **unit** *and* (as is customary) *will be denoted by* 1.

M–5. *Existence of a reciprocal.* *For each number a different from zero there is one and only one number x such that ax* $=$ *1. This number x is called the* **reciprocal** *of a and is denoted by* a^{-1} *(or* $1/a$*).*

Remarks. We observe that Axioms M–1 through M–4 are the parallels of Axioms A–1 through A–4 with addition replaced by multiplication. However M–5 is not the exact analogue of A–5 since there is the additional requirement $a \neq 0$. The reason for this is given below in Theorem 3 where it is shown that the result of multiplication of any number by zero is zero. In familiar terms, we say that division by zero is excluded.

Special axiom

D. *Distributive law*. For all numbers a, b, and c, we have

$$a(b + c) = ab + ac.$$

Remarks. The axioms of addition and multiplication and the distributive law are supposed to hold for all real numbers whether positive, negative, or zero. In fact, the axioms hold for many number systems of which the collection of real numbers is only one. For example, all the axioms stated so far hold for the system consisting of all *complex numbers*. Furthermore, there are many systems each consisting of only a finite number of elements which satisfy all the above axioms. Additional axioms are needed if we require the real numbers to be the *only* collection satisfying all the axioms. The additional required axioms are usually discussed in calculus.

Theorem 3. *If a is any number, then $a \cdot 0 = 0$.*

Proof. Let b be any number. Then $b + 0 = b$ and hence $a \cdot (b + 0) = a \cdot b$. From the distributive law, we conclude that

$$(a \cdot b) + (a \cdot 0) = (a \cdot b),$$

so that $a \cdot 0 = 0$ by Axiom A–4.

Theorem 4. *If a and b are numbers and $a \neq 0$, then there is one and only one number x such that $a \cdot x = b$. The number x is given by $x = ba^{-1}$.*

The proof is just like the proof of Theorem 1 with addition replaced by multiplication, 0 replaced by 1, and $-a$ replaced by a^{-1}. The details are left to the student. (See problem 2 at the end of Appendix 1.)

We now establish the familiar principle which underlies the solution of quadratic equations by factoring.

Theorem 5. (i) *We have $a \cdot b = 0$ if and only if $a = 0$ or $b = 0$ or both.*

(ii) We have $a \neq 0$ and $b \neq 0$ if and only if $a \cdot b \neq 0$.

Proof. We must prove two statements in each of the parts (i) and (ii). To prove (i), if $a = 0$ or $b = 0$ or both it follows from Theorem 3 that $a \cdot b = 0$. Going the other way, suppose $a \cdot b = 0$. Then there are two cases: either $a = 0$ or $a \neq 0$. If $a = 0$, the result follows. If $a \neq 0$, then we see that

$$a^{-1}(ab) = (a^{-1}a)b = 1 \cdot b = b = a^{-1} \cdot 0 = 0.$$

Hence $b = 0$ and the result follows in all cases.

(ii) Suppose $a \neq 0$ and $b \neq 0$. Then $a \cdot b \neq 0$ because $a \neq 0$ and $b \neq 0$ is the negation of the statement "$a = 0$ or $b = 0$ or both." Thus (i) applies. For the second part of (ii), suppose $a \cdot b \neq 0$. Then $a \neq 0$ and $b \neq 0$ for, if one of them were zero, Theorem 3 would apply to give $a \cdot b = 0$.

We define abc as the common value of $(ab)c$ and $a(bc)$. The student can prove the following lemmas which are similar to Lemmas 1 and 2.

Lemma 3. *If a, b, and c are numbers, then*

$$abc = acb = bac = bca = cab = cba.$$

Lemma 4. *If a, b, c, and d are numbers, then*

$$(ac) \cdot (bd) = (ab) \cdot (cd).$$

Theorem 6. (i) *If* $a \neq 0$, *then* $a^{-1} \neq 0$ *and* $[(a^{-1})^{-1}] = a$. (ii) *If* $a \neq 0$ *and* $b \neq 0$, *then* $(a \cdot b)^{-1} = (a^{-1}) \cdot (b^{-1})$.

The proof of this theorem is like the proof of Theorem 2 with addition replaced by multiplication, 0 replaced by 1, and $(-a)$, $(-b)$ replaced by a^{-1}, b^{-1}. The details are left to the student. (See problem 5 at the end of Appendix 1.) We note that if $a \neq 0$, then $a^{-1} \neq 0$ because $aa^{-1} = 1$ and $1 \neq 0$. Then Theorem 5 (ii) may be used with $b = a^{-1}$.

Using Theorem 3 and the distributive law, we easily prove the **laws of signs** stated as Theorem 7 below. We emphasize that the numbers a and b may be positive, negative, or zero.

Theorem 7. *If a and b are any numbers, then* (i) $a \cdot (-b) = -(a \cdot b)$, (ii) $(-a) \cdot b = -(a \cdot b)$, (iii) $(-a) \cdot (-b) = a \cdot b$.

Proof. (i) Since $b + (-b) = 0$, we find from the distributive law that

$$a[b + (-b)] = a \cdot b + a \cdot (-b) = 0.$$

On the other hand, the negative of $a \cdot b$ has the property: $a \cdot b + [-(a \cdot b)] = 0$.

Hence it follows from Axiom A–5 that $a \cdot (-b) = -(a \cdot b)$. Part (ii) follows from (i) by interchanging a and b. The proof of (iii) is left to the student. (See problem 6 at the end of Appendix 1.)

Corollary. *We have* $(-1) \cdot a = -a$.

We now show that the **laws of fractions,** as given in elementary algebra courses, follow from the axioms and theorems above.

Notation. We introduce the following symbols for $a \cdot b^{-1}$:

$$a \cdot b^{-1} = \frac{a}{b} = a/b = a \div b.$$

These symbols, representing an indicated division are called **fractions.** The **numerator** and **denominator** of a fraction are defined as usual. A fraction with *denominator* zero has no meaning.

Theorem 8. (i) *For every number* a, *we have* $a/1 = a$. (ii) *If* $a \neq 0$, *then* $a/a = 1$.

Proof. (i) We have $a/1 = a \cdot (1)^{-1} = a \cdot 1 = a$.

(ii) If $a \neq 0$, then $a/a = a \cdot a^{-1} = 1$, by definition.

Theorem 9. *If* $b \neq 0$ *and* $d \neq 0$, *then*

$$\left(\frac{a}{b}\right) \cdot \left(\frac{c}{d}\right) = \frac{a \cdot c}{b \cdot d}.$$

Proof. Using the notation for fractions, Lemma 4, and Theorem 6(ii), we find

$$\left(\frac{a}{b}\right) \cdot \left(\frac{c}{d}\right) = (a \cdot b^{-1}) \cdot (cd^{-1}) = (a \cdot c) \cdot (b^{-1}d^{-1}) = (a \cdot c)(bd)^{-1}$$
$$= \frac{a \cdot c}{b \cdot d}.$$

The proofs of Theorems 10 through 14 are left to the student. (See problems 10 through 14 at the end of Appendix 1.)

Theorem 10. *If* $b \neq 0$ *and* $c \neq 0$, *then*

$$\frac{a}{b} = \frac{a \cdot c}{b \cdot c}.$$

Theorem 11. *If* $c \neq 0$, *then*

$$\frac{a}{c} + \frac{b}{c} = \frac{(a + b)}{c}.$$

Theorem 12. *If $b \neq 0$, then $-b \neq 0$ and*

$$\frac{(-a)}{b} = \frac{a}{(-b)} = -\left(\frac{a}{b}\right).$$

Theorem 13. *If $b \neq 0$, $c \neq 0$, and $d \neq 0$, then $(c/d) \neq 0$ and*

$$\frac{(a/b)}{(c/d)} = \frac{a \cdot d}{b \cdot c} = \left(\frac{a}{b}\right) \cdot \left(\frac{d}{c}\right).$$

Theorem 14. *If $b \neq 0$ and $d \neq 0$, then*

$$\frac{a}{b} + \frac{c}{d} = \frac{ad + bc}{bd}.$$

PROBLEMS

1. Prove Lemma 1.

2. Prove Theorem 4.

3. Prove Lemma 3.

4. Prove Lemma 4.

5. Prove Theorem 6.

6. Complete the proof of Theorem 7.

7. Prove that $a(b + c + d) = ab + ac + ad$.

8. Assuming that $A + B + C + D$ means $(A + B + C) + D$, prove that $A + B + C + D = (A + B) + (C + D)$.

9. Assuming the result of problem 8, prove that

$$(a + b) \cdot (c + d) = ac + bc + ad + bd.$$

10. Prove Theorem 10.

11. Prove Theorem 11.

12. Prove Theorem 12.

13. Prove Theorem 14.

14. Prove Theorem 13. (*Hint.* Use Theorem 10 appropriately.)

15. Prove that if $b \neq 0$, $d \neq 0$, $f \neq 0$, then

$$\left(\frac{a}{b}\right) \cdot \left(\frac{c}{d}\right) \cdot \left(\frac{e}{f}\right) = \frac{(a \cdot c \cdot e)}{(b \cdot d \cdot f)}.$$

16. Prove that if $d \neq 0$, then

$$\frac{a}{d} + \frac{b}{d} + \frac{c}{d} = \frac{(a + b + c)}{d}.$$

APPENDIX 2
Natural Numbers. Sequences. Extensions.

Historically, the present-day concept of the real number system was built up by a sequence of successive enlargements. To begin with, the positive integers were extended to include positive rational numbers (quotients, or ratios, of integers); these were enlarged to include all the positive real numbers, and finally, all real numbers are obtained by the inclusion of negative numbers.

Since the system of axioms in Appendix 1 does not distinguish (or even mention) positive numbers, we shall discuss first the **natural numbers.** As we know, these turn out to be nothing but the positive integers.

Intuitively, the totality of natural numbers can be obtained by starting with 1 and then forming $1 + 1$, $(1 + 1) + 1$, $[(1 + 1) + 1] + 1$, and so on. We call $1 + 1$ the number 2, $(1 + 1) + 1$ is called 3, and in this way the collection of natural numbers is generated. Actually, it is possible to give a more logically satisfactory (but rather abstract) definition of a natural number which yields the same set. This process is frequently carried out in advanced mathematics courses. If we add to our system of axioms the axiom of inequality, then all the usual properties of natural numbers can be proved, including the principle of mathematical induction. Since it would take us too far afield to establish all the necessary results, we shall just assume that mathematical induction and various applications to counting are known and continue from that point.

In Appendix 1 we defined the sum and product of three numbers. We wish now to indicate extensions of these notions to more terms. For our purposes, it is sufficient to think of a **sequence** as a set of numbers arranged in a specific order. For example, the numbers

$$1, \quad 3, \quad -2, \quad 1, \quad \pi, \quad 5, \quad \sqrt{2}$$

form a sequence. The numbers in the sequence, separated by commas, are called its **terms.** The reader is undoubtedly familiar with arithmetic and geometric progressions which are examples of sequences. A sequence may be **finite,** in which case there is a first and a last term, or a sequence may be **infinite,** in which case the terms continue without terminating. We shall indicate sequences by notations such as

$$a_1, a_2, \ldots, a_n \qquad \text{or} \qquad a_1, a_2, \ldots, a_n, \ldots,$$

the first sequence being finite and having n terms; the second sequence is infinite with the dots indicating the intervening and the succeeding terms. When $n = 1$ the sequence indicated has only one term a_1. In our notation a_i denotes the **i**th **term.**

We define the sum and product of a sequence $a_1, a_2, \ldots, a_n$ as follows: If $n = 2$, the sum is $a_1 + a_2$ and the product is $a_1 \cdot a_2$, as usual. For $n > 2$, we have

$$a_1 + a_2 + a_3 = (a_1 + a_2) + a_3,$$

$$a_1 + a_2 + a_3 + a_4 = (a_1 + a_2 + a_3) + a_4,$$

and so on. A similar definition may be given for products. For completeness, if $n = 1$ (so that there is only one term a_1), we define the sum and product as a_1, itself.

It is possible to prove the **Extended Commutative Law** for sums and products, which states that:

The sum and product of a given finite sequence are independent of the order of its terms.

This has been proved for $n = 3$ in Lemma 1, Section A–1. The associative law for sums and products may be extended as follows:

The sum of a finite sequence can be obtained by separating the given sequence into several shorter sequences, adding the terms in each of these and then adding the results; a similar statement holds for products.

For example,

$$1 + 5 + 3 + 7 + 4 + 1 + 6 = (1 + 5 + 3) + (7 + 4) + (1 + 6).$$

Finally, the distributive law may be extended so that

$$a \cdot (b_1 + \cdots + b_n) = (b_1 + \cdots + b_n) \cdot a = ab_1 + \cdots + ab_n.$$

This was proved for $n = 3$ in problem 7, Section A–1. This in turn leads to the general rule for multiplying two sums:

To multiply two sums, multiply each term of one sum by each term of the other and add all the results.

See problem 9, Section A–1 and the following example:

$$(a_1 + a_2)(b_1 + b_2 + b_3) = a_1(b_1 + b_2 + b_3) + a_2(b_1 + b_2 + b_3)$$
$$= a_1b_1 + a_1b_2 + a_1b_3 + a_2b_1 + a_2b_2 + a_2b_3.$$

We understand the signed sum

$$a - b - c + d - e,$$

for example, to mean

$$a + (-b) + (-c) + d + (-e).$$

Theorem 2(ii) (the negative of a sum is the sum of the negatives) can be generalized to sequences of any finite number of terms. Using Theorem 2(i) $[-(-a) = a]$, we obtain the usual rule for the negative of a signed sum. Then, since $A - B = A + (-B)$ by definition, we obtain the usual rule for subtraction of signed sums. With the aid of the sign laws for multiplication and the extended distributive law, we get the usual rules for the multiplication of signed sums. The symbol x^n, for n a natural number, is defined as usual, and the familiar laws of exponents follow. From these and the preceding rules follow the established rules for adding, subtracting, and multiplying polynomials. The validity of these rules, as well as those for the division and factoring of polynomials, will be assumed in the discussion below.

Finally, it can be shown that any natural number n has one and only one representation of the form

$$n = d_0 \cdot 10^k + d_1 \cdot 10^{k-1} + \cdots + d_k$$

in which each d_i is 0 or a natural number ≤ 9 (i.e., a **digit**) and $d_0 \neq 0$. By verifying the rules for addition and multiplication of digits, we can then derive the rules for the addition and multiplication of natural numbers from the corresponding rules for polynomials (with $x = 10$). Of course we shall assume these rules.

We now define some frequently used terms.

DEFINITIONS. A real number is an **integer** if and only if it is either zero, a natural number, or the negative of a natural number. A real number r is said to be **rational** if and only if there are *integers* p and q, with $q \neq 0$, such that $r = p/q$.

It is clear that the sum and product of a finite sequence of either integers or rational numbers is again an integer or a rational number, respectively, and that the quotient of two rational numbers is again rational.

We now sketch briefly how the general rules apply in specific cases with which the student is familiar. We use the customary notations and assume the laws of exponents for exponents which are positive integers.

EXAMPLE 1. Add: $2x^3 - 3x^3 + 5x^3$.

Solution.

$$2x^3 - 3x^3 + 5x^3 = 2x^3 + (-3)x^3 + 5x^3$$
$$= [2 + (-3) + 5] \cdot x^3 = 4x^3 \qquad \text{(distributive law)}.$$

EXAMPLE 2. Multiply: $3x^3 \cdot (-5x^5)$.

Solution.

$3x^3 \cdot (-5x^5) = 3 \cdot x^3 \cdot (-5) \cdot x^5$ (extended commutative and associative laws).

$\qquad = 3 \cdot (-5) \cdot x^3 \cdot x^5 = [3 \cdot (-5)] \cdot (x^3 \cdot x^5) = (-15) \cdot x^8 = -15x^8$

EXAMPLE 3. Carry out the indicated operation:

$$(2x^3 - 3x^2 + x + 2) - (-x^3 - x^2 + 2x + 1).$$

Solution.

$2x^3 - 3x^2 + x + 2 - (-x^3 - x^2 + 2x + 1)$

$\quad = 2x^3 - 3x^2 + x + 2 + [-(-x^3 - x^2 + 2x + 1)]$ $[A - B = A + (-B)]$

$\quad = 2x^3 - 3x^2 + x + 2 + [x^3 + x^2 - 2x - 1]$ [negative of a sum is the sum of

$\qquad\qquad\qquad\qquad\qquad\qquad\qquad\qquad\qquad\qquad$ the negatives; $-(-a) = a$]

$\quad = [2x^3 + x^3] + [(-3)x^2 + x^2] + [x + (-2)x] + [2 + (-1)]$

$\quad = 3x^3 - 2x^2 - x + 1.$

EXAMPLE 4. Multiply: $(2x^2 - x - 3) \cdot (x^3 - 2x^2 + 3x - 4)$.

Solution. Each term of one sum must be multiplied by each term of the other. In order to ensure this, we arrange the work (essentially) as in multiplication in arithmetic:

$$
\begin{array}{r}
x^3 + (-2)x^2 + 3x + (-4) \\
2x^2 + (-1)x + (-3) \\
\hline
2x^5 + (-4)x^4 + 6x^3 + (-8)x^2 \\
(-1)x^4 + 2x^3 + (-3)x^2 + 4x \\
(-3)x^3 + 6x^2 + (-9)x + 12 \\
\hline
2x^5 - 5x^4 + 5x^3 - 5x^2 - 5x + 12 \quad \text{(Ans.)}
\end{array}
$$

Theorem 5 ($a \cdot b = 0$, etc.) can be extended to products of more than two terms. Then the rule for the multiplication of fractions (Theorem 9) can be extended as follows:

$$\frac{a_1}{b_1} \cdot \frac{a_2}{b_2} \cdots \frac{a_n}{b_n} = \frac{a_1 \cdot a_2 \cdots a_n}{b_1 \cdot b_2 \cdots b_n}.$$

Theorems 11 and 12, together with the definition of signed sums, etc., yield the following rule:

A signed sum of a finite sequence of fractions having a common denominator is equal to a single fraction of which the denominator is the common denominator and the numerator is the corresponding signed sum of the numerators.

When one adds only two fractions, Theorem 14 is useful. However, when more fractions are involved, it is best to find the least common denominator of all the fractions, express each fraction as a fraction with that denominator, and then add. We illustrate the procedure, giving a reason for each step and noting values which make a *denominator* zero; these must be excluded.

EXAMPLE 5. Express the following sum as a single fraction in its lowest terms (i.e., in which the numerator and denominator have no common factor). Give a reason for each step.

$$\frac{x}{8x^2 - 2} + \frac{5}{3 + 6x} - \frac{5}{3 - 6x} \qquad (x \neq \tfrac{1}{2} \text{ or } -\tfrac{1}{2})$$

Solution.

$$\frac{x}{8x^2 - 2} + \frac{5}{3 + 6x} - \frac{5}{3 - 6x} \qquad (x \neq \pm\tfrac{1}{2})$$

$$= \frac{x}{2(2x + 1)(2x - 1)} + \frac{5}{3(2x + 1)} - \frac{5}{-3(2x - 1)} \qquad \text{(factoring)}$$

$$= \frac{x}{2(2x + 1)(2x - 1)} + \frac{5}{3(2x + 1)} + \frac{5}{3(2x - 1)} \qquad \left[\frac{a}{-b} = -\frac{a}{b}, \right.$$

$$= \frac{3x}{6(2x + 1)(2x - 1)} + \frac{10(2x - 1)}{6(2x + 1)(2x - 1)} \qquad \left. -(-a) = a \right]$$

$$+ \frac{10(2x + 1)}{6(2x + 1)(2x - 1)} \qquad \left(\frac{a}{b} = \frac{ac}{bc}, c \neq 0 \right)$$

$$= \frac{3x + 10(2x - 1) + 10(2x + 1)}{6(2x + 1)(2x - 1)} \qquad \left(\frac{a_1}{b} + \cdots + \frac{a_n}{b} = \frac{a_1 + \cdots + a_n}{b} \right)$$

$$= \frac{43x}{6(2x + 1)(2x - 1)} \qquad \text{(simplifying numerator).}$$

Any set of objects satisfying the axioms of this section is called a **field.** The axioms given above imply only theorems concerned with the operations of addition, subtraction, multiplication, and division. They do not imply the *existence* of a number whose square is 2, for instance, since the set of all rational numbers is a field and there is no rational number whose square is 2.

PROBLEMS

In problems 1 through 4, subtract the second polynomial from the first. Put in enough steps to indicate the fundamental principles being used.

1. $x^2 - 2xy + 3y^2$; $2x^2 - 3xy - y^2$
2. $x^3 + x - 2x^2 - 2$; $x^2 - x^3 + 4$
3. $4x^3 - 3x^2 - 2$; $x^2 - 3x - 3$
4. $4xy - 3x^2y - 2y^2 - 4x - 2$; $7 - 2x^2y - xy - 2y^2$

In problems 5 through 8, multiply the two polynomials.

5. $2x - 3;\quad x^3 - 2x^2 - 3x + 1$

6. $x^2 - x - 2;\quad 2x^3 - x + 1$

7. $2x^2 - x + 3;\quad x^3 - 2x^2 - 3x + 3$

8. $x^3 - 3x^2 - 2x + 4;\quad 2x^4 - x^3 + 3x^2 - 1$

Reduce each of the following expressions to a single fraction in its lowest terms. Note excluded values and give a reason for each step as in Example 5.

9. $\dfrac{2}{a - b} + \dfrac{3}{b - a} + \dfrac{a}{a^2 - b^2}$ $(a \neq \pm b)$

10. $\dfrac{x + y}{(y - z)(y - x)} + \dfrac{y + z}{(z - x)(z - y)} + \dfrac{z + x}{(x - y)(x - z)}$

11. $\dfrac{x + 1}{2(x + 4)} - \dfrac{4x - 1}{3x - 3} + \dfrac{7}{6}$

12. $\dfrac{x^2 - x - 2}{x^2 + x - 6} - \dfrac{2}{9 - x^2} + 1 - \dfrac{5}{3 - x}$

13. $\dfrac{x^2 + 6x + 5}{x^2 - 1} \div \dfrac{x^2 + 3x + 2}{x^3 - 1}$

14. $\dfrac{3x^2 + 8x + 4}{2x^2 + 7x + 3} \cdot \dfrac{x^2 + 2x - 3}{9x^2 + 12x + 4} \cdot \dfrac{2x^2 - x - 1}{3x^2 - x - 2}$

15. $\dfrac{\left(\dfrac{2x + y}{x + y} - 1\right)}{\left(1 - \dfrac{y}{x + y}\right)}$

16. $\dfrac{\left(1 - \dfrac{2}{a} + \dfrac{1}{a^2}\right)}{\left(1 - \dfrac{1}{a^2}\right)}$

17. $\dfrac{\left(\dfrac{a^2 + b^2}{a^2 - b^2} - \dfrac{a^2 - b^2}{a^2 + b^2}\right)}{\left(\dfrac{a + b}{a - b} - \dfrac{a - b}{a + b}\right)}$

3 Trigonometry Review

We assume that the reader has successfully completed a course in trigonometry. However, no matter how well he has learned the formulas of trigonometry, he can forget them unless he has used them repeatedly and frequently. We shall therefore devote this appendix to a review of the basic notions of trigonometry and a compilation of many of the elementary formulas, for we shall need to use them from now on. In any case, the formulas serve as handy tools for future use, and the student who has difficulty remembering them should relearn them systematically.

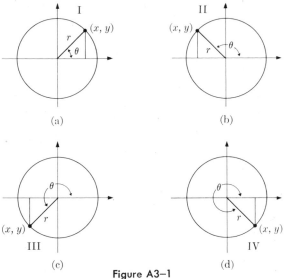

Figure A3–1

A circle of radius r is drawn and a Cartesian coordinate system constructed with origin at the center of this circle. An angle θ is drawn, starting from the positive direction of the x axis, and is measured counterclockwise. Angles in the first, second, third, and fourth quadrants are shown in Fig. A3–1. The basic definitions of the trigonometric functions are

$$\sin \theta = \frac{y}{r}, \qquad \cos \theta = \frac{x}{r},$$

where (x, y) are the coordinates of the point where the terminating side of the angle θ intersects the circle. These definitions depend only on the angle θ and not on the size of the circle since, by similar triangles, the ratios are independent of the size of r. The quantity r is always positive, while x and y have the sign that goes with the quadrant: x and y are positive in the first quadrant; x is negative, y is positive in the second quadrant, and so on.

The remaining trigonometric functions are defined by the relations

$$\tan \theta = \frac{\sin \theta}{\cos \theta}, \qquad \cot \theta = \frac{\cos \theta}{\sin \theta},$$

$$\sec \theta = \frac{1}{\cos \theta}, \qquad \csc \theta = \frac{1}{\sin \theta}.$$

We also can define them by the formulas

$$\tan \theta = \frac{y}{x}, \quad \cot \theta = \frac{x}{y}, \quad \sec \theta = \frac{r}{x}, \quad \csc \theta = \frac{r}{y}.$$

For angles larger than 360° the trigonometric functions repeat, and for this reason they are called *periodic* functions. This means that for all θ

$$\sin (360° + \theta) = \sin \theta, \qquad \cos (360° + \theta) = \cos \theta.$$

Negative angles are measured by starting from the positive x direction and going *clockwise*. Figure A3–2 shows a negative angle $-\theta$ and its corresponding positive angle θ. By congruent triangles it is easy to see that for *every* θ we have

$$\sin (-\theta) = -\sin \theta \qquad \text{and} \qquad \cos (-\theta) = \cos \theta,$$

which says that the sine function is an **odd** function of θ, while the cosine function is an **even** function of θ. It is easy to show that

$$\tan (-\theta) = -\tan \theta, \qquad \cot (-\theta) = -\cot \theta,$$
$$\sec (-\theta) = \sec \theta, \qquad \csc (-\theta) = -\csc \theta.$$

No matter what quadrant the angle θ is in, the Pythagorean theorem tells us that

$$x^2 + y^2 = r^2,$$

or, upon dividing by r^2,

$$\left(\frac{x}{r}\right)^2 + \left(\frac{y}{r}\right)^2 = 1.$$

This formula asserts that for every angle θ,

$$\cos^2 \theta + \sin^2 \theta = 1.$$

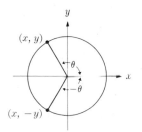

Figure A3–2

Let *ABC* be any triangle with sides *a*, *b*, *c*, as shown in Fig. A3–3. We have the **Law of Sines:**

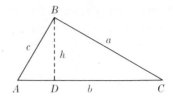

$$\frac{\sin A}{a} = \frac{\sin B}{b} = \frac{\sin C}{c},$$

and the **Law of Cosines:**

$$a^2 = b^2 + c^2 - 2bc \cos A,$$
$$b^2 = a^2 + c^2 - 2ac \cos B,$$
$$c^2 = a^2 + b^2 - 2ab \cos C.$$

Figure A3–3

The area of $\triangle ABC$ is: Area $= \frac{1}{2}ab \sin C = \frac{1}{2}bc \sin A = \frac{1}{2}ac \sin B$. The proofs of these statements should be reviewed. An ambitious student may want to re-discover these proofs for himself. As a help in establishing the law of sines, the student should drop the perpendicular from *B* to side *b* in Fig. A3–3.

The most common unit for measurement of angles is the degree. The circum-ference of a circle is divided into 360 equal parts, and the angle from the center to two adjacent subdivision points is **one degree.** Each degree is divided into 60 **minutes** and each minute into 60 **seconds.**

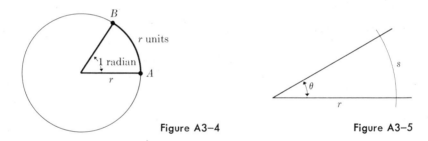

Figure A3–4   Figure A3–5

A second type of unit for measuring angles is the *radian*. A circle is drawn and an arc is measured which is equal in length to the radius, as in Fig. A3–4. Note that the arc $\overset{\frown}{AB}$ (*not* the chord) is *r* units in length. *The angle subtended from the center of the circle is* 1 **radian,** by definition. This angle is the same regardless of the size of the circle and so is a perfectly good unit for measurement. To measure any angle θ in radian units (Fig. A3–5), we measure the arc *s* intercepted by a circle of radius *r*. Then division by *r* gives the radian measure of the angle:

$$\theta = \frac{s}{r} \text{ radians.}$$

We are all familiar with the definition of the number π, which is

$$\pi = \frac{\text{length of circumference of circle}}{\text{length of diameter of circle}}.$$

If C is the length of the circumference and r is the radius, then

$$2\pi = C/r.$$

This is the same as saying that the radian measure of the angle we know as 360° is exactly 2π radians. We have the basic relation

$$\pi \text{ radians} = 180 \text{ degrees.}$$

The fundamental formulas upon which many of the trigonometric identities and relations are based are the **addition formulas for the sine and cosine.** These are

$$\sin(\phi + \theta) = \sin\phi\cos\theta + \cos\phi\sin\theta;$$
$$\sin(\phi - \theta) = \sin\phi\cos\theta - \cos\phi\sin\theta;$$
$$\cos(\phi + \theta) = \cos\phi\cos\theta - \sin\phi\sin\theta;$$
$$\cos(\phi - \theta) = \cos\phi\cos\theta + \sin\phi\sin\theta.$$

Proofs of these formulas should be reviewed. See problems 4 and 5 at the end of this section. Many of the formulas listed below are special cases of the above formulas. Before listing them, we shall indicate some special values of the three principal trigonometric functions in a way that somewhat reduces the strain on memorization. The tangent function tends to infinity as $\theta \to 90°$ and we sometimes say, speaking loosely, that "tan 90° is infinite."

Function \ Angle	0°	30°	45°	60°	90°	120°	135°	150°	180°
sin	$\dfrac{\sqrt{0}}{2}$	$\dfrac{\sqrt{1}}{2}$	$\dfrac{\sqrt{2}}{2}$	$\dfrac{\sqrt{3}}{2}$	$\dfrac{\sqrt{4}}{2}$	$\dfrac{\sqrt{3}}{2}$	$\dfrac{\sqrt{2}}{2}$	$\dfrac{\sqrt{1}}{2}$	$\dfrac{\sqrt{0}}{2}$
cos	$\dfrac{\sqrt{4}}{2}$	$\dfrac{\sqrt{3}}{2}$	$\dfrac{\sqrt{2}}{2}$	$\dfrac{\sqrt{1}}{2}$	$\dfrac{\sqrt{0}}{2}$	$-\dfrac{\sqrt{1}}{2}$	$-\dfrac{\sqrt{2}}{2}$	$-\dfrac{\sqrt{3}}{2}$	$-\dfrac{\sqrt{4}}{2}$
tan	$\sqrt{\tfrac{0}{4}}$	$\sqrt{\tfrac{1}{3}}$	$\sqrt{\tfrac{2}{2}}$	$\sqrt{\tfrac{3}{1}}$	*	$-\sqrt{\tfrac{3}{1}}$	$-\sqrt{\tfrac{2}{2}}$	$-\sqrt{\tfrac{1}{3}}$	$-\sqrt{\tfrac{0}{4}}$

From the relation

$$\sin^2\theta + \cos^2\theta = 1,$$

we obtain, when we divide by $\cos^2\theta$, the relation

$$\tan^2\theta + 1 = \sec^2\theta.$$

If we divide by $\sin^2\theta$, the identity

$$1 + \cot^2\theta = \csc^2\theta$$

results.

The **double-angle formulas** are obtained by letting $\theta = \phi$ in the addition formulas. We find that

$$\sin 2\theta = 2 \sin \theta \cos \theta;$$
$$\cos 2\theta = \cos^2 \theta - \sin^2 \theta = 2 \cos^2 \theta - 1 = 1 - 2 \sin^2 \theta.$$

Since $\tan (\phi + \theta) = \sin (\phi + \theta)/\cos (\phi + \theta)$, we have

$$\tan (\phi + \theta) = \frac{\tan \phi + \tan \theta}{1 - \tan \phi \tan \theta}; \qquad \tan (\phi - \theta) = \frac{\tan \phi - \tan \theta}{1 + \tan \phi \tan \theta}.$$

The double-angle formula becomes

$$\tan 2\theta = \frac{2 \tan \theta}{1 - \tan^2 \theta}.$$

The **half-angle formulas** are derived from the double-angle formulas by writing $\theta/2$ for θ. We obtain

$$\sin^2 \frac{\theta}{2} = \frac{1 - \cos \theta}{2}, \qquad \cos^2 \frac{\theta}{2} = \frac{1 + \cos \theta}{2};$$

$$\tan \frac{\theta}{2} = \frac{1 - \cos \theta}{\sin \theta} = \frac{\sin \theta}{1 + \cos \theta}.$$

The following relations may be verified from the addition formulas:

$$\sin \phi + \sin \theta = 2 \sin \left(\frac{\phi + \theta}{2}\right) \cos \left(\frac{\phi - \theta}{2}\right);$$

$$\sin \phi - \sin \theta = 2 \cos \left(\frac{\phi + \theta}{2}\right) \sin \left(\frac{\phi - \theta}{2}\right);$$

$$\cos \phi + \cos \theta = 2 \cos \left(\frac{\phi + \theta}{2}\right) \cos \left(\frac{\phi - \theta}{2}\right);$$

$$\cos \phi - \cos \theta = -2 \sin \left(\frac{\phi + \theta}{2}\right) \sin \left(\frac{\phi - \theta}{2}\right).$$

The addition formulas, with one of the angles taken as a special value, yield the following useful relations. Starting with the formula

$$\sin \left(\frac{\pi}{2} - \theta\right) = \sin \frac{\pi}{2} \cos \theta - \cos \frac{\pi}{2} \sin \theta \qquad \left(\frac{\pi}{2} = 90°\right),$$

and noting that $\sin 90° = 1$, $\cos 90° = 0$, we obtain

$$\sin \left(\frac{\pi}{2} - \theta\right) = \cos \theta.$$

In a similar way, we have

$$\cos\left(\frac{\pi}{2} - \theta\right) = \sin \theta;$$

$$\sin\left(\frac{\pi}{2} + \theta\right) = \cos \theta, \qquad \cos\left(\frac{\pi}{2} + \theta\right) = -\sin \theta;$$

$$\cos(\pi - \theta) = -\cos \theta, \qquad \sin(\pi - \theta) = \sin \theta;$$

$$\cos(\pi + \theta) = -\cos \theta, \qquad \sin(\pi + \theta) = -\sin \theta.$$

PROBLEMS

1. Prove the Law of Sines.
2. Prove the Law of Cosines.
3. Establish the formula: Area $= \frac{1}{2}ab \sin C$, for the area of a triangle.

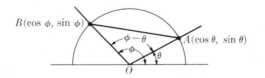

$B(\cos \phi, \sin \phi)$ $\phi - \theta$ $A(\cos \theta, \sin \theta)$ ϕ θ O

                                                                    Figure A3–6

4. Given the diagram in Fig. A3–6 (circle has radius $= 1$). Apply the Law of Cosines to triangle ABO to obtain a formula for $\cos(\phi - \theta)$. Then use the distance formula for the length $|AB|$ to derive

$$\cos(\phi - \theta) = \cos \phi \cos \theta + \sin \phi \sin \theta.$$

5. Derive formulas for $\sin 3\theta$, $\cos 3\theta$ in terms of $\sin \theta$ and $\cos \theta$.
6. Derive formulas for $\sin 4\theta$, $\cos 4\theta$ in terms of $\sin \theta$ and $\cos \theta$.
7. Write out the derivation of the formulas for $\sin \theta/2$ and $\cos \theta/2$ in terms of $\cos \theta$.
8. Derive formulas for $\sin \frac{1}{4}\theta$ and $\cos \frac{1}{4}\theta$ in terms of $\cos \theta$.
9. Without using tables, find the values of the following:

$$\sin 15°, \quad \cos 15°, \quad \sin 22\tfrac{1}{2}°, \quad \cos 22\tfrac{1}{2}°, \quad \tan 75°, \quad \sec 75°.$$

10. Without using tables, find the values of the following:

$$\tan 22\tfrac{1}{2}°, \quad \sin 7\tfrac{1}{2}°, \quad \cos 7\tfrac{1}{2}°, \quad \sin 37\tfrac{1}{2}°, \quad \tan 37\tfrac{1}{2}°.$$

11. Write formulas in terms of $\sin \theta$ and $\cos \theta$ for

$$\sin\left(\frac{3\pi}{2} - \theta\right), \quad \cos\left(\frac{3\pi}{2} - \theta\right), \quad \sin\left(\frac{3\pi}{2} + \theta\right), \quad \cos\left(\frac{3\pi}{2} + \theta\right),$$

$$\sin(2\pi - \theta), \quad \cos(2\pi - \theta).$$

12. Since $\pi = 3.14^+$ approximately, let

$$x = 0, \quad \frac{\pi}{6}, \quad \frac{\pi}{4}, \quad \frac{\pi}{3}, \quad \frac{\pi}{2}, \quad \frac{2\pi}{3}, \quad \frac{3\pi}{4}, \quad \frac{5\pi}{6}, \quad \pi, \quad \frac{7\pi}{6}, \quad \text{etc.,}$$

and plot the curve $y = \sin x$ on a Cartesian coordinate system. Extend the graph from -2π to $+4\pi$.

13. Work problem 12 again for $y = \cos x$.

14. Work problem 12 again for $y = \tan x$.

15. Derive (from the addition formulas) the identity

$$\sin \phi + \sin \theta = 2 \sin \left(\frac{\phi + \theta}{2} \right) \cos \left(\frac{\phi - \theta}{2} \right).$$

16. Work problem 15 again for

$$\cos \phi + \cos \theta = 2 \cos \left(\frac{\phi + \theta}{2} \right) \cos \left(\frac{\phi - \theta}{2} \right).$$

17. In terms of $\sin \theta$ and $\cos \theta$, find

$$\sin \left(\frac{\pi}{6} - \theta \right), \qquad \cos \left(\frac{\pi}{3} + \theta \right), \qquad \cos \left(\frac{5\pi}{6} - \theta \right).$$

18. In terms of $\sin \theta$ and $\cos \theta$, find

$$\cos \left(\frac{\pi}{3} + \frac{\pi}{4} - \theta \right), \qquad \sin \left(\frac{\pi}{12} - \theta \right), \qquad \sin \left(\frac{\pi}{4} - \frac{\pi}{24} - \theta \right).$$

19. Sketch, in Cartesian coordinates, the graph of $y = \cot x$.

20. Sketch, in Cartesian coordinates, the graph of $y = \sec x$.

21. Sketch, in Cartesian coordinates, the graph of $y = \csc x$.

Determinants

In this section we present an introduction to determinants of the second and third order. A determinant of the second order is denoted by

$$\begin{vmatrix} a & b \\ c & d \end{vmatrix}$$

in which a, b, c, and d are numbers. A determinant of the third order is denoted by

$$\begin{vmatrix} a & b & c \\ d & e & f \\ g & h & j \end{vmatrix}$$

in which a, b, ... , j are numbers.

Determinants are actually functions of the elements which appear. For a second order determinant, we define

$$\begin{vmatrix} a & b \\ c & d \end{vmatrix} = ad - bc.$$

For example,

$$\begin{vmatrix} 2 & 1 \\ -5 & 6 \end{vmatrix} = 12 - (-5) = 17.$$

A third order determinant is defined in terms of second order determinants by the formula:

$$\begin{vmatrix} a & b & c \\ d & e & f \\ g & h & j \end{vmatrix} = c \begin{vmatrix} d & e \\ g & h \end{vmatrix} - f \begin{vmatrix} a & b \\ g & h \end{vmatrix} + j \begin{vmatrix} a & b \\ d & e \end{vmatrix}.$$

This formula is known as an **expansion in terms of the last column.** For example,

$$\begin{vmatrix} 2 & 1 & -1 \\ 0 & 3 & 2 \\ 1 & 4 & 3 \end{vmatrix} = -1 \begin{vmatrix} 0 & 3 \\ 1 & 4 \end{vmatrix} - 2 \begin{vmatrix} 2 & 1 \\ 1 & 4 \end{vmatrix} + 3 \begin{vmatrix} 2 & 1 \\ 0 & 3 \end{vmatrix}$$

$$= -(-3) - 2(8 - 1) + 3(6) = 7.$$

Theorem 1 (Cramer's rule for $n = 2$). *Given the two equations*

$$ax \mid by - e,$$
$$cx + dy = f, \tag{1}$$

if the determinant of the coefficients

$$\begin{vmatrix} a & b \\ c & d \end{vmatrix}$$

is not zero, then the two simultaneous equations (1) *have a unique solution given by*

$$x = \frac{\begin{vmatrix} e & b \\ f & d \end{vmatrix}}{\begin{vmatrix} a & b \\ c & d \end{vmatrix}}, \qquad y = \frac{\begin{vmatrix} a & e \\ c & f \end{vmatrix}}{\begin{vmatrix} a & b \\ c & d \end{vmatrix}}.$$

The proof is left to the reader. (See problem 1 at the end of the appendix.)

Theorem 2. *Two sets of numbers* $\{a_1, b_1, c_1\}$ *and* $\{a_2, b_2, c_2\}$, *neither all zero, are proportional if and only if all three determinants*

$$\begin{vmatrix} a_1 & b_1 \\ a_2 & b_2 \end{vmatrix}, \qquad \begin{vmatrix} a_1 & c_1 \\ a_2 & c_2 \end{vmatrix}, \qquad \begin{vmatrix} b_1 & c_1 \\ b_2 & c_2 \end{vmatrix}$$

are zero.

Proof. If the sets are proportional, then there is a number k such that

$$\bullet \qquad a_2 = ka_1, \qquad b_2 = kb_1, \qquad c_2 = kc_1. \tag{2}$$

It is immediate that the three determinants vanish. Now suppose all determinants vanish. We wish to prove the sets are proportional. Suppose $c_1 \neq 0$, for instance. Then we have

$$\begin{vmatrix} a_1 & c_1 \\ a_2 & c_2 \end{vmatrix} = a_1 c_2 - a_2 c_1 = 0 \Leftrightarrow a_2 = \left(\frac{c_2}{c_1}\right) a_1,$$

$$\begin{vmatrix} b_1 & c_1 \\ b_2 & c_2 \end{vmatrix} = b_1 c_2 - b_2 c_1 = 0 \Leftrightarrow b_2 = \left(\frac{c_2}{c_1}\right) b_1.$$

We define $k = c_2/c_1$ and we see that (2) holds. If $c_1 = 0$, then either $a_1 \neq 0$ or $b_1 \neq 0$ and a similar argument leads to the same conclusion.

It is clear that the value of a determinant depends on the arrangement of the elements which enter it. It is simpler to state theorems about determinants if we introduce a systematic notation for the entries. The standard method consists of a double subscript for each element, the first subscript identifying the row in which the element is situated, and the second subscript identifying the column. That is, we denote the element in the ith row and jth column by the symbol a_{ij}. The evaluation of a third order determinant may now be written

$$\begin{vmatrix} a_{11} & a_{12} & a_{13} \\ a_{21} & a_{22} & a_{23} \\ a_{31} & a_{32} & a_{33} \end{vmatrix} = a_{13} \begin{vmatrix} a_{21} & a_{22} \\ a_{31} & a_{32} \end{vmatrix} - a_{23} \begin{vmatrix} a_{11} & a_{12} \\ a_{31} & a_{32} \end{vmatrix} + a_{33} \begin{vmatrix} a_{11} & a_{12} \\ a_{21} & a_{22} \end{vmatrix}.$$

Using the definition of second order determinants, we obtain for the value of a third order determinant:

$$\begin{vmatrix} a_{11} & a_{12} & a_{13} \\ a_{21} & a_{22} & a_{23} \\ a_{31} & a_{32} & a_{33} \end{vmatrix} = \begin{matrix} a_{11}a_{22}a_{33} + a_{12}a_{31}a_{23} + a_{13}a_{21}a_{32} \\ - a_{11}a_{32}a_{23} - a_{12}a_{21}a_{33} - a_{13}a_{22}a_{31}. \end{matrix} \tag{3}$$

It is a simple matter to verify that the interchange of each a_{ij} with a_{ji} in the right-hand side of (3) does not alter it in any way. Therefore we conclude the following result for third order determinants.

Theorem 3. *If a new determinant is obtained from a given one by an interchange of its rows and columns, the value of the new determinant is equal to the value of the given one.*

DEFINITIONS. Given a determinant

$$D \equiv \begin{vmatrix} a_{11} & a_{12} & a_{13} \\ a_{21} & a_{22} & a_{23} \\ a_{31} & a_{32} & a_{33} \end{vmatrix}, \tag{4}$$

we define the quantity D_{ij} *as the second order determinant obtained from* D *by crossing out the* ith *row and* jth *column in* D. For example,

$$D_{21} = \begin{vmatrix} a_{12} & a_{13} \\ a_{32} & a_{33} \end{vmatrix}.$$

The **cofactor** of the element a_{ij} in D is defined to be the number A_{ij} given by

$$A_{ij} = (-1)^{i+j} D_{ij}.$$

For example, the cofactor of a_{13} is

$$A_{13} = (-1)^{1+3} D_{13} = (+1) \begin{vmatrix} a_{21} & a_{22} \\ a_{31} & a_{32} \end{vmatrix} = a_{21}a_{32} - a_{31}a_{22}.$$

Theorem 4. *If D denotes the determinant in* (4), *we have*

$$D = a_{11}A_{11} + a_{12}A_{12} + a_{13}A_{13} = a_{11}A_{11} + a_{21}A_{21} + a_{31}A_{31}, \tag{5a}$$

$$D = a_{21}A_{21} + a_{22}A_{22} + a_{23}A_{23} = a_{12}A_{12} + a_{22}A_{22} + a_{32}A_{32}, \tag{5b}$$

$$D = a_{31}A_{31} + a_{32}A_{32} + a_{33}A_{33} = a_{13}A_{13} + a_{23}A_{23} + a_{33}A_{33}. \tag{5c}$$

Proof. We verify the first formula on the right:

$$a_{11}A_{11} + a_{21}A_{21} + a_{31}A_{31}$$

$$= a_{11} \begin{vmatrix} a_{22} & a_{23} \\ a_{32} & a_{33} \end{vmatrix} - a_{21} \begin{vmatrix} a_{12} & a_{13} \\ a_{32} & a_{33} \end{vmatrix} + a_{31} \begin{vmatrix} a_{12} & a_{13} \\ a_{22} & a_{23} \end{vmatrix}$$

$$= a_{11}(a_{22}a_{33} - a_{32}a_{23}) - a_{21}(a_{12}a_{33} - a_{32}a_{13}) + a_{31}(a_{12}a_{23} - a_{22}a_{13})$$

$$= D.$$

Theorem 3 and a computation may be used to verify the remaining formulas. The details are left to the student. (See problem 2 at the end of this appendix.)

DEFINITIONS. The first sums in (5a), (5b), and (5c) are called the **expansions of D according to the first, second, and third rows,** respectively. The second sums in (5a), (5b), and (5c) are called the **expansions of D according to the first, second, and third columns,** respectively.

EXAMPLE 1. Evaluate the following determinant by expanding it according to (i) the second column, and (ii) the third row:

$$\begin{vmatrix} 1 & -2 & 3 \\ 2 & 1 & -1 \\ -2 & -1 & 2 \end{vmatrix}.$$

Solution. (i) Expanding according to the second column, we obtain

$$\begin{vmatrix} 1 & -2 & 3 \\ 2 & 1 & -1 \\ -2 & -1 & 2 \end{vmatrix} = -(-2) \begin{vmatrix} 2 & -1 \\ -2 & 2 \end{vmatrix} + 1 \cdot \begin{vmatrix} 1 & 3 \\ -2 & 2 \end{vmatrix} - (-1) \begin{vmatrix} 1 & 3 \\ 2 & -1 \end{vmatrix}$$

$$= 2(4 - 2) + (2 + 6) + (-1 - 6) = 5.$$

(ii) Expanding according to the third row, we obtain

$$\begin{vmatrix} 1 & -2 & 3 \\ 2 & 1 & -1 \\ -2 & -1 & 2 \end{vmatrix} = (-2) \begin{vmatrix} -2 & 3 \\ 1 & -1 \end{vmatrix} - (-1) \begin{vmatrix} 1 & 3 \\ 2 & -1 \end{vmatrix} + 2 \begin{vmatrix} 1 & -2 \\ 2 & 1 \end{vmatrix}$$

$$= (-2)(-1) + 1 \cdot (-7) + 2 \cdot 5 = 5.$$

The definitions and theorems given for determinants of the second and third order may be extended to determinants of order n, where n is any positive integer. A determinant of order n has n rows and n columns. The evaluation of such determinants, a process beyond the scope of this appendix, is most easily accomplished by an induction technique. We now state a number of general properties of determinants, valid for any order, which are most useful in such evaluations. The proofs depend on the expansion given in Theorem 4 which we have proved only for $n = 3$. Actually the appropriate extension of Theorem 4 is valid for determinants of any order. The proof is usually given in courses devoted to matrix theory.

Theorem 5. *Let D be a determinant and suppose that each element of the kth row a_{kj} is the sum of two numbers: $a_{kj} = a'_{kj} + a''_{kj}$. Then $D = D' + D''$ where D' is the determinant obtained from D by inserting a'_{kj} instead of a_{kj} for each element of the kth row. Similarly, D'' is obtained by inserting a''_{kj} for a_{kj}. The result also holds if the elements of a column are treated analogously.*

Proof. We expand D according to the elements of the kth row:

$$D = a_{k1}A_{k1} + a_{k2}A_{k2} + a_{k3}A_{k3}$$

$$= (a'_{k1} + a''_{k1})A_{k1} + (a'_{k2} + a''_{k2})A_{k2} + (a'_{k3} + a''_{k3})A_{k3}$$

$$= (a'_{k1}A_{k1} + a'_{k2}A_{k2} + a'_{k3}A_{k3}) + (a''_{k1}A_{k1} + a''_{k2}A_{k2} + a''_{k3}A_{k3})$$

$$= D' + D''.$$

Theorem 4 shows that the same analysis works for columns.

Theorem 5 shows that, for example,

$$\begin{vmatrix} a_{11} & a_{12} & a_{13} \\ a_{21} & a_{22} & a_{23} \\ a'_{31} + a''_{31} & a'_{32} + a''_{32} & a'_{33} + a''_{33} \end{vmatrix} = \begin{vmatrix} a_{11} & a_{12} & a_{13} \\ a_{21} & a_{22} & a_{23} \\ a'_{31} & a'_{32} & a'_{33} \end{vmatrix} + \begin{vmatrix} a_{11} & a_{12} & a_{13} \\ a_{21} & a_{22} & a_{23} \\ a''_{31} & a''_{32} & a''_{33} \end{vmatrix}.$$

Theorem 6. *If each element $a_{kj} = ca'_{kj}$ for k fixed and $j = 1, 2,$ and 3, then $D = cD'$ where D' is obtained from D as in Theorem 5.*

The proof is similar to that of Theorem 5. In words, Theorem 6 states that if all the elements of a row are multiplied by a constant, then the determinant is multiplied by that constant.

Theorem 7. *If D' is obtained from D by interchanging any two rows, then $D' = -D$. The same result holds for the interchange of two columns.*

Proof. The result is a direct consequence of formula (3). The details are left to the student.

Theorem 8. *If any two rows (or columns) of a determinant D are proportional, then D = 0.*

Proof. If two rows are identical, then interchanging them has no effect. On the other hand, Theorem 7 shows that $D' = -D$. Thus $D' = -D = D$ and $D = 0$. If two rows are proportional, one is c times the other, and by Theorem 6, $D = cD''$ where D'' has two identical rows. Hence $D = 0$. The proof for columns is similar.

The next lemma and theorem are very useful in evaluating determinants, especially those of high order.

Lemma 1. *If D' is obtained from D by multiplying the kth row by the constant c and adding the result to the ith row, where $i \neq k$, then D' = D. The same result holds for columns.*

Proof. The elements of the ith row of D', denoted by a'_{ij}, have the form

$$a'_{ij} = a_{ij} + ca_{kj}.$$

Using Theorem 5, we find

$$D' = D + cD''$$

where D'' is obtained by replacing the ith row of D by the kth row. But then D'' has two identical rows (the ith and kth), so that $D'' = 0$. Hence

$$D' = D.$$

Theorem 9. *If D' is obtained from D by multiplying the kth row by c_i and adding the result to the ith row for all $i \neq k$ in turn, then D' = D. The same is true for columns.*

Proof. Each step in the process is one for which the lemma applies leaving the value unchanged.

The next example shows how to use Theorem 9 to simplify the determinant before applying the expansion theorem.

EXAMPLE 2. Simplify by using Theorem 9 and use the expansion theorem to evaluate the determinant

$$D = \begin{vmatrix} 2 & 1 & 3 \\ -1 & 2 & 2 \\ 2 & -3 & 1 \end{vmatrix}.$$

Solution. Multiplying the first row by -2 and adding to the second row, and then multiplying it by 3 and adding to the third row, we obtain

$$D = \begin{vmatrix} 2 & 1 & 3 \\ -5 & 0 & -4 \\ 8 & 0 & 10 \end{vmatrix}.$$

We can expand the new determinant according to the second column, thus getting

$$D = -1 \cdot \begin{vmatrix} -5 & -4 \\ 8 & 10 \end{vmatrix} = (-1) \cdot (-18) = 18.$$

The following theorem includes and supplements the expansion theorem (Theorem 4).

Theorem 10. *If D is any third order determinant, then*

(a) $a_{i1}A_{k1} + a_{i2}A_{k2} + a_{i3}A_{k3} = \begin{cases} D \text{ if } k = i, \\ 0 \text{ if } k \neq i, \end{cases}$

(b) $a_{1j}A_{1k} + a_{2j}A_{2k} + a_{3j}A_{3k} = \begin{cases} D \text{ if } k = j, \\ 0 \text{ if } k \neq j. \end{cases}$

Proof. The cases $k = i$ in (a) and $k = j$ in (b) restate Theorem 4. If $k \neq i$ in (a), then we see from the expansion theorem that the left-hand side in (a) is the expansion of a determinant D' obtained from D by replacing the kth row by the ith row. But then D' has two identical rows and so is zero. The proof of (b) is the same.

Theorem 11 (Cramer's rule for $n = 3$). *If the determinant D*

$$D = \begin{vmatrix} a_{11} & a_{12} & a_{13} \\ a_{21} & a_{22} & a_{23} \\ a_{31} & a_{32} & a_{33} \end{vmatrix}$$

of the coefficients of the three equations

$$\begin{aligned} a_{11}x + a_{12}y + a_{13}z &= b_1, \\ a_{21}x + a_{22}y + a_{23}z &= b_2, \\ a_{31}x + a_{32}y + a_{33}z &= b_3, \end{aligned} \tag{6}$$

is not zero, then the equations have one and only one solution given by

$$x = \frac{D_1}{D}, \qquad y = \frac{D_2}{D}, \qquad z = \frac{D_3}{D}, \tag{7}$$

where

$$D_1 = \begin{vmatrix} b_1 & a_{12} & a_{13} \\ b_2 & a_{22} & a_{23} \\ b_3 & a_{32} & a_{33} \end{vmatrix}, \quad D_2 = \begin{vmatrix} a_{11} & b_1 & a_{13} \\ a_{21} & b_2 & a_{23} \\ a_{31} & b_3 & a_{33} \end{vmatrix}, \quad D_3 = \begin{vmatrix} a_{11} & a_{12} & b_1 \\ a_{21} & a_{22} & b_2 \\ a_{31} & a_{32} & b_3 \end{vmatrix}.$$

Proof. (a) Suppose the equations (6) hold. If we multiply the first, second, and third equations by A_{11}, A_{21}, and A_{31}, respectively, and then add, we get

$$Dx + (a_{12}A_{11} + a_{22}A_{21} + a_{32}A_{31})y + (a_{13}A_{11} + a_{23}A_{21} + a_{33}A_{31})z$$
$$= (b_1A_{11} + b_2A_{21} + b_3A_{31}) \qquad (8)$$

By Theorem 10, the coefficients of y and z are zero. The right-hand side of the above equation is the expansion of D_1 according to its first column. Hence $x = D_1/D$. Multiplying the equations (6) by A_{12}, A_{22}, A_{32} and proceeding similarly, we get the result $y = D_2/D$. The value for z is obtained in the same way.

(b) Now assume that equations (7) hold and that $D \neq 0$. Then (8) holds; similarly for equations for y and z. Therefore, by addition,

$$D(a_{11}x + a_{12}y + a_{13}z) = b_1(a_{11}A_{11} + a_{12}A_{12} + a_{13}A_{13})$$
$$+ b_2(a_{11}A_{21} + a_{12}A_{22} + a_{13}A_{23})$$
$$+ b_3(a_{11}A_{31} + a_{12}A_{32} + a_{13}A_{33})$$
$$= Db_1.$$

The last equality is valid because of Theorem 10. In a similar way, we see that the remaining equations of (6) are satisfied.

EXAMPLE 3. Use Cramer's rule to solve the following equations:

$$3x - 2y + 4z = 5,$$
$$x + y + 3z = 2,$$
$$-x + 2y - z = 1.$$

Solution. To simplify and evaluate D, we multiply the second row by 2 and add it to the first row and then multiply the second row by -2 and add it to the third row. We obtain

$$D = \begin{vmatrix} 3 & -2 & 4 \\ 1 & 1 & 3 \\ -1 & 2 & -1 \end{vmatrix} = \begin{vmatrix} 5 & 0 & 10 \\ 1 & 1 & 3 \\ -3 & 0 & -7 \end{vmatrix}.$$

Now expanding according to the second column is easy. We get

$$D = 1 \cdot \begin{vmatrix} 5 & 10 \\ -3 & -7 \end{vmatrix} = -5.$$

Proceeding similarly for D_1, D_2, and D_3, we find

$$D_1 = \begin{vmatrix} 5 & -2 & 4 \\ 2 & 1 & 3 \\ 1 & 2 & -1 \end{vmatrix} = \begin{vmatrix} 9 & 0 & 10 \\ 2 & 1 & 3 \\ -3 & 0 & -7 \end{vmatrix} = 1 \cdot \begin{vmatrix} 9 & 10 \\ -3 & -7 \end{vmatrix} = -33,$$

$$D_2 = \begin{vmatrix} 3 & 5 & 4 \\ 1 & 2 & 3 \\ -1 & 1 & -1 \end{vmatrix} = \begin{vmatrix} 8 & 5 & 9 \\ 3 & 2 & 5 \\ 0 & 1 & 0 \end{vmatrix} = -1 \cdot \begin{vmatrix} 8 & 9 \\ 3 & 5 \end{vmatrix} = -13,$$

$$D_3 = \begin{vmatrix} 3 & -2 & 5 \\ 1 & 1 & 2 \\ -1 & 2 & 1 \end{vmatrix} = \begin{vmatrix} 5 & -2 & 9 \\ 0 & 1 & 0 \\ -3 & 2 & -3 \end{vmatrix} = 1 \cdot \begin{vmatrix} 5 & 9 \\ -3 & -3 \end{vmatrix} = 12.$$

Therefore $x = \frac{33}{5}$, $y = \frac{13}{5}$, $z = -\frac{12}{5}$. The student should verify that these numbers satisfy the three equations.

Corollary. *If there are numbers x, y, and z, not all zero, which satisfy Eqs.* (6) *with* $b_1 = b_2 = b_3 = 0$, *then* $D = 0$.

PROBLEMS

1. Prove Theorem 1.

2. Complete the proof of Theorem 4.

In problems 3 through 5, evaluate the given determinant by expanding it according to (a) the second row, (b) the third column.

3. $\begin{vmatrix} 3 & 2 & -1 \\ -1 & 0 & 1 \\ 2 & 1 & -2 \end{vmatrix}$ 4. $\begin{vmatrix} 1 & 0 & 3 \\ 2 & -1 & -2 \\ 1 & 3 & 2 \end{vmatrix}$ 5. $\begin{vmatrix} 2 & 3 & 1 \\ 1 & 2 & -2 \\ -2 & 1 & 3 \end{vmatrix}$

In problems 6 through 8, simplify the determinant, using Theorem 9, and then evaluate it by the expansion theorem.

6. $\begin{vmatrix} 2 & -1 & 3 \\ 3 & -1 & 2 \\ -1 & 2 & 3 \end{vmatrix}$ 7. $\begin{vmatrix} 1 & -1 & -2 \\ 2 & 3 & -2 \\ -1 & -3 & 2 \end{vmatrix}$ 8. $\begin{vmatrix} 2 & 1 & -2 \\ -1 & 3 & 2 \\ -3 & 1 & -2 \end{vmatrix}$

In problems 9 through 11, solve by Cramer's Rule and check

9. $2x - y + 3z = 1$ 10. $2x - y + z = -3$ 11. $x + 3y - 2z = 4$
     $3x + y - z = 2$ $x + 3y - 2z = 0$ $-2x + y + 3z = 2$
     $x + 2y + 3z = -6$ $x - y + z = -2$ $2x + 4y - z = -1$

Proofs of Theorems 6, 12, and 13 of Chapter 11

We saw in Chapter 11, Section 2, that the vectors $\mathbf{i}$, $\mathbf{j}$, and $\mathbf{k}$ corresponding to any given coordinate system in space are linearly independent. Consequently Theorem 6 is a special case of the following Theorem 6':

Theorem 6'. *Suppose that* $\mathbf{u}_1$, $\mathbf{v}_1$, *and* $\mathbf{w}_1$ *are linearly independent and suppose that*

$$
\begin{aligned}
\mathbf{u}_2 &= a_{11}\mathbf{u}_1 + a_{12}\mathbf{v}_1 + a_{13}\mathbf{w}_1, \\
\mathbf{v}_2 &= a_{21}\mathbf{u}_1 + a_{22}\mathbf{v}_1 + a_{23}\mathbf{w}_1, \\
\mathbf{w}_2 &= a_{31}\mathbf{u}_1 + a_{32}\mathbf{v}_1 + a_{33}\mathbf{w}_1,
\end{aligned}
\qquad
D = \begin{vmatrix} a_{11} & a_{12} & a_{13} \\ a_{21} & a_{22} & a_{23} \\ a_{31} & a_{32} & a_{33} \end{vmatrix}. \tag{1}
$$

Then the set $\{\mathbf{u}_2, \mathbf{v}_2, \mathbf{w}_2\}$ *is linearly dependent* $\Leftrightarrow D = 0$.

Proof. (a) Suppose the set is linearly dependent. Then there are constants c_1, c_2, and c_3, not all zero, such that

$$
c_1\mathbf{u}_2 + c_2\mathbf{v}_2 + c_3\mathbf{w}_2 = \mathbf{0}. \tag{2}
$$

If we substitute (1) into (2), we obtain

$$
(c_1 a_{11} + c_2 a_{21} + c_3 a_{31})\mathbf{u}_1 + (c_1 a_{12} + c_2 a_{22} + c_3 a_{32})\mathbf{v}_1 \\
+ (c_1 a_{13} + c_2 a_{23} + c_3 a_{33})\mathbf{w}_1 = \mathbf{0}. \tag{3}
$$

Since $\mathbf{u}_1$, $\mathbf{v}_1$, and $\mathbf{w}_1$ are linearly independent, their coefficients must all vanish. That is, we must have

$$
\begin{aligned}
a_{11}c_1 + a_{21}c_2 + a_{31}c_3 &= 0, \\
a_{12}c_1 + a_{22}c_2 + a_{32}c_3 &= 0, \\
a_{13}c_1 + a_{23}c_2 + a_{33}c_3 &= 0.
\end{aligned} \tag{4}
$$

But if (4) holds with c_1, c_2, and c_3 not all zero, the determinant D' of the coefficients must vanish according to the Corollary (p. 335) to Cramer's Rule (Theorem 11, Appendix 4). But D' is obtained from D by interchanging rows and columns. Accordingly, $D = D' = 0$.

(b) Now suppose $D = 0$. If all the cofactors A_{ij} are zero, any two rows of D and hence any two of the vectors $\mathbf{u}_2$, $\mathbf{v}_2$, and $\mathbf{w}_2$ are proportional and the set is linearly dependent. Otherwise, some $A_{pq} \neq 0$. By interchanging the order of the vectors, if necessary, we may assume that $p = 3$. From the expansion theorem (Theorem 10, Appendix 4), we conclude that

$$a_{11}A_{1q} + a_{21}A_{2q} + a_{31}A_{3q} = 0,$$
$$a_{12}A_{1q} + a_{22}A_{2q} + a_{32}A_{3q} = 0, \tag{5}$$
$$a_{13}A_{1q} + a_{23}A_{2q} + a_{33}A_{3q} = 0.$$

Since $A_{3q} \neq 0$, we can solve equations (5) for the a_{3j}, obtaining

$$a_{31} = ka_{11} + la_{21}, \qquad a_{32} = ka_{12} + la_{22}, \qquad a_{33} = ka_{13} + la_{23}, \\ k = -A_{1q}/A_{3q}, \qquad l = -A_{2q}/A_{3q}. \tag{6}$$

In this case it follows from (6) and (1) that

$$\mathbf{w}_2 = k\mathbf{u}_2 + l\mathbf{v}_2,$$

and the vectors $\mathbf{u}_2$, $\mathbf{v}_2$, and $\mathbf{w}_2$ are linearly dependent.

Theorem 12. *Suppose that $\mathbf{u}$ and $\mathbf{v}$ are any vectors, that $\{\mathbf{i}, \mathbf{j}, \mathbf{k}\}$ is a right-handed coordinate triple, and that t is any number. Then*

(i) $\mathbf{v} \times \mathbf{u} = -\mathbf{u} \times \mathbf{v}$,

(ii) $(t\mathbf{u}) \times \mathbf{v} = t(\mathbf{u} \times \mathbf{v}) = \mathbf{u} \times (t\mathbf{v})$,

(iii) $\mathbf{i} \times \mathbf{j} = -\mathbf{j} \times \mathbf{i} = \mathbf{k}$,
   $\mathbf{j} \times \mathbf{k} = -\mathbf{k} \times \mathbf{j} = \mathbf{i}$,
   $\mathbf{k} \times \mathbf{i} = -\mathbf{i} \times \mathbf{k} = \mathbf{j}$,

(iv) $\mathbf{i} \times \mathbf{i} = \mathbf{j} \times \mathbf{j} = \mathbf{k} \times \mathbf{k} = \mathbf{0}$.

Proofs. (i) By definition, $|\mathbf{v} \times \mathbf{u}| = |\mathbf{u} \times \mathbf{v}|$ and $\mathbf{v} \times \mathbf{u}$ and $\mathbf{u} \times \mathbf{v}$ are both orthogonal to both $\mathbf{u}$ and $\mathbf{v}$ (or are both zero if $\mathbf{u}$ and $\mathbf{v}$ are proportional). Thus $\mathbf{v} \times \mathbf{u} = \pm\mathbf{u} \times \mathbf{v}$. If we let $\mathbf{w} = \mathbf{u} \times \mathbf{v}$, then $\{\mathbf{u}, \mathbf{v}, \mathbf{w}\}$ and $\{\mathbf{v}, \mathbf{u}, -\mathbf{w}\}$ are right-handed (see Theorem 11, Chapter 11), so $\mathbf{v} \times \mathbf{u}$ must equal $-\mathbf{w}$.

(ii) If $t = 0$ or $\mathbf{u}$ and $\mathbf{v}$ are proportional, (ii) certainly holds. Otherwise, let us set $\mathbf{w} = \mathbf{u} \times \mathbf{v}$. Then (ii) follows since all the terms in (ii) have the same magnitude, all are orthogonal to both $\mathbf{u}$ and $\mathbf{v}$, and $\{t\mathbf{u}, \mathbf{v}, t\mathbf{w}\}$ and $\{\mathbf{u}, t\mathbf{v}, t\mathbf{w}\}$ are right-handed by Theorem 11, Chapter 11.

(iv) This follows, since we must have $\mathbf{i} \times \mathbf{i} = -\mathbf{i} \times \mathbf{i} = \mathbf{0}$, etc.

(iii) To prove (iii), we note that, since $\{\mathbf{i}, \mathbf{j}, \mathbf{k}\}$ is right-handed, $\theta = \pi/2$, $|\mathbf{i}| = |\mathbf{j}| = |\mathbf{k}| = 1$, and $\mathbf{k}$ is orthogonal to both $\mathbf{i}$ and $\mathbf{j}$, it follows that $\mathbf{i} \times \mathbf{j} = \mathbf{k}$.

That $\mathbf{j} \times \mathbf{i} = -\mathbf{k}$ follows from this and from (i). Since $\{\mathbf{i}, \mathbf{j}, \mathbf{k}\}$ is a coordinate triple, it follows as above that $\mathbf{j} \times \mathbf{k} = \pm\mathbf{i}$. Setting

$$\mathbf{u}_1 = \mathbf{i}, \quad \mathbf{v}_1 = \mathbf{j}, \quad \mathbf{w}_1 = \mathbf{k}, \quad \mathbf{u}_2 = \mathbf{j}, \quad \mathbf{v}_2 = \mathbf{k}, \quad \mathbf{w}_2 = \mathbf{i},$$

we see that

$$\mathbf{u}_2 = 0 \cdot \mathbf{u}_1 + 1 \cdot \mathbf{v}_1 + 0 \cdot \mathbf{w}_1,$$
$$\mathbf{v}_2 = 0 \cdot \mathbf{u}_1 + 0 \cdot \mathbf{v}_1 + 1 \cdot \mathbf{w}_1,$$
$$\mathbf{w}_2 = 1 \cdot \mathbf{u}_1 + 0 \cdot \mathbf{v}_1 + 0 \cdot \mathbf{w}_1.$$

Since $\{\mathbf{i}, \mathbf{j}, \mathbf{k}\}$ was given as right-handed, it follows from the discussion in Chapter 11, Section 4, that $\{\mathbf{u}_2, \mathbf{v}_2, \mathbf{w}_2\}$ is right-handed since

$$D = \begin{vmatrix} 0 & 1 & 0 \\ 0 & 0 & 1 \\ 1 & 0 & 0 \end{vmatrix} = +1.$$

The proof that $\mathbf{k} \times \mathbf{i} = \mathbf{j}$ is similar.

Theorem 13. (Distributive law). *If* $\mathbf{u}$, $\mathbf{v}$, *and* $\mathbf{w}$ *are any vectors,*

(i) $\mathbf{u} \times (\mathbf{v} + \mathbf{w}) = (\mathbf{u} \times \mathbf{v}) + (\mathbf{u} \times \mathbf{w})$ *and*

(ii) $(\mathbf{v} + \mathbf{w}) \times \mathbf{u} = (\mathbf{v} \times \mathbf{u}) + (\mathbf{w} \times \mathbf{u}).$

Proof. Part (ii) follows from part (i) and part (i) of Theorem 12, for

$$(\mathbf{v} + \mathbf{w}) \times \mathbf{u} = -[\mathbf{u} \times (\mathbf{v} + \mathbf{w})] = -[(\mathbf{u} \times \mathbf{v}) + (\mathbf{u} \times \mathbf{w})]$$
$$= [-(\mathbf{u} \times \mathbf{v})] + [-(\mathbf{u} \times \mathbf{w})] = (\mathbf{v} \times \mathbf{u}) + (\mathbf{w} \times \mathbf{u}).$$

It is clear that (i) holds if $\mathbf{u} = \mathbf{0}$. Otherwise, let $\{\mathbf{i}', \mathbf{j}', \mathbf{k}'\}$ be a right-handed coordinate triple such that $\mathbf{u} = |\mathbf{u}|\mathbf{i}'$ (i.e., $\mathbf{i}'$ is the unit vector in the direction of $\mathbf{u}$). Suppose that

$$\mathbf{v} = a_1\mathbf{i}' + b_1\mathbf{j}' + c_1\mathbf{k}', \quad \mathbf{u} \times \mathbf{v} = \mathbf{V} = A_1\mathbf{i}' + B_1\mathbf{j}' + C_1\mathbf{k}'.$$

We first find A_1, B_1, C_1 in terms of a_1, b_1, and c_1.
Since $\mathbf{V}$ is orthogonal to both $\mathbf{u}$ and $\mathbf{v}$, we must have

$$\mathbf{V} \cdot \mathbf{u} = A_1|\mathbf{u}| = 0, \quad \mathbf{V} \cdot \mathbf{v} = A_1a_1 + B_1b_1 + C_1c_1 = 0.$$

Thus

$$A_1 = 0 \quad \text{and} \quad b_1B_1 + c_1C_1 = 0 \quad \text{so that} \quad B_1 = -kc_1, \quad C_1 = kb_1 \qquad (7)$$

for some k. Moreover,

$$|\mathbf{V}| = |\mathbf{u}| \cdot \sqrt{a_1^2 + b_1^2 + c_1^2} \sin \theta$$

and

$$\mathbf{u} \cdot \mathbf{v} = |\mathbf{u}| \cdot \sqrt{a_1^2 + b_1^2 + c_1^2} \cos \theta = |\mathbf{u}|a_1.$$

Since $0 \le \theta \le \pi$ and $\cos \theta = a_1/|\mathbf{v}|$, it follows that

$$\sin \theta = \sqrt{b_1^2 + c_1^2}/|\mathbf{v}|.$$

Thus

$$|\mathbf{V}| = |k| \cdot \sqrt{b_1^2 + c_1^2} = |\mathbf{u}| \cdot \sqrt{b_1^2 + c_1^2} \qquad \text{so} \quad k = \pm|\mathbf{u}|.$$

Finally $\{\mathbf{u}, \mathbf{v}, \mathbf{V}\}$ must be right-handed, so that

$$\begin{vmatrix} |\mathbf{u}| & 0 & 0 \\ a_1 & b_1 & c_1 \\ 0 & -kc_1 & kb_1 \end{vmatrix} = k|\mathbf{u}| \cdot (b_1^2 + c_1^2) > 0$$

(unless $\mathbf{v} = \mathbf{0}$ or $\mathbf{v}$ is proportional to $\mathbf{u}$). Hence $k = +1$ and

$$\mathbf{V} = |\mathbf{u}| \cdot (-c_1\mathbf{j}' + b_1\mathbf{k}'). \tag{8}$$

The result in (8) evidently holds also if $\mathbf{v} = \mathbf{0}$ or is proportional to $\mathbf{u}$.
  In like manner, if we let

$$\mathbf{w} = a_2\mathbf{i}' + b_2\mathbf{j}' + c_2\mathbf{k}', \qquad \mathbf{W} = \mathbf{u} \times \mathbf{w}, \qquad \mathbf{X} = \mathbf{u} \times (\mathbf{v} + \mathbf{w}),$$

we see that

$$\mathbf{W} = |\mathbf{u}|(-c_2\mathbf{j}' + b_2\mathbf{k}'), \qquad \mathbf{X} = |\mathbf{u}| \cdot [-(c_1 + c_2)\mathbf{j}' + (b_1 + b_2)\mathbf{k}']$$

from which (i) follows.

Proofs of Theorems 8, 13 (ii), and Lemma 2 of Chapter 12

Proof of Theorem 8. We wish to show that Equations (3) and (4) on pages 286 and 287 imply that $e = f = g = h = 0$. We combine (3) and (4) to get:

$$(a_1r + a_2s + e)^2 + (b_1r + b_2s + f)^2$$
$$+ (c_1r + c_2s + g)^2$$
$$+ (d_1r + d_2s + h)^2 = r^2 + s^2, \tag{1}$$

$$(a_1r + a_2s + e - a_1)^2 + (b_1r + b_2s + f - b_1)^2$$
$$+ (c_1r + c_2s + g - c_1)^2$$
$$+ (d_1r + d_2s + h - d_1)^2 = r^2 + s^2 - 2r + 1, \tag{2}$$

$$(a_1r + a_2s + e - a_2)^2 + (b_1r + b_2s + f - b_2)^2$$
$$+ (c_1r + c_2s + g - c_2)^2$$
$$+ (d_1r + d_2s + h - d_2)^2 = r^2 + s^2 - 2s + 1. \tag{3}$$

Squaring out (2) and subtracting it from (1), we get (taking into account that $a_1^2 + b_1^2 + c_1^2 + d_1^2 = 1$):

$$a_1(a_1r + a_2s + e) + b_1(b_1r + b_2s + f)$$
$$+ c_1(c_1r + c_2s + g)$$
$$+ d_1(d_1r + d_2s + h) = r. \tag{4}$$

Similarly, squaring out (3) and subtracting from (1) yields

$$a_2(a_1r + a_2s + e) + b_2(b_1r + b_2s + f)$$
$$+ c_2(c_1r + c_2s + g)$$
$$+ d_2(d_1r + d_2s + h) = s. \tag{5}$$

From (4) and (5) we obtain (taking (1) in the statement of Theorem 8 into account)

$$a_1e + b_1f + c_1g + d_1h = 0$$
$$a_2e + b_2f + c_2g + d_2h = 0. \tag{6}$$

Now we square out the left side of (1) writing $a_1r + a_2s + e = [(a_1r + a_2s) + e]$, etc. When we employ (1) of Theorem 8 and (6), the result is

$$e^2 + f^2 + g^2 + h^2 = 0$$

which implies that $e = f = g = h = 0$.

Proof of Theorem 13(ii). By hypothesis the quadrulpes A_k, B_k, C_k, D_k, $k = 1, 2$ are not proportional. We may assume the determinant $A_1B_2 - A_2B_1$ is not zero. If one of the other determinants is not zero, the proof is similar. We set $D = A_1B_2 - A_2B_1$ and write

$$A_1x + B_1y = -C_1z - D_1v - E_1$$
$$A_2x + B_2y = -C_2z - D_2v - E_2.$$

These equations may be solved for x and y in terms of the remaining quantities. We get

$$x = \alpha_1z + \alpha_2v + \alpha_3$$
$$y = \beta_1z + \beta_2v + \beta_3 \tag{7}$$

where

$$\alpha_1 = \frac{B_1C_2 - B_2C_1}{D}, \qquad \alpha_2 = \frac{B_1D_2 - B_2D_1}{D}, \qquad \alpha_3 = \frac{B_1E_2 - B_2E_1}{D}$$

$$\beta_1 = \frac{A_2C_1 - A_1C_2}{D}, \qquad \beta_2 = \frac{A_2D_1 - A_1D_2}{D}, \qquad \beta_3 = \frac{A_2E_1 - A_1E_2}{D}.$$

We now set $x_0 = \alpha_3$, $y_0 = \beta_3$, $z_0 = 0$, $v_0 = 0$ and observe that equations (7) are in the form

$$x - x_0 = \alpha_1(z - z_0) + \alpha_2(v - v_0), \qquad y - y_0 = \beta_1(z - z_0) + \beta_2(v - v_0). \tag{8}$$

Now we wish to choose c_1, c_2, d_1, and d_2 so that if we set

$$z - z_0 = c_1r + c_2s, \qquad v - v_0 = d_1r + d_2s \tag{9}$$

and substitute these values in (8) we will get

$$x - x_0 = a_1r + a_2s, \qquad y - y_0 = b_1r + b_2s, \tag{10}$$

in which case we recognize that (9) and (10) together yield the parametric equations of a plane.

If we substitute (9) into (8) the result must be the same as (10). We get the conditions

$$a_1 = \alpha_1c_1 + \alpha_2d_1, \qquad b_1 = \beta_1c_1 + \beta_2d_1$$
$$a_2 = \alpha_1c_2 + \alpha_2d_2, \qquad b_2 = \beta_1c_2 + \beta_2d_2.$$

We set

$$e = \frac{-(\alpha_1\alpha_2 + \beta_1\beta_2)}{1 + \alpha_2^2 + \beta_2^2}$$

and we make the following choices:

$$c_1 = [(\alpha_1 + e\alpha_2)^2 + (\beta_1 + e\beta_2)^2 + e^2 + 1]^{-1/2}, \qquad c_2 = 0$$
$$d_1 = ec_1, \qquad d_2 = (1 + \alpha_2^2 + \beta_2^2)^{-1/2}.$$

Then it may be verified that

$$a_1^2 + b_1^2 + c_1^2 + d_1^2 = 1,$$
$$a_2^2 + b_2^2 + c_2^2 + d_2^2 = 1,$$
$$a_1a_2 + b_1b_2 + c_1c_2 + d_1d_2 = 0.$$

Hence equations (9) and (10) are the parametric equations of a plane. From the way they were derived, each point satisfying these equations satisfies (8), and conversely. However, (8) is equivalent to (7), which, in turn, is equivalent to the equations (10) in the statement of Theorem 13(ii).

Proof of Lemma 2.   (i) We define

$$a = \begin{vmatrix} b_1 & c_1 & d_1 \\ b_2 & c_2 & d_2 \\ b_3 & c_3 & d_3 \end{vmatrix}, \qquad b = -\begin{vmatrix} a_1 & c_1 & d_1 \\ a_2 & c_2 & d_2 \\ a_3 & c_3 & d_3 \end{vmatrix},$$

$$c = \begin{vmatrix} a_1 & b_1 & d_1 \\ a_2 & b_2 & d_2 \\ a_3 & b_3 & d_3 \end{vmatrix}, \qquad d = -\begin{vmatrix} a_1 & b_1 & c_1 \\ a_2 & b_2 & c_2 \\ a_3 & b_3 & c_3 \end{vmatrix}. \tag{11}$$

Then by expanding the following determinant with respect to its top row, and recalling that any determinant with two identical rows must be zero, we find

$$\begin{vmatrix} a_k & b_k & c_k & d_k \\ a_1 & b_1 & c_1 & d_1 \\ a_2 & b_2 & c_2 & d_2 \\ a_3 & b_3 & c_3 & d_3 \end{vmatrix} = aa_k + bb_k + cc_k + dd_k = 0, \qquad k = 1, 2, 3.$$

The algebraic identity

$$\begin{vmatrix} b_1 & c_1 & d_1 \\ b_2 & c_2 & d_2 \\ b_3 & c_3 & d_3 \end{vmatrix}^2 + \begin{vmatrix} a_1 & c_1 & d_1 \\ a_2 & c_2 & d_2 \\ a_3 & c_3 & d_3 \end{vmatrix}^2 + \begin{vmatrix} a_1 & b_1 & d_1 \\ a_2 & b_2 & d_2 \\ a_3 & b_3 & d_3 \end{vmatrix}^2$$

$$+ \begin{vmatrix} a_1 & b_1 & c_1 \\ a_2 & b_2 & c_2 \\ a_3 & b_3 & c_3 \end{vmatrix}^2 = \begin{vmatrix} g_{11} & g_{12} & g_{13} \\ g_{21} & g_{22} & g_{23} \\ g_{31} & g_{32} & g_{33} \end{vmatrix}$$

holds, where $g_{kl} = a_k a_l + b_k b_l + c_k c_l + d_k d_l$. From the hypotheses on the quadruples a_k, b_k, c_k, d_k, we have

$$\begin{vmatrix} g_{11} & g_{12} & g_{13} \\ g_{21} & g_{22} & g_{23} \\ g_{31} & g_{32} & g_{33} \end{vmatrix} = \begin{vmatrix} 1 & 0 & 0 \\ 0 & 1 & 0 \\ 0 & 0 & 1 \end{vmatrix} = 1 = a^2 + b^2 + c^2 + d^2,$$

the last equality coming from the definition of a, b, c, and d in (11).

(ii) We assume that $d \neq 0$, the proof in other cases being similar. We define a_k, b_k, c_k, d_k in turn as follows: We begin by setting $a_1 = b_1 = 0$. Then

$$aa_1 + bb_1 + cc_1 + dd_1 = 0 \quad \text{and} \quad a_1^2 + b_1^2 + c_1^2 + d_1^2 = 1$$

$$\Leftrightarrow d_1 = -\frac{c}{d}c_1 \quad \text{and} \quad c_1^2 + d_1^2 = 1.$$

We set $e = \sqrt{c^2 + d^2}$ and then $c_1 = -(d/e)h$, $d_1 = (c/e)h$. We take $h = 1$, so

$$a_1, b_1, c_1, d_1 = 0, 0, -\frac{d}{e}, \frac{c}{e}.$$

To choose a_2, b_2, c_2, d_2, we begin by setting $a_2 = 0$. Then $aa_2 + bb_2 + cc_2 + dd_2 = 0$, $a_1a_2 + b_1b_2 + c_1c_2 + d_1d_2 = 0$ and $a_2^2 + b_2^2 + c_2^2 + d_2^2 = 1$ imply that

$$c_2 = \frac{cd_2}{d}, \quad bdb_2 + e^2d_2 = 0, \quad b_2^2 + c_2^2 + d_2^2 = 1$$

$$\Leftrightarrow c_2 = -\frac{bc}{e^2}b_2, \quad d_2 = -\frac{bd}{e^2}b_2 \quad \text{and} \quad b_2^2 + c_2^2 + d_2^2 = 1$$

$$\Leftrightarrow b_2 = \frac{he}{\sqrt{b^2 + e^2}}, \quad c_2 = -\frac{hbc}{e\sqrt{b^2 + e^2}}, \quad d_2 = -\frac{hbd}{e\sqrt{b^2 + e^2}},$$

$$h = \pm 1.$$

We may set $h = 1$, so that

$$a_2, b_2, c_2, d_2 = 0, \frac{e^2}{D}, -\frac{bc}{D}, -\frac{bd}{D}, \quad \text{where } D = e\sqrt{b^2 + e^2}.$$

To choose a_3, b_3, c_3, d_3, we must have

$$c_1c_3 + d_1d_3 = b_2b_3 + c_2c_3 + d_2d_3 = aa_3 + bb_3 + cc_3 + dd_3 = 0,$$
$$a_3^2 + b_3^2 + c_3^2 + d_3^2 = 1$$

$$\Leftrightarrow -cd_3 + cd_3 = 0 = e^2b_3 - bcc_3 - bdd_3, \quad a_3^2 + b_3^2 + c_3^2 + d_3^2 = 1,$$
$$aa_3 + bb_3 + cc_3 + dd_3 = 0$$

$$\Leftrightarrow c_3 = \frac{cd_3}{d}, \quad b_3 = \frac{bcdc_3 + bd^2d_3}{de^2} = \frac{bd_3}{d}, \quad a_3^2 + b_3^2 + c_3^2 + d_3^2 = 1,$$

$$aa_3 + bb_3 + cc_3 + dd_3 = 0$$

$$\Leftrightarrow aa_3 + \frac{d_3(b^2 + c^2 + d^2)}{d} = 0, \quad b_3 = \frac{bd_3}{d}, \quad c_3 = \frac{cd_3}{d}, \quad d_3 = \frac{ad_3}{d},$$

$$a_3^2 + b_3^2 + c_3^2 + d_3^2 = 1$$

$$\Leftrightarrow d_3 = \frac{ad}{b^2 + e^2} a_3, \qquad c_3 = \frac{aca_3}{b^2 + e^2}, \qquad b_3 = \frac{aba_3}{b^2 + e^2},$$

$$a_3^2 + b_3^2 + c_3^2 + d_3^2 = 1$$

$$\Leftrightarrow \left[1 + \frac{a^2(b^2 + e^2)}{(b^2 + e^2)^2}\right] a_3^2 = 1 \Leftrightarrow a_3^2 = \frac{b^2 + e^2}{a^2 + b^2 + e^2}.$$

With a_3 determined by this last equation, the line above it determines b_3, c_3, and d_3.

Answers to Odd-numbered Problems

CHAPTER 1

▶ Section 1-2 (p. 10)

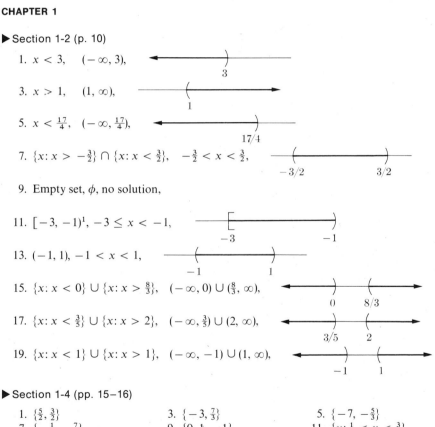

1. $x < 3$, $(-\infty, 3)$,

3. $x > 1$, $(1, \infty)$,

5. $x < \frac{17}{4}$, $(-\infty, \frac{17}{4})$,

7. $\{x: x > -\frac{3}{2}\} \cap \{x: x < \frac{3}{2}\}$, $-\frac{3}{2} < x < \frac{3}{2}$,

9. Empty set, ϕ, no solution,

11. $[-3, -1)^1$, $-3 \le x < -1$,

13. $(-1, 1)$, $-1 < x < 1$,

15. $\{x: x < 0\} \cup \{x: x > \frac{8}{3}\}$, $(-\infty, 0) \cup (\frac{8}{3}, \infty)$,

17. $\{x: x < \frac{3}{5}\} \cup \{x: x > 2\}$, $(-\infty, \frac{3}{5}) \cup (2, \infty)$,

19. $\{x: x < 1\} \cup \{x: x > 1\}$, $(-\infty, -1) \cup (1, \infty)$,

▶ Section 1-4 (pp. 15–16)

1. $\{\frac{5}{2}, \frac{3}{2}\}$ 3. $\{-3, \frac{7}{3}\}$ 5. $\{-7, -\frac{5}{3}\}$

7. $\{-\frac{1}{4}, -\frac{7}{8}\}$ 9. $\{0, 1, -1\}$ 11. $\{x: \frac{1}{2} < x < \frac{3}{2}\}$

13. $\{x: x < -\frac{1}{3}\} \cup \{x: x > 3\}$ 15. $\{x: x > -\frac{5}{4}\} \cup \{x: x < -3\}$

17. $\{x: x > 0\}$ 19. $\{x: x \le -\frac{1}{4}\} \cup \{x: x \ge \frac{7}{2}\}$

21. $\{x: x \le 2\} \cup \{x \ge 10\}$

▶ Section 1-5 (pp. 19–20)

1. $-1 < x < 3$ 3. $-1 < x < -\frac{1}{2}$ 5. $(-\infty, -\frac{1}{3}) \cup (0, 1)$

7. $-3 < x < 3$ 9. $-\sqrt{10} < x < \sqrt{10}$ 11. empty set

13. $(-3, -2) \cup (2, 3)$ 15. $(-\infty, 0) \cup (\frac{1}{2}, \infty)$

17. $(-\infty, 1) \cup (2, 6)$ 19. $(-1, 1) \cup (4, \infty)$ 21. $(-2, 0)$

23. $(-\frac{5}{2}, -\frac{51}{42}) \cup (\frac{1}{2}, \infty)$ 25. $(-\infty, -\frac{1}{2}) \cup (1, \infty)$

CHAPTER 2

▶ Section 2-1 (pp. 24–25)

1. $\{(x, y): y = 2x\}$ 3. $\{(x, y) \ y = \frac{x}{2} + \frac{5}{2}\}$ 5. $\{(x, y): x = -2\}$

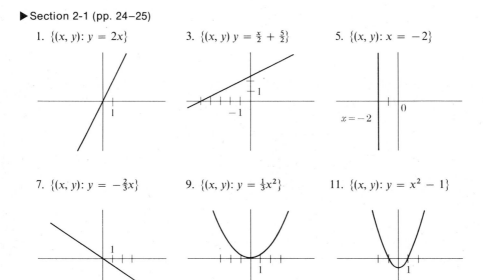

7. $\{(x, y): y = -\frac{2}{3}x\}$ 9. $\{(x, y): y = \frac{1}{3}x^2\}$ 11. $\{(x, y): y = x^2 - 1\}$

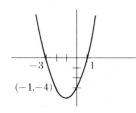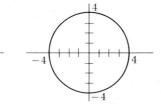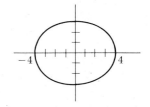

13. $\{(x, y): y = (x + 1)^2 - 4\}$ 15. $\{(x, y): x^2 + y^2 = 16\}$ 17. $\{(x, y): x^2/16 + y^2/9 = 1\}$

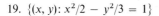

19. $\{(x, y): x^2/2 - y^2/3 = 1\}$ 21. $\{(x, y): y = -(x + 1)^2 + 4\}$

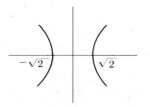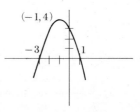

23. $\{(x, y): y = x(x^2 - 1)\}$

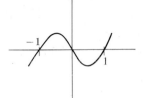

25. $\{(x, y): (y - x)(y + 3x) = 0\}$

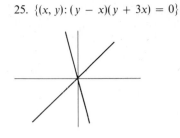

27. $\{(x, y): x^2 + 2xy + y^2 + 2x - 2y - 3 = 0\}$

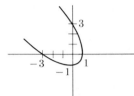

29. $\{(x, y): (x - 1)^2 + y^2 = -8\}$ but sum of squares $\not< 0$ $\therefore$ ϕ

▶ Section 2-2 (pp. 29–30)

1.

x	-4	-3	-2	-1	0	1	2	$a + 2$
$f(x)$	1	-3	-5	-5	-3	1	7	$a^2 + 7a + 7$

3.

x	-4	-3	-2	-1	0	1	-1000	1000
$f(x)$	$\frac{5}{11}$	$\frac{4}{9}$	$\frac{3}{7}$	$\frac{2}{5}$	$\frac{1}{3}$	0	$\frac{1001}{2003}$	$\frac{999}{1997}$

; no

5.

x	-1000	-3	-2	-1	0	1	2	3	1000
$f(x)$	$-\frac{4000}{4000001}$	$-\frac{12}{37}$	$-\frac{8}{17}$	$-\frac{4}{5}$	0	$\frac{4}{5}$	$\frac{8}{17}$	$\frac{12}{37}$	$\frac{4000}{4000001}$

$$f(-x) = \frac{4(-x)}{4(-x)^2 + 1} = -\frac{4x}{4x^2 + 1} = -f(x)$$

7.

x	-3	-2	-1	0	1
$f(x)$	0	1	$\sqrt{2}$	$\sqrt{3}$	2

; Domain: $x \geq -3$

9.

x	-2	-1	0	1	2	3	4
$f(x)$	-1	0	1	2	1	0	-1

11. Domain: $x \neq \pm 2,$

x	-3	-1	0	1	3
$f(x)$	0	$\frac{2}{3}$	$\frac{3}{4}$	$\frac{4}{3}$	$\frac{6}{5}$

13. Domain: $(-\infty, 2] \cup [4, \infty)$

15. $x + \dfrac{h}{2}$

17. $6x^2 + 6xh + 2h^2$

19. $\dfrac{-2x + 2 - h}{(x - 1)^2(x + h - 1)^2}$

▶Section 2-3 (p. 37)

Intercepts	*Symmetry*	*As functions*
1. $(0, -9), (\pm 3, 0)$	y axis	
3. $(0, 0)$	x axis	$f_1(x) = 2\sqrt{x}, f_2(x) = -2\sqrt{x}$
5. $(1, 0)$	x axis	$f_1(x) = \sqrt{2x - 2}, f_2(x) = -\sqrt{2x - 2}$
7. $(0, \pm 2), (-4, 0)$	x axis	$f_1(x) = \sqrt{x + 4}, f_2(x) = -\sqrt{x + 4}$
9. $(0, \pm\sqrt{3}), (\pm 2, 0)$	x, y axes, origin	$f_1(x) = \frac{1}{2}\sqrt{12 - 3x^2}, f_2(x) = -\frac{1}{2}\sqrt{12 - 3x^2}$
11. $(0, \pm 2)$	origin	$f_1(x) = -2x + 2\sqrt{x^2 + 1},$
		$f_2(x) = -2x - 2\sqrt{x^2 + 1}$
13. $(0, \pm 3)$	x, y axes, origin	$f_1(x) = \sqrt{x^2 + 9}, f_2(x) = -\sqrt{x^2 + 9}$
15. $(0, \pm\sqrt{5}), (\pm\sqrt{\frac{5}{2}}, 0)$	origin	$f_1(x) = -x + \sqrt{5 - x^2}, f_2(x) = -x - \sqrt{5 - x^2}$
17. $(0, 1), (-2^{1/3}, 0)$	none	
19. $(\frac{2}{3}, 0)$	none	$f_1(x) = -1 + \sqrt{3x - 1}, f_2(x) = -1 - \sqrt{3x - 1}$
21. $(0, 0)$	x, y axes, origin	$f_1(x) = 2x^2, f_2(x) = -2x^2$
23. $(0, 0), (2, 0)$	x axis	$f_1(x) = \sqrt{-x}, f_2(x) = -\sqrt{-x},$
		$f_3(x) = \sqrt{x - 2}, f_4(x) = -\sqrt{x - 2}$
25. $(0, 0)$	x, y axes, origin	$y = \pm\frac{x}{2}\sqrt{x^2 + 4}$
29. $(0, -1)$	none	

▶Section 2-4 (p. 39)

1. $(0, 0), (2, 4)$ 3. $(-1, -4), (2, 2)$

5. $(1, 2), (21, -8)$ 7. $(\frac{1}{2}, 4), (2, 1)$

9. $(2, 3), (2, -3), (-2, 3), (-2, -3),$ 11. $(3, 4), (3, -4), (-3, 4), (-3, -4)$

13. $(\sqrt{3}, \sqrt{2}), (\sqrt{3}, -\sqrt{2}), (-\sqrt{3}, \sqrt{2}), (-\sqrt{3}, -\sqrt{2}),$

15. $(0, 0), (\frac{1}{9}, \pm\sqrt{\frac{2}{3}})$ 17. $(-\frac{7}{2}, -1), (-\frac{3}{2}, 1)$

CHAPTER 3

▶Section 3-2 (pp. 44–45)

1. $|AB| = \sqrt{10}, |BC| = \sqrt{5}, |AC| = \sqrt{5}$

3. $|AB| = 2\sqrt{17}, |BC| = \sqrt{26}, |AC| = 3\sqrt{2}$

5. $(3, 5)$,   7. $(-\frac{1}{2}, \frac{3}{2})$,   9. $B(-4, 5)$

11. median through $A = \sqrt{26}$ 13. $|AB| = \sqrt{20}, |BC| = 5, |AC| = 5,$

       median through $B = \sqrt{17}$ since $|BC| = |AC|,$

       median through $C = \sqrt{41}$ the triangle is isosceles

15. $|AB| = \sqrt{13}, |AC| = \sqrt{26}, |BC| = \sqrt{13}$; isosceles right triangle

17. $|AB| = 3\sqrt{2}, = |BC| = |CD| = |DA|$ and $|AC| = 6, |AC|^2 = |AB|^2 + |BC|^2$ $\therefore \angle B$ is a right angle and the quadrilateral is a square

19. midpoint $AC = (\frac{5}{2}, 1) = $ midpoint (BD)

21. $|AB| = \frac{1}{2}\sqrt{b^2 + c^2} = |CD|, |BC| = \frac{1}{2}\sqrt{(a - d)^2 + e^2} = |DA|$ $\therefore$ $ABCD$ is a parallelogram

23. $x^2 - 2x + y^2 - 2y - 2 = 0$ 25. $y^2 - 8x + 2y + 17 = 0$

27. $x^2 + y^2 - 10x + 2y + 22 = 0$ 29. $x^2 + y^2 - 8x + 4y = 0$

▶Section 3-4 (pp. 50–51)

1. parallel   3. parallel   5. neither

7. not on line
13. trapezoid
19. none

9. on same line
15. none
21. none

11. $Q(4, \frac{5}{3})$
17. parallelogram
23. rhombus

27. $C\left(a, \frac{a\sqrt{3}}{2}\right)$, $D\left(\frac{a}{2}, a\sqrt{3}\right)$, $E\left(-\frac{a}{2}, a\sqrt{3}\right)$, $F\left(-a, \frac{a\sqrt{3}}{2}\right)$; diagonal $= 2a$

▶ Section 3-5 (pp. 57–58)

1. $y - 2x - 1 = 0$
7. $y = 5$
13. $x - 2y - 2 = 0$
19. $5x - 2y - 8 = 0$
25. $2x + 3y - 13 = 0$
31. $7x + 5y + 4 = 0$

3. $3x + 2y - 5 = 0$
9. $y = 1$
15. $y = -2x$
21. $m = -\frac{3}{2}, b = 3$
27. $2x + 3y - 13 = 0$
33. $x - y - 3 = 0$

5. $7x + 5y + 1 = 0$
11. $y = 3$
17. $4x + 3y + 12 = 0$
23. $a = -\frac{7}{3}, b = -\frac{7}{2}$
29. $2x + 5y - 9 = 0$

▶ Section 3-6 (pp. 64–65)

1. $x = 2 + t/\sqrt{26}, y = -1 + 5t/\sqrt{26}$
5. $x = 2 + t/\sqrt{2}, y = -3 + t/\sqrt{2}$
9. $x = 1 - t/\sqrt{2}, y = -t/\sqrt{2}$
13. $(-3, -7)$
19. $\frac{2}{3}$

3. $x = 1 + t/\sqrt{2}, y = -2 + t/\sqrt{2}$
7. $x = 1 + 3t/\sqrt{13}, y = -1 - 2t/\sqrt{13}$
11. $(6, 3)$
15. $(\frac{8}{3}, -\frac{5}{3})$
21. $\frac{3}{7}$

17. $(-6, -8)$
23. $(-3, 7)$

▶ Section 3-7 (pp. 69–70)

1. 2
7. 4
13. $21/\sqrt{41}, \frac{21}{2}$
19. $\left(\dfrac{12(1 + \sqrt{13}}{-17}, \dfrac{42 - 9\sqrt{13}}{17}\right)$
23. $x - 3y + 5 = 0$
$3x + y - 1 = 0$
29. 0

3. $\frac{11}{13}$
9. 2
15. $18/\sqrt{10}, 18$

25. 0

31. $-\sqrt{5}$

5. $3/\sqrt{5}$
11. $8/\sqrt{41}$
17. $(0, 7), (0, -3)$
21. $x + 2y - 3 - 2\sqrt{5} = 0$
$x + 2y - 3 + 2\sqrt{5} = 0$
27. $-1 - \sqrt{3}$

▶ Section 3-8 (p. 72)

1. $2x + y + k = 0$
7. $x + y = a$
13. $11x + 12y - 43 = 0$
19. $(-1 \pm \frac{1}{4}\sqrt{86})x + (-4 \pm \frac{1}{4}\sqrt{86})y + 5 \mp \sqrt{86} = 0$
21. $y = 2x \pm 2$

3. $y = m(x - 3)$
9. $2x + y = 2a$
15. $2x - y - 1 = 0$

5. $y - 2 = m(x + 1)$
11. $5x - ax + ay = 5a - a^2$
17. $5x - y - 3 = 0$

▶ Section 3-9 (pp. 75–76)

1. 3
7. $-\frac{1}{2}$
13. $8, \frac{2}{3}, 2$

3. 3
9. $-\frac{2}{3}$

5. undefined, $\phi = 90°$
11. $2, B = 90°, \frac{1}{2}$
15. $x + 3y - 4 = 0$
$3x + y - 4 = 0$

17. $(\sqrt{3} - 1)x - y(1 + \sqrt{3}) - \sqrt{3} - 5 = 0, (1 + \sqrt{3})x + y(1 - \sqrt{3}) - \sqrt{3} + 5 = 0$
19. $x - 7y + 6 = 0, 7x + y - 18 = 0$
21. $21x + 3y + 10 = 0, x - 7y + 40 = 0$
23. $(3 \pm \sqrt{13})x - 2y + 2 = 0$
27. $x + (-3 - \sqrt{10})y + 6 + \sqrt{10} = 0$
29. $y = -\frac{1}{3}x + b, y = 3x + b$

▶Section 3-11 (pp. 81–82)

1. Set above L_1 and L_2, Intersection $(\frac{1}{3}, \frac{7}{3})$ does not belong to S
3. $P(\frac{2}{3}, \frac{2}{3})$, $P \notin S$, S to right of L_1, includes L_1, to right of L_2
5. $P(-\frac{3}{13}, \frac{28}{13})$, $P \in S$, S to right of L_1, below L_2, includes L_1 and L_2
7. $P_1(\frac{3}{2}, 2)$, $P_2(\frac{25}{14}, -\frac{4}{14})$, $P_3(-\frac{1}{2}, -2)$, $P_1 \notin S$, $P_2 \notin S$, $P_3 \notin S$, S below L_1, above L_2, left of L_3
9. S points between L_1 and L_2
11. $P_1(\frac{1}{2}, -7)$, $P_2(\frac{1}{2}, 0)$, $P_3(\frac{17}{6}, \frac{14}{3})$, $P_4(\frac{23}{4}, -7)$; $P_1, P_2, P_3 \in S$, $P_4 \notin S$, S inside quadrilateral including sides L_1, L_2, L_3

▶Section 3-12 (p. 84)

1. $P_1(5, -2)$, $P_2(2, 1)$, $P_3(2, 2)$, $P_4(6, 2)$, $P_5(6, -\frac{5}{2})$; S inside and includes boundaries of $P_1P_2P_3P_4P_5$
3. $P_1(-1, 1)$, $P_2(0, 1)$, $P_3(0, \frac{4}{3})$; S inside and boundaries of $P_1P_2P_3$
5. yes at $(3, 6)$ 7. $(0, 1)$

CHAPTER 4

▶Section 4-2 (pp. 92–93)

1. $-3\mathbf{i} - 3\mathbf{j}$ 3. $3\mathbf{i} + 4\mathbf{j}$ 5. $-3\mathbf{i} - 4\mathbf{j}$
7. $-4\mathbf{i}$ 9. $\mathbf{u} = \frac{12}{13}\mathbf{i} - \frac{5}{13}\mathbf{j}$ 11. $\mathbf{u} = (2/\sqrt{13})\mathbf{i} + (3/\sqrt{13})\mathbf{j}$
13. $B(1, 8)$ 15. $A(3, 7)$ 17. $A(-\frac{9}{2}, 3)$, $B(-\frac{3}{2}, 1)$
19. $\mathbf{v} = 2\mathbf{i} + 2\mathbf{j}$, $\mathbf{v} = -2\mathbf{i} + 2\mathbf{j}$ 21. $\mathbf{i} + \mathbf{j}$

▶Section 4-3 (pp. 97–98)

1. $|\mathbf{v}| = 5$, $|\mathbf{w}| = 5$, $\cos \theta = \frac{24}{25}$, $\text{proj}_\mathbf{w} \mathbf{v} = \frac{24}{5}$
3. $|\mathbf{v}|$ 25, $|\mathbf{w}|$ 13, $\cos \theta -\frac{36}{325}$, $\text{proj}_\mathbf{w} \mathbf{v} -\frac{36}{13}$
5. $|\mathbf{v}| \sqrt{13}$, $|\mathbf{w}|$ 5, $\cos \theta$ $(6/5\sqrt{13})$, $\text{proj}_\mathbf{w} \mathbf{v} \frac{6}{5}$
7. $|\mathbf{v}| \sqrt{13}$, $|\mathbf{w}| \sqrt{10}$, $\cos \theta$ $(3/\sqrt{130})$, $\text{proj}_\mathbf{w} \mathbf{v}$ $(3/\sqrt{10})$
9. $\cos \theta = 1/\sqrt{26}$, $\cos \alpha = 11/\sqrt{221}$ 11. $\cos \theta = 1/\sqrt{2}$, $\cos \alpha = 1/\sqrt{2}$
13. $\cos \theta = 19/(5\sqrt{26})$, $\cos \alpha = -9/\sqrt{370}$ 15. 1
17. $\pm \dfrac{1}{\sqrt{2}}$ 19. $-\dfrac{45 \pm 25\sqrt{3}}{39}$
23. $|\overrightarrow{BC}| = 2\sqrt{10}$, $\text{proj}_{\overrightarrow{BC}} \overrightarrow{AB} = -8/\sqrt{10}$, $\text{proj}_{\overrightarrow{BC}} \overrightarrow{AC} = 12/\sqrt{10}$
25. $\text{proj}_{\overrightarrow{CB}} \overrightarrow{AC} = -\frac{35}{6}$, $\text{proj}_{\overrightarrow{CB}} \overrightarrow{AB} = \frac{19}{6}$
27. $\overrightarrow{DE} = \frac{3}{4}\overrightarrow{AB} + \frac{1}{2}\overrightarrow{BC}$ 29. $\overrightarrow{EF} = \frac{1}{2}\overrightarrow{AB}$

CHAPTER 5

▶Section 5-1 (pp. 101–102)

1. $x^2 + y^2 = 16$ 3. $x^2 + y^2 - 4x - 6y + 4 = 0$
5. $x^2 + y^2 - 2x - 6y + 2 = 0$ 7. $x^2 + y^2 - 2x - 4y - 20 = 0$
9. $x^2 + y^2 - 4x - 6y + 9 = 0$ 11. $x^2 + y^2 - 6x - 4y - \frac{36}{25} = 0$
13. $x^2 + y^2 + 8x + 8y + 16 = 0$ 15. circle, $C(2, -1)$, $r = \sqrt{5}$
17. point $(-3, 2)$, $r = 0$ 19. circle, $C(-\frac{3}{2}, \frac{5}{2})$, $r = 2\sqrt{2}$
21. circle, $C(\frac{5}{4}, -\frac{3}{4})$, $r = \frac{3}{2}$ 25. $x^2 + y^2 - 4x - 2y - 20 = 0$
27. $x^2 + y^2 - 2x + 3y - 3 = 0$ 29. $x^2 + y^2 - 15x/7 - 19y/7 - \frac{1624}{196} = 0$

31. $x^2 + y^2 + 2y - 19 = 0$ 33. $(\frac{3}{5}, -\frac{4}{5}), (3, 4)$
35. no intersection

▶ Section 5-2 (pp. 106–107)

1. $x - 2y + 5 = 0$ 3. $8x + 9y - 26 = 0$
5. $2x - y - 4 = 0$ 7. $8x - 5y + 38 = 0$
9. $3x + 5y + 35 = 0, 5x - 3y + 13 = 0$ 11. P inside circle, no lines
13. $(x - h)^2 + y^2 = 9$ 15. $(x - h)^2 + (y - 6 + h)^2 = 1$
17. $x^2 + y^2 - 2hx - 2hy + h^2 = 0$
19. $(x - 7 + 2k)^2 + (y - k)^2 = (13 - k)^2/13$
21. $(-2, 0), (0, 2)$ 23. $(2, 1), (1, -1)$

CHAPTER 6

▶ Section 6-1 (pp. 113–114)

1. $F(1, 0), x = -1,$ 3. $F(3, 0), x = -3$
5. $F(\frac{9}{8}, 0), x = -\frac{9}{8}$ 7. $y^2 = -16x$ 9. $x^2 = 8y$
11. $x^2 = 10y$ 13. $y^2 = -\frac{25}{2}x$
15. $F(-1, \frac{17}{4}), V(-1, 4), y = \frac{15}{4}, x = -1$ 17. $F(-1, \frac{71}{8}), V(-1, 9), y = \frac{73}{8}, x = -1$
19. $F(-\frac{25}{4}, 1), V(-6, 1), x = -\frac{23}{4}, y = 1$ 21. $F(\frac{1}{2}, -1), V(\frac{3}{2}, -1), x = \frac{5}{2}, y = -1$
23. $y^2 = 20(x - 5)$ 25. $x^2 = -16(y - 3)$
27. $16x^2 - 24xy + 9y^2 - 130x - 90y + 100 = 0$
29. $x = \frac{9}{2}\sqrt{6}$ 31. $y^2 = 18(x + \frac{1}{2})$
33. $x = 2700/(64\sqrt{3})$ ft., maximum $y = \frac{225}{64}$
35. $x^2 + 2xy + y^2 - 16x - 14y + 34 = 0$, parabola

▶ Section 6-2 (pp. 119–120)

1. $2x + 3y + 3 = 0$ 3. $6x + y - 4 = 0$
5. $2x + 3y - 4 = 0$ 7. $x + 8y + 9 = 0$
9. $5x + 4y - 24 = 0, 5x - 4y + 9 = 0$
11. $(-5 \pm \sqrt{21})x - 2y - 23 \pm 5\sqrt{21} = 0$
13. $(5 \pm \sqrt{21})x - 2y - 23 \mp 5\sqrt{21} = 0$
15. $(1 \pm \sqrt{37})x + 6y + 3 \mp 3\sqrt{37} = 0$
17. $11x + y - 12 = 0, 3x + y - 4 = 0$
19. no solution   21. $8x + 32y - 17 = 0$
23. $81y^2 - 56x - 224y + 168 = 0$ 27. $(50, \frac{80}{9}), (100, \frac{320}{9})$
29. $50' \times 100'$

▶ Section 6-3 (pp. 124–125)

	a	b	c	Foci	Vertices	e
1.	3	2	$\sqrt{5}$	$(\pm\sqrt{5}, 0)$	$(\pm 3, 0)$	$\sqrt{5}/3$
3.	6	4	$2\sqrt{5}$	$(0, \pm 2\sqrt{5})$	$(0, \pm 6)$	$\sqrt{5}/3$
5.	2	$\sqrt{3}$	1	$(0, \pm 1)$	$(0, \pm 2)$	$\frac{1}{2}$
7.	$\sqrt{5}$	$\sqrt{3}$	$\sqrt{2}$	$(0, \pm\sqrt{2})$	$(0, \pm\sqrt{5})$	$\sqrt{10}/5$
9.	$\sqrt{26}/2$	$\sqrt{39}/3$	$\sqrt{\frac{13}{6}}$	$(0, \pm\sqrt{\frac{13}{6}})$	$(0, \pm\sqrt{26}/2)$	$1/\sqrt{3}$

11. $16x^2 + 25y^2 = 400$ 13. $100x^2 + 64y^2 = 6400$ 15. $25x^2 + 16y^2 = 400$
17. $15x^2 + 7y^2 = 247$ 19. $3x^2 + 5y^2 = 120$ 21. $16x^2 + 7y^2 = 176$
23. $3x^2 - 36x + 4y^2 = 0$ 25. $94.581 \times 10^6, 91.419 \times 10^6$

▶Section 6-4 (pp. 128–130)

1. $25x^2 + 16y^2 = 400$
3. $6x^2 + 8y^2 = 384$
5. $16x^2 + 7y^2 = 252$
7. $3x^2 + 4y^2 = 192$
9. $21x^2 + 25y^2 - 152x + 256 = 0$, $C(\frac{76}{21}, 0)$, minor axis $x = \frac{76}{21}$, major axis $y = 0$
11. $-2x + \sqrt{10}\,y = 10$
13. $12x + 9\sqrt{5}y - 10\sqrt{5} = 0$
21. $2(1 \mp 2\sqrt{58})x + (16 \pm \sqrt{58})y = 66$
25. 50 ft.

▶Section 6-5 (p. 134)

	a	b	c	Foci	Vertices	e
1.	2	3	$\sqrt{13}$	$(\pm\sqrt{13}, 0)$	$(\pm 2, 0)$	$\sqrt{13}/2$
3.	4	3	5	$(\pm 5, 0)$	$(\pm 4, 0)$	$\frac{5}{4}$
5.	2	$\sqrt{3}$	$\sqrt{7}$	$(\pm\sqrt{7}, 0)$	$(\pm 2, 0)$	$\sqrt{7}/2$
7.	$\sqrt{3}$	$\sqrt{5}$	$2\sqrt{2}$	$(\pm 2\sqrt{2}, 0)$	$(\pm\sqrt{3}, 0)$	$2\sqrt{2}/\sqrt{3}$
9.	$\sqrt{2}$	2	$\sqrt{6}$	$(\pm\sqrt{6}, 0)$	$(\pm\sqrt{2}, 0)$	$\sqrt{3}$

11. $16x^2 - 9y^2 = 144$
13. $5y^2 - 4x^2 = 180$
15. $2x^2 - y^2 = 7$
17. $(-2 + 2\sqrt{41})y^2 - 2(11 - \sqrt{41})x^2 = 12\sqrt{41} - 52$
19. $4x^2 - 5y^2 = 19$
21. $8x^2 - y^2 = 8$
23. $y^2 - 8x^2 + 64x = 120$
25. $7x^2 - 9y^2 = 63$

▶Section 6-6 (pp. 139–140)

	a	b	c	V	F	$e = c/a$	Asymptotes	Directrices a/e
1.	2	4	$2\sqrt{5}$	$(\pm 2, 0)$	$(\pm 2\sqrt{5}, 0)$	$\sqrt{5}$	$y = \pm 2x$	$x = \pm 2/\sqrt{5}$
3.	1	3	$\sqrt{10}$	$(\pm 1, 0)$	$(\pm\sqrt{10}, 0)$	$\sqrt{10}$	$y = \pm 3x$	$x = \pm\sqrt{10}/10$
5.	6	8	10	$(\pm 6, 0)$	$(\pm 10, 0)$	$\frac{5}{3}$	$y = \pm\frac{4}{3}x$	$x = \pm\frac{18}{5}$
7.	3	3	$\sqrt{18}$	$(0, \pm 3)$	$(0, \pm 3\sqrt{2})$	$\sqrt{2}$	$y = \pm x$	$y = \pm\frac{3}{2}\sqrt{2}$

9. $9x^2 - 4y^2 = 144$
11. $5x^2 - 4y^2 = 405$
13. $4x^2 - y^2 = 4$
15. $16x^2 - 9y^2 = 256$
17. $3y^2 - x^2 = 27$ or $4y^2 - 3x^2 = 21$
19. $x = \frac{2}{3}(4\sqrt{10} + 5)$, $y = \frac{1}{3}(13 + 5\sqrt{10})$
23. $25x^2 - 16y^2 = 400$
27. P: no; Q: no; R: $y = \pm\sqrt{\frac{2}{11}}(x - 4)$; S: no

▶Section 6-7 (p. 145)

1. $P'(1, 7)$; $Q'(-3, 1)$; $R'(\frac{5}{2}, -3)$; $S'(6, -2)$
3. $x' = x - 20$, $y' = y + 47$; $7x' + 3y' = 0$, $5x' + 2y' = 0$
5. ellipse: $C(-2, -1)$; $V(3, -1), (-7, -1)$; $F(1, -1), (-5, -1)$; $D: x = \frac{19}{3}, x = -\frac{31}{3}$
7. hyperbola: $C(2, 1)$; $V(5, 1), (-1, 1)$; $F(2 \pm \sqrt{13}, 1)$; $D: x = 2 \pm 9/\sqrt{13}$;
asymptotes: $y - 1 = \pm\frac{2}{3}(x - 2)$
9. ellipse: $C(-2, -1)$; $V(-2, 2), (-2, -4)$; $F(-2, -1 \pm \sqrt{5})$; $D: y = -1 \pm 9/\sqrt{5}$
11. ellipse: $C(1, 1)$; $V(1 \pm \frac{3}{2}, 0)$; $F(1 \pm \sqrt{5}/2, 0)$; $D: x = 1 \pm 9/(2\sqrt{5})$
13. ellipse: $C(1, 2)$; $V(1, 4), (1, 0)$; $F(1, 3), (1, 1)$; $D: y = 2 \pm 4$
15. hyperbola: $C(-2, 1)$; $V(-2, 3), (-2, -1)$; $F(-2, 1 \pm \sqrt{10})$; $D: y = 1 \pm 4/\sqrt{10}$;
asymptotes: $y - 1 = \pm(2/\sqrt{6})(x + 2)$
17. two straight lines: $y - 1 = \pm(\sqrt{3}/2)(x - 2)$
19. $7x^2 + 16y^2 + 14x - 64y - 41 = 0$
21. $y^2 - 5x^2 - 10x - 2y - 20 = 0$
23. $x^2 - 2x - 6y + 22 = 0$
25. ellipse: $3x^2 + 4y^2 + 12x - 16y + 16 = 0$
27. circle: $x^2 + y^2 + 10x + 12y + 29 = 0$

▶ Section 6-8 (p. 154)

1. two lines: $y = 2x \pm 3$
3. hyperbola: $x'^2 - y'^2 = 10$; $V(1, 3), (-1, -3)$
5. parabola: $y'^2 = (8/\sqrt{2})x'$; $F(1, 1)$; $D: x + y + 2 = 0$
7. ellipse: $4x'^2 + 17y'^2 = 884$; $V(2\sqrt{17}, 3\sqrt{17}), (-2\sqrt{17}, -3\sqrt{17})$
9. hyperbola: $(1 + \sqrt{5})x'^2 - (\sqrt{5} - 1)y'^2 = 20$; $V(\pm\sqrt{3\sqrt{5} - 5}, \pm\sqrt{2\sqrt{5}})$
11. hyperbola: $(\sqrt{13} - 3)y'^2 - (\sqrt{13} + 3)x'^2 = 10$; $V\left(\mp\sqrt{\dfrac{5}{\sqrt{13}}}, \pm\sqrt{\dfrac{5}{\sqrt{13}}}, \dfrac{\sqrt{\sqrt{13} + 3}}{2}\right)$
13. two lines: $4y''^2 - x''^2 = 0$; $y - 3x + 1 = 0, -x + 3y - 5 = 0$
15. parabola: $y''^2 = 4\sqrt{5}x''$; $F(-1, 0)$; $V(-3, -1)$; $D: y + 2x + 12 = 0$
17. point: $(x'', y'') = (0, 0)$; $(x, y) = (2, \frac{3}{2})$
19. ellipse: $4(x'')^2 + 9(y'')^2$; $C(4, 2)$; $V(4 + \frac{3}{2}\sqrt{3}, \frac{7}{2}), (4 - \frac{3}{2}\sqrt{3}, \frac{1}{2})$
21. hyperbola: $3(x'')^2 - (y'')^2 = 12$; $C(0, -\sqrt{3})$; $V(1, 0), (-1, -2\sqrt{3})$

CHAPTER 7

▶ Section 7-1 (pp. 159–160)

	Intercepts	Symmetries	Domain	Range	Asymptotes
1.	$(0, 9), (\pm 3, 0)$	y axis	all x	$y \le 9$	none
3.	$(0, \pm 2), (\pm\sqrt{3}, 0)$	x, y axes, origin	$\|x\| \le \sqrt{3}$	$\|y\| \le 2$	none
5.	$(\pm\sqrt{3}, 0)$	x, y axes, origin	$\|x\| \ge \sqrt{3}$	all y	$y = \pm(2/\sqrt{3})x$
7.	$(0, \pm\sqrt{12}), (\pm\sqrt{3}, 0)$	origin	$\|x\| \le 2$	$\|y\| \le 4$	none
9.	none	x axis	$x > 2$	$y \ne 0$	$x = a$ $y = 0$
11.	none	x, y axes, origin	$\|x\| > 3$	$y \ne 0$	$x = \pm 3$ $y = 0$
13.	$(0, \pm\sqrt{2})$	x, y axes, origin	all x	$\|y\| \le \sqrt{2}$ $y \ne 0$	$y = 0$
15.	none	none	$x \ne 0, 2,$	$\{y \le -16\} \cup \{y > 0\}$	$x = 0, 2$ $y = 0$
17.	$(0, 0)$	origin	all x	$y \ne \pm 3$	$x = 0$ $y = \pm 3$
19.	$(0, 0)$	origin	all x	$y \ne \pm 3$	$x = 0$ $y = \pm 3$
21.	$(0, 0)$	x, y axes, origin	all x	all y	none
23.	none	x axis	$x \ne 0$	$\|y\| \ge \sqrt{\frac{7}{8}}$	$x = 0$ $y = \pm 1$
25.	$(2, 0), (3, 0)$	no symmetry	$x = 2$ $x = 3$	all y	

▶ Section 7-2 (pp. 163–164)

1. S: points to left of $y^2 = 2x$
3. S: outside, to right of $x = 9 - y^2$
5. S: points between and including branches of the hyperbola $4x^2 - y^2 = 4$

7. trapezoid: boundary and inside $x = -1, x = 2, y = 0, y = x + 2$
9. triangle bounded by $y = 2, x = 1, y = x - 2$
11. area bounded by $y = x + 1, x = 1, y = x^2 - 1$
13. strip inside circle $x^2 + y^2 = 9$ and between $-1 \le x \le 1$ including boundaries
15. area between $-2 \le x \le 2$ and $x^2 - 8 \le y \le -x$
17. $\{(x, y): -1 \le y \le 1, y^2 < x < 2 - y\}$
19. $\{(x, y): 0 \le x \le 3, x^2 \le y \le 12x\}$
21. $\{(x, y): 0 \le y \le \frac{3}{4}, y^2 \le x \le y\}$
23. $\{(x, y): 1 \le x \le 2, 2/x^2 \le y \le 4 - x\}$
25. $\{(x, y): 1 \le y \le 2, y/3 \le x \le \sqrt{y}\}$

CHAPTER 8

▶ Section 8-1 (pp. 168–169)

1. amplitude $= 3$; per $= 6\pi$

3. amp $= 2$; per $= 6$

5. amp $= 5$; per $= 5\pi$

7. amp $= 3$; per $= \dfrac{4\pi}{3}$; shift: $x = -\dfrac{\pi}{3}$

9. amp $= 3$; per $= \dfrac{2\pi}{3}$

11. amp $= 3$; per $= \dfrac{2\pi}{3}$; shift: $x = -\dfrac{\pi}{12}$

13. amp $= 4$; per $= \frac{8}{3}$

15. amp $= 1$; per $= \pi$

17. no amplitude; per $= \dfrac{2\pi}{3}$

19. per $= \dfrac{2\pi}{3}$

21. per $= 1$; shift: $x = \frac{1}{2}$

23. amp $= \sqrt{2}$; per $= \pi$; shift: $x = -\dfrac{\pi}{8}$

35. $y = \sqrt{a^2 + b^2} \sin\left(nx + \arcsin\dfrac{b}{\sqrt{a^2 + b^2}}\right)$

▶ Section 8-2 (p. 173)

1. $g(x) = \dfrac{x - 3}{2}$

3. $y_1 = g_1(x) = -2 + \sqrt{x - 1}$
   $y_2 = g_2(x) = -2 - \sqrt{x - 1}$

5. $g_1(x) = -\frac{3}{2} + \frac{1}{2}\sqrt{25 - 4x}$
   $g_2(x) = -\frac{3}{2} - \frac{1}{2}\sqrt{25 - 4x}$

7. $g(x) = \dfrac{1}{1 - x}$

9. $g_1(x) = 1/\sqrt{x}, g_2(x) = -1/\sqrt{x}$

11. $g_1(x) = -\sqrt[4]{-x}, (-\infty < x < 0)$
    $g_2(x) = +\sqrt[4]{-x}, (-\infty < x < 0)$

13. $g(x) = 2 + \sqrt[3]{x}, (-\infty < x < \infty)$

15. $g(x) = \dfrac{2x + 1}{2} - \frac{1}{2}\sqrt{4x + 1} \ (0 \le x < \infty)$

▶ Section 8-3 (p. 176)

1. (a) $\dfrac{\pi}{6}$ (b) $\dfrac{\pi}{6}$

3. (a) $\dfrac{\pi}{6}$ (b) $-\dfrac{\pi}{3}$

▶ Section 8-4 (pp. 180–181)

17. (a) 4 (b) 2 (c) -3

19. (a) 0 (b) -2 (c) 1

25. (a) $\dfrac{\log 23}{\log 7}$ (b) $\dfrac{\log 13}{-\log 3}$ (c) $\dfrac{\log 16}{\log 7}$

27. 2.8926

29. -2.0959

31. 1.8983

CHAPTER 9

▶Section 9-1 (p. 186)

1. $y = -\frac{4}{3}x$ 　　　　　 3. $y = \dfrac{2}{x}$ 　　　　 5. $\dfrac{(x-1)^2}{9} + \dfrac{(y+2)^2}{9} = 1$

7. $\left(\dfrac{x}{4}\right)^{2/3} + \left(\dfrac{y}{4}\right)^{2/3} - 1$ 　　　 9. $\dfrac{(x+2)^2}{25} - \dfrac{(y-3)^2}{9} = 1$

11. $y^2 - 2xy + x^2 + 6x - 10y + 16 = 0$

13. $x^2 + y^2 = 25$ 　　 but $(0, -5)$ not in locus

15. $y = 3/x$

17.

t	2	4	6	8	10
x	800	1600	2400	3200	4000
y	536	944	1224	1376	1400

$T = 18.75$ sec, $R = 7500$ ft.

19. $x = a\theta - b\sin\theta,\quad 0 \le \theta \le \pi/2$
　　$y = a - b\cos\theta,\quad 0 < \theta \le \pi/2$

21. $x = a\cos\theta + a\theta\sin\theta$
　　$y = a\sin\theta - a\theta\cos\theta$

23. $y = \dfrac{8a^3}{x^2 + 4a^2},\quad y = 0$ is asymptote

▶Section 9-2 (pp. 189–190)

1. $(\frac{3}{2}\sqrt{2}, \frac{3}{2}\sqrt{2}), (-2, 2\sqrt{3}), (-4, 0), (2, 0), (3, 0)$

3. $(-2, 0), (\frac{3}{2}\sqrt{2}, -\frac{3}{2}\sqrt{2}), (\sqrt{3}, -1), (2\sqrt{3}, -2), (0, 0)$

5. $\left(4, \dfrac{\pi}{3}\right), \left(2, \dfrac{\pi}{2}\right), \left(2, \dfrac{5\pi}{6}\right), \left(2, -\dfrac{\pi}{2}\right), (3, 0)$

7. $\left(2\sqrt{2}, \dfrac{3\pi}{4}\right), \left(2\sqrt{3}, -\dfrac{\pi}{6}\right), \left(2, \dfrac{\pi}{6}\right), \left(4, \dfrac{\pi}{3}\right), \left(4, \dfrac{2\pi}{3}\right)$

9. circle radius 5, center $(0, 0)$; line through pole, inclination $\dfrac{\pi}{3}$ or slope $\sqrt{3}$; line through origin, slope $\dfrac{1}{\sqrt{3}}$ or inclination $\dfrac{\pi}{6}$; $r = $ constant and $\theta = $ constant intersect at right angles

11. $\sqrt{13}$

▶Section 9-3 (pp. 194–195)

1. x axis 　　　　　 3. no symmetry 　　　 5. x axis

7. x axis 　　　　　 9. y axis 　　　　　 11. x axis

13. y axis 　　　　　 15. x axis, pole 　　 17. pole, y axis

19. x axis 　　　　　 21. pole, x axis 　　 23. pole, x, y axes

25. y axis 　　　　　 27. pole, x, y axes 　 29. x axis

31. y axis 　　　 33. $\left(\dfrac{6}{\sqrt{13}}, \arctan \frac{3}{2}\right)$, origin 　 35. $\left(\sqrt{2}, \dfrac{\pi}{4}\right), \left(\sqrt{2}, -\dfrac{\pi}{4}\right)$

37. origin, $\left(\sqrt{3}, \dfrac{\pi}{3}\right), \left(-\sqrt{3}, -\dfrac{\pi}{3}\right)$

39. origin, $\left(\dfrac{\sqrt{3}}{2}, \pm\dfrac{\sqrt{\pi}}{6}\right), \left(\dfrac{\sqrt{3}}{2}, \pm\dfrac{\pi}{3}\right), \left(\dfrac{\sqrt{3}}{2}, \pm\dfrac{2\pi}{3}\right), \left(\dfrac{\sqrt{3}}{2}, \pm\dfrac{5\pi}{6}\right)$

▶Section 9-4 (p. 197)

1. $r \cos \theta = 4$ 3. $\theta = \dfrac{\pi}{4}$ 5. $r \cos \left(\theta - \dfrac{\pi}{3} \right) = 1$

7. $r^2 = \dfrac{10}{\sin 2\theta}$ 9. $r = 2 \tan \theta \sec \theta$ 11. $r = 4 \cos \theta - 2 \sin \theta$

13. $x^2 + y^2 = 16$ 15. $x^2 + y^2 = 5x$ 17. $x + 2 = 0$

19. $x - \sqrt{3}y = 8$ 21. $x^2 - y^2 = 6$ 23. $(x^2 + y^2)^2 = 2x^2 - 2y^2$

25. $x^2 = 2y$ 27. $9(x^2 + y^2) = (4 + 2x)^2$ 29. $(x^2 + y^2)^2 = 9x^2y - 3y^3$

31. $(x^2 + y^2 - 2x)^2 = x^2 + y^2$

▶Section 9-5 (pp. 200–201)

1. $r \cos \left(\theta - \dfrac{\pi}{3} \right) = 2$ 3. $r \sin \theta = 4$

5. $r \cos \theta = -2$ 7. $r \cos \left(\theta - \dfrac{3\pi}{4} \right) = \dfrac{5\pi}{4}$ 9. $r^2 - 6r \cos \left(\theta - \dfrac{\pi}{4} \right) = 7$

11. $r = 6 \cos \left(\theta - \dfrac{\pi}{6} \right)$ 13. $r^2 - 6r \cos \left(\theta + \dfrac{\pi}{3} \right) = 25$

15. $r + 4\sqrt{2} \cos \left(\theta - \dfrac{\pi}{4} \right) = 0$ 17. $r = \dfrac{3}{1 - \cos \theta}$

19. $r = \dfrac{-6}{4 + 3 \sin \theta}$ 21. $r = \dfrac{-4}{1 - \sin \theta}$

23. $d = |r_1 \cos (\theta_1 - \alpha) - p|$ 25. $r \cos \theta = -3, e = 1$

27. $e = \frac{1}{2}, r \cos \left(\theta - \dfrac{\pi}{2} \right) = -3$ or $r \sin \theta = -3$

29. $e = \frac{1}{2}, r \cos \theta = 4$ 31. $e = 1, r \sin \left(\theta - \dfrac{\pi}{6} \right) = \frac{3}{2}$

CHAPTER 10

▶Section 10-1 (pp. 205–206)

1. $|AB| = \sqrt{38}, |BC| = \sqrt{19}, |AC| = \sqrt{17}$, no special triangle

3. $|AB| = 3, |BC| = 3, |AC| = \sqrt{26}$, isosceles triangle

5. $|AB| = \sqrt{14}, |BC| = \sqrt{14}, |AC| = \sqrt{30}$, isosceles triangle

7. $M = (2, 3, \frac{5}{2})$,   9. $M = (-\frac{1}{2}, -1, 2)$

11. |median through $A| = \frac{1}{2}\sqrt{61}$, |median through $B| = \frac{1}{2}\sqrt{10}$, |median through $C| = \frac{1}{2}\sqrt{61}$

13. |median through $A| = \frac{1}{2}\sqrt{74}$, |median through $B| = \frac{1}{2}\sqrt{26}$, |median through $C| = \frac{1}{2}\sqrt{74}$

15. $Q(5, 3, -\frac{5}{2}), P_2(4, 8, -6)$

17. not same line   19. not same line

21. line parallel to y axis through point $(1, -2, 0)$

23. one octant including boundary plane $x = 0$

27. $8x - 6y - 4z + 5 = 0$, plane $\perp$ and bisecting P_1P_2

▶Section 10-2 (pp. 211–212)

1. direction numbers $-2, -1, 5$; direction cosines $-2/\sqrt{30}, -1/\sqrt{30}, 5/\sqrt{30}$

3. direction numbers $2, 7, 1$; direction cosines $2/(3\sqrt{6}), 7/(3\sqrt{6}), 1/(3\sqrt{6})$

5. $P(1 + 5k, 2 + 3k, 5 + 2k)$ 7. $P(-3 + 5k, 0, 4)$
9. ABC on line   11. ABC not on a line
13. $P_1P_2 \parallel Q_1Q_2$ 15. $P_1P_2 \nparallel Q_1Q_2$ 17. $P_1P_2 \perp Q_1Q_2$
19. $-\sqrt{2}/3$ 21. $41/(3\sqrt{190})$ 23. $4\sqrt{10}$

▶ Section 10-3 (pp. 215–216)

1. parametric: $x = 4 - 2t, y = 2 - 3t, z = 3 - 6t$
   symmetric: $(x - 4)/2 = (y - 2)/-3 = (z - 3)/-6, P_3(6, 5, 9), P_4(0, -4, -9)$
3. parametric: $x = -1 + 3t, y = 1 + 2t, z = 2 - 3t$
   symmetric: $(x + 1)/3 = (y - 1)/2 = (z - 2)/-3, P_3(-4, -1, 5), P_4(5, 5, -4)$
5. $L: x = -1 - 3t, y = 2t, z = 1 + t$
7. $L: x = -2t, y = 4 + 3t, z = -t$ 9. $L: x = -2, y = 3 + 2t, z = -1$
11. $L_1 \perp L_2$ 13. L_1 not $\perp L_2$
15. med A, M_{BC}: $(x - 2)/2 = (y - 4)/-\frac{3}{2} = z/3$
    med B, M_{AC}: $(x - 4)/-3 = (y - 3)/0 = (z - 1)/\frac{3}{2}$
    med C, M_{AB}: $x/3 = (y - 2)/\frac{3}{2} = (z - 5)/-\frac{9}{2}$
17. $(0, -\frac{9}{7}, \frac{17}{7}), (-\frac{9}{2}, 0, \frac{1}{2}), (-\frac{17}{3}, \frac{1}{3}, 0)$
19. $(x - 3)/-1 = (y - 1)/3 = (z - 5)/1$
21. $(x - \frac{5}{4})/1 = (y - \frac{13}{4})/5 = z/8$

▶ Section 10-4 (pp. 219–220)

1. $x - 4y - 2z + 4 = 0$ 3. $x - 4y + 7 = 0$
5. $x - 2y + 1 = 0$ 7. $5x - 9y - z + 16 = 0$
9. $4x - 2y - 3z - 11 = 0$ 11. $2x - 3y + 10 = 0$
13. $(x - 1)/1 = (y + 2)/2 = (z - 3)/3$ 15. $(x + 2)/2 = (y + 1)/1 = z/0$
17. $x - 3y - 2z = 0$ 19. $3x - y + 2z - 5 = 0$
21. $(x - 3)/4 = (y - 2)/3 = (z + 1)/-2$ 23. $x/4 = (y - 1)/2 = (z + 2)/-1$
25. $x - 2y + 2z + 4 = 0$ 27. $5x - 2y + 3z + 7 = 0$
29. $x - 2y + 2z + 6 = 0$ 33. $(x - 2)/7 = (y + 1)/-3 = (z - 3)/5$
35. $(x - \frac{3}{2})/2 = (y - \frac{11}{4})/3 = z/4, P_1(\frac{3}{2}, \frac{11}{4}, 0), P_2(2, \frac{7}{2}, 1)$

▶ Section 10-5 (pp. 224–226)

1. $\frac{8}{21}$ 3. $\sqrt{\frac{14}{17}}$
5. $x = \frac{11}{5} + t, y = -5t, z = -\frac{7}{5} + 8t$ 7. $x = 3, y = -1 - 2t, z = t$
9. $x = -1, y = 3, z = 2$ 11. $(-\frac{6}{5}, -\frac{3}{5}, 0)$ 13. 1
15. $6/\sqrt{29}$ 17. $x + 10y - 17z + 25 = 0$
19. $13x - 14y - 8z - 15 = 0$ 21. $x + z - 1 = 0$
23. point $(\frac{25}{57}, \frac{91}{57}, \frac{31}{57})$ 25. no intersection
27. $L: x = 2 + t, y = 3 - 4t, z = -1 - 5t$
29. $x = 4 - 22t, y = 29t, z = 2 + t$ 31. $\sqrt{6}a/3$
33. (a) $\sqrt{14}$ (b) $4/\sqrt{6}$ (c) $\cos \theta = 1/\sqrt{3}$

▶ Section 10-6 (pp. 229–230)

1. $x^2 + y^2 + z^2 - 2x - 4y - 11 = 0$ 3. $x^2 + y^2 + z^2 - 12x - 6y + 4z = 0$
5. sphere, $C(2, -1, 0), r = 2$ 7. no locus, $C(1, 1, -2), r =$ imaginary
9. sphere, $C(-1, 3, -2), r = 2$ 11. sphere, $C(6, -\frac{5}{2}, -\frac{7}{2}), r = \frac{3}{2}\sqrt{6}$
13. plane $\parallel$ to xz plane
15. plane $\parallel$ to x axis intersecting yz plane in $x = 0, 2y + z = 3$

17. parabolic cylinder $\parallel$ to z axis, parabola in xy plane
19. elliptic cylinder $\parallel$ to x axis   21. circular cylinder $\parallel$ to x axis, $r = 3$
23. circular cylinder, axis $\parallel$ to x axis, axis through point $(0, 1, 0)$, $r = 1$
25. $\{(x, y, z): x = 3 \text{ and } y^2 + z^2 = 16\}$, circle $r = 4$, $C(3, 0, 0)$, parallel to yz plane
27. no intersection
29. (a) $\{(x, y, z): x = 2, y = 1\}$ (b) intersection of two planes, yes

▶ Section 10-7 (p. 236)

  1. ellipsoid   3. ellipsoid of revolution
  5. elliptic hyperboloid of two sheets   7. elliptic hyperboloid of one sheet
  9. elliptic paraboloid   11. circular paraboloid   13. circular cone
15. hyperboloid of one sheet   17. hyperbolic paraboloid

▶ Section 10-8 (p. 240)

  1. sphere, $C(\frac{3}{2}, -2, 4)$, $r = \sqrt{89}/2$ 3. elliptic hyperboloid of one sheet
  5. elliptic paraboloid   7. hyperbolic paraboloid   9. right circular cone
11. hyperbolic paraboloid   13. ellipsoid   15. paraboloid

▶ Section 10-9 (pp. 243–244)

  1. (a) $(3\sqrt{2}, \pi/4, 7)$ (b) $(4\sqrt{5}, \arctan 2, 2)$ (c) $(\sqrt{13}, \arctan(-\frac{3}{2}), 1)$
  3. (a) $(2\sqrt{3}, \pi/4, \arccos 1/\sqrt{3})$ (b) $(2\sqrt{3}, -\pi/4, \arccos(-1/\sqrt{3}))$ (c) $(2\sqrt{2}, 2\pi/3, \pi/4)$
  5. (a) $(4, \pi/3, 0)$ (b) $(1, 2\pi/3, -\sqrt{3})$ (c) $(\frac{7}{2}, \pi/2, 7\sqrt{3}/2)$
  7. $r^2 + z^2 = 9$, sphere   9. $r^2 = 4z$, paraboloid
11. $r^2 = z^2$, right circular cone   13. $r^2 \cos 2\theta = 4$, hyperbolic cylinder
15. $r = 4 \sin \theta$, circular cylinder   17. $\rho = 4 \cos \phi$, sphere
19. $\phi = \pm \pi/4$, right circular cone
21. $\rho = \pm 2/(1 \mp \cos \phi)$, paraboloid of revolution

▶ Section 10-10 (p. 247)

  1. tetrahedron; vertices: $(0, 0, 0)$, $(0, 0, 2)$, $(0, 4, 0)$, $(4, 0, 0)$
  3. tetrahedron; vertices: $(0, 0, 0)$, $(0, 0, -2)(0, 2, 0)$, $(4, 0, 0)$; bounding planes excluded
    except $x + 2y - 2z - 4 = 0$
  5. infinite polygon, $z \geq 0$; vertices of base $(0, 0, 0)$, $(2, 0, 0)(2, \frac{3}{2}, 0)$, $(\frac{5}{3}, 2, 0)$, $(0, 2, 0)$
  7. $S: S_1 \cup S_2: S_1 = \{(x, y, z); 3 \leq x \leq 5, \text{ and } z \geq 2\}$
             $S_2 = \{(x, y, z): x \leq 3, x + 1 \leq y \leq 4, z \geq 1 - x\}$

CHAPTER 11

▶ Section 11-1 (p. 253)

  1. $\mathbf{v} = \mathbf{i} + \mathbf{j} + 2\mathbf{k}$ 3. $\mathbf{v} = -\mathbf{i} + \mathbf{j} + 2\mathbf{k}$ 5. $\mathbf{v} = 2\mathbf{i} - \mathbf{j} + 6\mathbf{k}$
  7. $\mathbf{u} = \frac{2}{3}\mathbf{i} - \frac{2}{3}\mathbf{j} - \frac{1}{3}\mathbf{k}$ 9. $\mathbf{u} = (3/\sqrt{29})\mathbf{i} - (2/\sqrt{29})\mathbf{j} - (4/\sqrt{29})\mathbf{k}$
11. $B(3, -2, -1)$ 13. $A(3, -2, -2)$ 15. $A(\frac{5}{2}, -1, -1)$, $B(\frac{7}{2}, -3, 3)$
17. $A(\frac{7}{3}, -1, \frac{2}{3})$, $B(\frac{10}{3}, -1, \frac{8}{3})$ 19. $\mathbf{i} - \mathbf{j} - 4\mathbf{k}$

▶Section 11-2 (pp. 256–257)

1. independent

3. dependent; $\mathbf{w} = \frac{6}{11}\mathbf{u} - \frac{1}{11}\mathbf{v}$

5. $\mathbf{u}, \mathbf{v}, \mathbf{w}$ independent; $\mathbf{r} = \frac{8}{3}\mathbf{u} + \frac{4}{3}\mathbf{v} + \frac{16}{3}\mathbf{w}$

7. independent; $\mathbf{r} = 7\mathbf{u} - 2\mathbf{v} + 13\mathbf{w}$

9. $\mathbf{r} = \frac{3}{8}\mathbf{u} - \frac{9}{8}\mathbf{v} + \frac{1}{4}\mathbf{w}$

11. $\mathbf{r} = 9\mathbf{u} - 3.2\mathbf{v} - 7.2\mathbf{w}$

▶Section 11-3 (pp. 260–261)

1. $-\frac{1}{6}$

3. $-\sqrt{\frac{6}{11}}$

5. $-2/\sqrt{66}$

7. $-6/\sqrt{14}$

9. 0

11. -160

13. -8

15. $\mathbf{v} = \frac{2}{3}\mathbf{i} + \frac{2}{3}\mathbf{j} - \frac{1}{3}\mathbf{k}$

17. $\mathbf{v} = (4/\sqrt{29})\mathbf{i} + (2/\sqrt{29})\mathbf{j} - (3/\sqrt{29})\mathbf{k}$

19. $k = -\frac{1}{5}, h = -\frac{1}{5}$

21. $k = -\frac{5}{11}, h = -\frac{5}{3}$

25. $g - -\frac{5}{3}h$

27. $g = \frac{9}{4}h$

29. $h = 1, g = -1$

▶Section 11-4 (pp. 266–267)

1. $-6\mathbf{i} - 4\mathbf{k}$

3. $-6\mathbf{j} - 4\mathbf{k}$

5. $3\mathbf{i} + 8\mathbf{j} + 7\mathbf{k}$

7. $\frac{1}{2}\sqrt{110}; 6x - 5y + 7z + 7 = 0$

9. $\sqrt{53}/2; 6x + 4y + z - 23 = 0$

11. $\sqrt{62}/2; 6x + 5y + z - 28 = 0$

13. $-5/\sqrt{3}$

15. $(x - 2)/2 = (y - 1)/5 = (z + 2)/4$

17. $(x + 3)/0 = (y - 2)/2 = (z - 1)/1$

19. $(x - 1)/0 = (y + 3)/2 = (z + 2)/1$

21. $(x - 1)/-5 = (y + 1)/-7 = (z - 2)/-8$

23. $2x - y - 2z - 2 = 0$

25. $2x + y - z + 3 = 0$

27. $2x + 3y - 1 = 0$

29. $x - 7y - 5z - 5 = 0$

31. $x + y + 3z - 8 = 0$

33. $(x - 1)/-10 = (y - 2)/1 = (z - 3)/16$

▶Section 11-5 (pp. 270–271)

1. $V = 1$

3. plane; $x + 2y - 5 = 0$ 7. $11\mathbf{i} - 10\mathbf{j} + 13\mathbf{k}$

9. $-11\mathbf{i} - 7\mathbf{j} - 2\mathbf{k}$

13. $(x - 2)/160 = (y + 1)/-45 = (z - 3)/37$

15. $-3\mathbf{v}$

17. $[(\mathbf{t} \times \mathbf{u}) \cdot \mathbf{w}]\mathbf{v} - [(\mathbf{t} \times \mathbf{u}) \cdot \mathbf{v}]\mathbf{w}$

19. $\mathbf{v} = \dfrac{p\mathbf{a} - (\mathbf{a} \times \mathbf{b})}{|\mathbf{a}|^2}$

CHAPTER 12

▶Section 12-1 (pp. 276–277)

1. $|AB| = \sqrt{18}, |AC| = \sqrt{10}, |BC| = \sqrt{30}$

3. $|AB| = \sqrt{7}, |AC| = \sqrt{24}, |BC| = \sqrt{35}$

5. $M = (2, 0, 0, 1), |MA| = \sqrt{7} = |MB|$ 7. $M = (\frac{3}{2}, 0, 1, \frac{5}{2}), |MA| = \sqrt{5/2} = |MB|$

9. $(4, -3, 1, 0)$

11. $(0, -1, 2, 0)$

13. $(3, -4, 5, 5)$

15. $3, 1, 2, -3; 3/\sqrt{23}, 1/\sqrt{23}, 2/\sqrt{23}, -3/\sqrt{23}$

17. $-3, 1, 4, -1; -3/\sqrt{27}, 1/\sqrt{27}, 4/\sqrt{27}, -1/\sqrt{27}$

19. $(2, 1, 4, 1)$

21. $(3, 1, 3, 2)$

23. not same line

25. same line

▶ Section 12-2 (pp. 283–284)

1. $\dfrac{x-2}{1} = \dfrac{y-1}{-2} = \dfrac{z+1}{2} = \dfrac{v-3}{1}$

3. $\dfrac{x+1}{3} = \dfrac{y}{2} = \dfrac{z-3}{-1} = \dfrac{v-2}{4}$

5. $x = 1+t,\ y = -1-t,\ z = -2+t,\ v = 1+t$

7. $x = 2+4t,\ y = 1-2t,\ z = 1+5t,\ v = -2+2t$

9. $\dfrac{x-2}{1} = \dfrac{y-1}{1} = \dfrac{z+2}{-1} = \dfrac{v+1}{-1}$ 11. $\dfrac{x-1}{2} = \dfrac{y+1}{-4} = \dfrac{z-2}{5} = \dfrac{v-3}{-2}$

13. $\dfrac{x-2}{-1} = \dfrac{y-1}{1} = \dfrac{z+1}{3} = \dfrac{v-3}{-4}$ 15. $\dfrac{x-1}{0} = \dfrac{y-2}{-3} = \dfrac{z+1}{0} = \dfrac{v+2}{5}$

17. $L_1 \perp L_2$ 19. L_1 not $\perp L_2$

21. $\frac{3}{2}$ 23. $-\frac{4}{3}$

25. $t = -\frac{1}{2};\ L_2:\dfrac{x-3}{9} = \dfrac{y+1}{3} = \dfrac{z-2}{5} = \dfrac{v-1}{11}$

27. $t = -\frac{5}{3};\ L_2:\dfrac{x+1}{17} = \dfrac{y-2}{-22} = \dfrac{z+3}{45} = \dfrac{v-1}{-28}$

▶ Section 12-3 (pp. 294–295)

1. $\frac{9}{14}$ 3. $-\frac{2}{3}$

7. $x + z - 1 = 0;\ y - z + 3v - 2 = 0$ 9. $x - 6z + 4v + 4 = 0;\ y + z - 2 = 0$

11. $x = 2 + 2hr + 3ks,\ y = 1 - 3hr + ks,\ z = -1 - 2hr + 2ks,\ v = 1 + hr + ks;$
    $h = 1/3\sqrt{2},\ k = 1/\sqrt{15}$

13. $x = -2 + 3hr + 2ks,\ y = -2 - hr + 2ks,\ z = -2 - 2hr + 3ks,\ v = 2 + 2hr + ks$
    if $h = k = 1/3\sqrt{2}$

15. $x = 1 + 2hr + 2ks,\ y = -1 - 3hr + 3ks,\ z = 1 + 2hr + 8ks,\ v = -1 + hr - 11ks,$
    $h = 1/3\sqrt{2},\ k = 1/\sqrt{198}$

17. $x = 2 + 2hr + 3ks,\ y = -2 - hr - 60ks,\ z = 2 - 2hr + 10ks,$
    $v = -2 + 2hr - 23ks$ if $h = 1/\sqrt{13},\ k = \frac{1}{13}$

19. $x = -3hr + 2ks,\ y = 2hr + 3ks,\ z = hr,\ v = ks,\ h = 1/\sqrt{14},\ k = 1/\sqrt{14}$

21. $x = -hr + 2ks,\ y = 2hr + ks,\ z = hr,\ v = ks,\ h = k = 1/\sqrt{6}$

▶ Section 12-4 (pp. 306–308)

1. $\dfrac{x-2}{2} = \dfrac{y+1}{2} = \dfrac{z-3}{-1} = \dfrac{v-2}{-3}$ 3. $\dfrac{x-1}{2} = \dfrac{y+2}{-1} = \dfrac{z-2}{2} = \dfrac{v-1}{1}$

5. $2(x-2) + 3(y-0) - 2(z-1) - 4(v+1) = 0$

7. $3(x+2) + 2(y-1) - 1(z) - 2(v-2) = 0$

9. $2(x+2) + (y-3) - 3(z-1) - 2(v-2) = 0$

11. $2x - y + 2z - v - 4 = 0$ 13. point of intersection $(\frac{7}{2}, \frac{35}{4}, -\frac{11}{2}, \frac{1}{4})$

15. H contains L 17. $P:(\frac{9}{4}, \frac{1}{2}, -\frac{1}{4}, \frac{7}{2})$

21. $9x - 8y - 14z - 7v + 53 = 0$ 23. $6x - 17y - 13z + 4v + 77 = 0$

25. $3/\sqrt{42}$

APPENDIX 2

1. $-x^2 + xy + 4y^2$

5. $2x^4 - 7x^3 + 11x - 3$

3. $4x^3 - 4x^2 + 3x + 1$

7. $2x^5 - 5x^4 - x^3 + 3x^2 - 12x + 9$

9. $-\dfrac{b}{a^2 - b^2}, a \neq \pm b$

11. $\dfrac{2x^2 - 9x - 23}{6(x - 1)(x + 4)}; x \neq 1, -4$

13. $\dfrac{(x + 5)(x^2 + x + 1)}{(x + 1)(x + 2)}; x \neq \pm 1, -2$

15. $1; x \neq 0, -y$

17. $\dfrac{ab}{a^2 + b^2}, a \neq \pm b, a \neq 0, b \neq 0$

APPENDIX 3

5. $\sin 3\theta = 3 \sin \theta - 4 \sin^3 \theta, \cos 3\theta = 4 \cos^3 \theta - 3 \cos \theta$

9. $\sin 15° = \dfrac{\sqrt{6} - \sqrt{2}}{4}, \cos 15° = \dfrac{\sqrt{6} + \sqrt{2}}{4}, \sin 22\frac{1}{2}° = \dfrac{\sqrt{2 - \sqrt{2}}}{2},$

$\cos 22\frac{1}{2}° = \dfrac{\sqrt{2 + \sqrt{2}}}{2}, \tan 75° = 2 + \sqrt{3}, \sec 75° = \sqrt{6} + \sqrt{2}$

11. $-\cos \theta, -\sin \theta, -\cos \theta, \sin \theta, -\sin \theta, \cos \theta$

17. $\dfrac{1}{2} \cos \theta - \dfrac{\sqrt{3}}{2} \sin \theta, \dfrac{1}{2} \cos \theta - \dfrac{\sqrt{3}}{2} \sin \theta, -\dfrac{\sqrt{3}}{2} \cos \theta + \dfrac{1}{2} \sin \theta$

APPENDIX 4

3. $D = -2$

5. $D = 24$

7. $D = 8$

9. $x = 1, y = -2, z = -1$

11. $x = -\frac{73}{19}, y = \frac{21}{19}, z = -\frac{43}{19}$

Index

Absolute value, 12
Addition, axioms of, 1, 2, 309
  of ordinates, 167
  of vectors, 88, 250
Amplitude (of a periodic function), 166
Angle, between two lines, 72, 209, 290
  between two planes, 221
  between two vectors, 93, 257
  bisector, 72
  direction, 207
Arccosine function, 175
Archimedes, spiral of, 194
Arcsine function, 174
Arctangent function, 175
Associative law, of addition, 2, 309
  of addition of vectors, 91, 252
  for the multiplication of
    vectors by scalars, 91, 252
  of multiplication, 2, 311
Asymptotes, horizontal, 157
  of a hyperbola, 136
  vertical, 157
Attitude numbers, 217
Axes, rotation of, 145
  translation of, 140, 237
Axiom of inequality, 3
Axioms of addition and multiplication,
    1–2, 309–312
Axis, conjugate, 132
  major, 121
  minor, 121
  of a parabola, 109
  polar, 187
  transverse, 132

Base of a directed segment, 85, 248

Cardioid, 193
Cartesian coordinate system, on a line, 58
  in a plane, 40
  in space, 202, 203
Center, of a circle, 99
  of an ellipse, 121
  of a hyperbola, 132
Central rectangle of a hyperbola, 133
Circle, 99
  center of, 99
  equation of, 99–100
    in polar coordinates, 198
  radius of, 99
  tangents to, 102
  through three points, 100–101
Closed interval, 5
Closed property, of addition, 1, 309
  of multiplication, 2, 311
Cofactor, 331
Colatitude, 243
Common logarithms, 180
Commutative law, of addition, 2, 309
  of addition of vectors, 91, 252
  of multiplication, 2, 311
Cone, elliptic, 235
Conic sections, 153
Conjugate axis of a hyperbola, 132
Convex polygon, 83
Convex set, 83
Coordinate plane, 202
Coordinates, 21, 202
  Cartesian, 40, 58, 202
  cylindrical, 241
  in a plane, 22
  in space, 202
  left-handed system of, 202

on a line, 58
polar, 187
right-handed system of, 202
spherical, 242
transformation of, 140
Cosine function, 166, 322
Cosines, direction, 207, 276
Cramer's rule, for $n = 2$, 330
    for $n = 3$, 335
Cross product of two vectors, 261
Cycloid, 185
Cylinder, 228
Cylindrical coordinates, 241
Cylindrical surface, 228
    generator of, 228

Decreasing function, 172
Determinants, expansion of, 329, 332
    of the second order, 329
    of the third order, 329
Directed distance along a directed line, 59
Directed line, 59, 206
Directed line segment, 59, 248
    base of, 85, 248
    equivalent, 85, 248
    head of, 85, 248
    magnitude of, 85, 248
Direction angles, 207
Direction cosines, 207, 276
Direction numbers, 208, 273
Directrix, ellipse, 125
    hyperbola, 135
    parabola, 108
Distance, between two points in the
        plane, 40
    between two points in space, 204
    directed, 59
    from a point to a line, 65
    from a point to a plane, 223
Distributive law, 2
    for the scalar product of two vectors,
        95, 258
    for the vector product of two vectors,
        263
Domain, of a function, 26
    of a relation, 31
Dot product of two vectors, 95, 257

Eccentricity, of an ellipse, 122
    of a hyperbola, 132
Edge (of a polyhedral domain), 246
Element (of a set), 7
Ellipse, center of, 121
    directrices of, 125
    eccentricity of, 122
    equation of, 121
        in polar coordinates, 199
    foci of, 120
    latus rectum of, 129
    major axis of, 121
    minor axis, 121
    reflected wave property of, 128
    tangent line to, 126, 128
    vertices, 121
Ellipsoid, 232
    of revolution, 232
Elliptic cone, 235
Elliptic hyperboloid, of one sheet, 233
    of two sheets, 233
Elliptic paraboloid, 234
Equation, of a circle, 99
        in polar coordinates, 198–199
    of a cylindrical surface, 228
    of an ellipse, 121
        in polar coordinates, 199
    of the first degree, 209, 55,
    of a hyperbola, 131
        in polar coordinates, 199
    of a line, 51
        in polar coordinates, 197
    of a parabola, 109–112
        in polar coordinates, 199
    of a plane in space, 216–217
    of the second degree, general, 147
    solution set of, 23
    of a sphere, 227
Equations, of certain loci, 43–44
    of a line 212–214, 278
    parametric, 61, 182, 212–213, 278
        of a line, 61, 212, 278
Equivalent directed segments, 85, 249
Euclidean space of four dimensions, 272
Even function, 323
Existence, of a negative, 2, 309
    of a reciprocal, 2, 311

of a unit, 2, 311
of a zero, 2, 309
Exponential curves, 178
Exponential functions, 178
Extended commutative law, 317

Faces (of a polyhedral domain), 246
Family, of circles, 102
of lines, 70
Focus, ellipse, 120
hyperbola, 130
parabola, 108
Free vector, 87
Function, 26
decreasing, 172
defined implicitly, 31
domain of, 26
even, 275
exponential, 178
increasing, 172
inverse, 170
inverse trigonometric, 173–176
linear, 55
logarithmic, 178
odd, 323
periodic, 166
range of, 26
transcendental, 165
trigonometric, 165, 322–323

Graph (of an equation), 23

h of the way from P_1 to P_2, 62
Half-open interval, 5
Half-plane, 80
Half-space, 204
Head (of a directed line segment), 85, 248
Hyperbola, 130
asymptotes of, 136
center of, 132
central rectangle of, 133
conjugate axis of, 132
directrices of, 135
eccentricity of, 132
equation of, 131
in polar coordinates, 199
foci of, 130

latus rectum of, 139
tangent line to, 137–138
transverse axis of, 132
vertices of, 132
Hyperbolic paraboloid, 234
Hyperplanes, 296
angle between, 305
parallel, 303
perpendicular to a line, 302

Image, 171
Inclination, 45
of a directed line, 60
Increasing function, 172
Inequalities, 3, 79, 160–164, 244–247
in three variables, 244–247
in two variables, 79, 160–164
Inequality, axiom of, 3
Infinite strip, 161
Initial line, 187
Inner product (of two vectors), 95, 257
Integer, 318
Intercepts, 34, 230
Intersection of sets, 8, 80
Interval, 5
closed, 5
half-open, 5
open, 5
Inverse function, 170
Inverse relation, 169
Inverse trigonometric functions, 173–176
arccos function, 175
arcsin function, 174
arctan function, 175

Latus rectum, ellipse, 129
hyperbola, 139
Law of cosines, 319
Law of sines, 319
Least period of a function, 166
Left-handed coordinate system, 202
Left-handed ordered triple of vectors, 262
Lemniscate, 194
Length of a vector, 87,
Limaçon, 191
Linear dependence, 253
Linear equation, 55, 217

Linear function, 55
Linear independence, 253
Linear inequalities, 79, 244
Linear programming, 83
Linearly dependent set (of vectors), 253
Linearly independent set (of vectors), 253
Lines, angle between, 72, 209
    Cartesian coordinates on, 58, 213
    directed, 59, 206
    equations of, 52, 212, 278
        in polar coodinates, 197
    inclinations of, 45, 60
    in four space, 273
    parallel, 48, 278
    parametric equations of, 61, 212
    perpendicular, 49, 278
    perpendicular to a plane, 216–217
    slopes of, 46
    tangent to a circle, 102
    tangent to an ellipse, 126, 128
    tangent to a hyperbola, 137–138
    tangent to a parabola, 115
Loci in polar coordinates, 190
Locus (of an equation), 23
Logarithm function, 178
Logarithmic spiral, 194
Logarithms, to the base a, 178
    common, 180
    natural, 180

Magnitude of a directed segment, 85, 248
Major axis of an ellipse, 121
Mapping, 171
Midpoint of a segment, 41–42
Minor axis of an ellipse, 121

Nappe (of a cone), 153
Natural logarithms, 180
Natural number, 316
Negative number, 3
Number plane (R_2), 21
    space (R_3), 202
    space (R_4), 272

Oblate spheroid, 232
Odd function, 323
Optical property of a parabola, 119

Ordered pair, 21
Ordered triple of vectors, 262
    left-handed, 262
    oppositely oriented, 262
    right-handed, 261
    similarly oriented, 262
Ordinates, addition of, 167
Orthogonal (vectors), 87, 258

Parabola, 108
    axis of, 109
    directrix of, 108
    equation of, 109–112
        in polar coordinates, 199
    focus of, 108
    "optical property" of, 119
    reflecting property of, 117–118
    tangent line to, 115
    vertex of, 108
Paraboloid, 117, 234
    elliptic, 234
    hyperbolic, 234
    of revolution, 234
Parallel lines, 48, 278
Parallel planes, 218
Parallel vectors, 257
Parallelepiped, 204
Parameter, 182
Parametric equations, in general, 61, 182
    of a line, 61, 212–213
Period (of a function), 166
Perpendicular lines, 49, 278
Perpendicular planes, 221
Perpendicular vectors, 87, 250
Plane, attitude numbers of, 217
    equation of, 216–217
    parallel to a plane, 218
    perpendicular to a line, 216–217
    through three points, 218
Planes in four space, 285
Point (in the number plane), 21
Point of division formula, 63, 214
Point-slope form, 52
Polar axis, 187
Polar coordinates, 187
Pole, 187

Polyhedral domain, 204
    edge of, 246
    face of, 246
    vertex of, 246
Positive numbers, 3
Projection of **v** along **w**, 94, 259
Prolate spheroid, 232
Proportional numbers, 208, 274
Proportional quadruples, 291
Proportional vectors, 253
Pythagorean theorem, 40

Quadric surface, 230

Radian, 324
Radical axis, 105
Range, of a function, 26
    of a relation, 31
Rational number, 318
Real number, 1
Reflected wave property of the ellipse,
    128
Reflecting property of the parabola, 117,
    118
Relation, 31
    domain of a, 31
    from R_1 to R_1, 31
    inverse, 169
    range of a, 31
Representative of a vector, 87, 250
Right circular cone, 153
Right-handed coordinate system, 202
Right-handed ordered triple of vectors,
    262
Rotation of axes, 145
Ruled surface, 236

Saddle-shaped surface, 235
Scalar product (of two vectors), 95, 257
Section of a surface by a plane, 231
Sequence, 316
    finite, 316
    infinite, 316
    ith term of, 316
    sum of, 316
Sine function, 165

Slope, of a line, 46
    of parallel lines, 48
    of perpendicular lines, 49
Solution set, of an equation, 23
    of inequalities, 6, 79, 160–164,
        244–247
Sphere, 227
    center of, 227
    equation of, 227
    radius of, 227
Spherical coordinates, 242
Spheroid, 232
    oblate, 232
    prolate, 232
Spiral, 194
    logarithmic, 194
    of Archimedes, 194
Subtraction of vectors, 91
Symmetric, with respect to the origin, 35,
    191
    with respect to the x axis, 35, 191, 232
    with respect to the xy plane, 232
    with respect to the y axis, 35, 191,

Tangent, of the angle between two lines,
    73
    function, 166, 275
    line, to a circle, 102
        to an ellipse, 126, 128
        to a hyperbola, 137, 138
        to a parabola, 115
Three-dimensional number space, 202
Traces of a surface, 231
Transcendental functions, 165
Transformation of coordinates, 140
Translation of axes, 140, 237
Transverse axis of a hyperbola, 132
Trigonometric functions, 165, 322, 323
Union of two sets, 7
Unit vector, 87, 250
Unit vectors, **i** and **j**, 89
    i, **j** and **k**, 250

Vector, 87, 250
    free, 87
    length of, 87, 250
    multiplied by a scalar, 88, 250

projection of, on a vector, 94, 259
representative of, 87, 250
unit, 87, 250
zero, 87, 250
Vector product of two vectors, 261
Vectors, addition of, 88, 250
    angle between, 93, 257
    cross product of, 261
    dot product of, 95, 257
    inner product of, 95, 257
    linearly dependent, 253
    linearly independent, 253
    orthogonal, 87, 250

parallel, 257
perpendicular, 87, 250
proportional, 253
scalar product of, 95, 257
vector product of, 261
Vertex, of an ellipse, 121
    of a hyperbola, 132
    of a parabola, 108
    of a polyhedral domain, 246

Work done (by a force), 259

Zero vector, 87, 250